Algebra and Applications

Volume 30

Algebra and Applications aims to publish well-written and carefully refereed monographs with up-to-date expositions of research in all fields of algebra, including its classical impact on commutative and noncommutative algebraic and differential geometry, K-theory and algebraic topology, and further applications in related domains, such as number theory, homotopy and (co)homology theory through to discrete mathematics and mathematical physics.

Particular emphasis will be put on state-of-the-art topics such as rings of differential operators, Lie algebras and super-algebras, group rings and algebras, Kac-Moody theory, arithmetic algebraic geometry, Hopf algebras and quantum groups, as well as their applications within mathematics and beyond. Books dedicated to computational aspects of these topics will also be welcome.

To Nelia.
To the memory of Andrzej Skowroński and Daniel Simson, great mathematicians and great friends.

Preface

The representation theory of associative algebras has developed rapidly since the 1970s, perhaps due to the depth within the problems and conjectures posed early on in the theory, perhaps for the role these problems play in the machinery of mathematics. This book intends to collect some of the most relevant advances obtained surrounding the tame and wild dichotomy problem, and what are now known as Brauer-Thrall conjectures. With these problems as guideline, we consider topics as integral quadratic forms, lattices of ideals, Auslander-Reiten theory, Bass's theorems on (semi-) perfect rings, Galois coverings and smash products, Coxeter spectral analysis, Weyl groups, and post-projective components, among others, all applied to the representation theory of wide classes of associative algebras.

A part of my professional career has been dedicated to the understanding of some classes of algebras, from a representation theoretical perspective, in view of the problems mentioned above, which has undoubtedly biased the selection of topics of the text. The task has led me to collaborate with some of the most influential figures in the development of the theory, cherished collaborations that morphed into friendships over time.

Throughout the text, A will denote a finite dimensional K-algebra over an algebraically closed field K. The main reason for taking these hypotheses is simplicity. For instance, gaining a uniform context not depending on whether we consider left or right modules by means of a duality functor

$$D = \mathrm{Hom}_K(-, K) : A\text{−Mod} \to A^{op}\text{−Mod},$$

taking left modules to right modules and preserving length. Moreover, finite length modules over A are finite dimensional, and therefore, the category A−mod is contained in the category of vector spaces K−mod. Finally, the consideration of an algebraically closed field sets in advance a nice hypothesis that will be used in many instances (for dichotomy theorems, for the existence of quivers, for the existence of roots of unity and other situations).

We say that A is of *finite representation type* if there are only finitely many indecomposable A-modules (up to isomorphism). The algebra A is *tame* if for every number n, almost every indecomposable A-module of dimension n is isomorphic to a module belonging to a finite number of 1-parameter families, where a 1-parameter family is of the form $S \otimes_L -$ for S a simple module and L a finitely generated $K[x] - K[x]$-bimodule. Finally, A is *wild* if mod $_A$ contains the representation theory of $K\langle x, y\rangle$, the free associative algebra in two indeterminates, that is, there is a faithful functor of the form $F = M \otimes_{K\langle x,y\rangle}$—which preserves isomorphism-classes. Drozd's theorem states that every finitely generated K-algebra is of one of these types.

The first explicit recognition that infinite representation type splits in two different classes arises in representations of groups: in 1954, Highman showed that the Klein group has infinitely many representations in characteristic 2 and Heller and Reiner classified them; in contrast, Krugljak showed in 1963 that solving the classification problem of groups of type (p, p) with $p \geq 3$ implies the classification of the representations of any group of the same characteristic, a task that was recognized as 'wild'. Donovan and Freislich conjectured at the middle of the 1970s that algebras split in *tame* and *wild* types, which was finally showed by Yuri Drozd in 1980.

The class of representation infinite algebras can be divided as follows: In 1957, J. Jans (then a student of Thrall) showed that a non-distributive algebra is strongly unbounded, i.e., that there exist infinitely many d such that there are infinitely many isomorphism classes of indecomposables of dimension d. Furthermore, he mentions two conjectures of Brauer and Thrall: The first says that A is representation-finite if there is a bound on the dimensions of indecomposables (BT1), and the second says that otherwise A is strongly unbounded (BT2).

The first conjecture was solved by Roiter [1] in 1968 by proving a somewhat stronger theorem. (Indeed, assume there are infinitely many isomorphism classes of indecomposable modules M_i, take the direct sum $M = \bigoplus_i M_i$. By Krull-Remak-Schmidt-Azumaya Theorem, this module M is not of finite type. Thus we obtain indecomposable modules of arbitrarily large finite length as submodules of this particular module M.) For the generalization of BT1 to artinian rings in 1972, Auslander invented almost split sequences. Jans proves BT2 in his paper for algebras that are not distributive. There is the long article [2] Nazarova and Roiter aiming at a proof of BT2, but the first complete proof was only given by Bautista in 1983. The proof of the second conjecture required some of the new concepts of representation theory introduced after 1968 and also an intensive study of representation-finite and distributive minimal representation-infinite algebras.

Integral quadratic forms, their roots and reflections, are classical instruments in representation theory, since the seminal work by Gabriel determining those hereditary basic algebras that are representation finite, via quiver algebras and their representations. In this case, the representation type of the algebra is determined by the positive-definiteness of its Euler quadratic form, an integral quadratic form containing some of the homological information of the algebra. Weaker arithmetical properties of the Tits form of an algebra, also known as geometrical form for its interpretation as (bounds of) dimensions of some varieties of modules associated with the algebra, have also been used as criterion to test finite-representation type and tameness of certain classes of algebras, most notoriously, strongly or weakly simply connected algebras (for instance, by Bongartz, Geiss, Brüstle, Skowroński and the author). For a more complete treatment of integral quadratic forms and their roots, the reader is referred to the book by Barot, Jiménez and the author [3].

In Chap. 1, we present some of the classical problems that triggered a rapid development in the representation theory of algebras: the Brauer-Thrall conjectures and Drozd's dichotomy theorem. After some preparatory concepts on quadratic forms, we introduce fundamental concepts and examples on algebras, categories, and algebraic varieties. Chapter 2 contains elementary (but fundamental) results in representation theory, namely the lemmas of Fitting, Harada-Sai, and Yoneda. We present Auslander's functorial approach on the existence of almost split sequences and use these results, together with Bass's characterization of perfect rings, to give a proof of the first Brauer-Thrall conjecture. Chapter 3 starts with the definition and characterization of distributive algebras, and some comments on other propositions under the label of Brauer-Thrall conjectures. The rest of the chapter treats the existence of post-projective components through an algorithmic approach and uses Tits quadratic form to determine the representation type of triangular algebras. We also give criteria to determine weak positivity and non-negative of integral quadratic forms. Chapter 4 deals with the Coxeter transformation associated with the Cartan matrix of a basic algebra. We relate Coxeter spectral properties to the representation theoretical structure of an algebra and seek conditions in the automorphism group of an algebra to determine spectral properties of the correspondent Coxeter transformation. Chapter 5 is dedicated to fundamental concepts on symmetries, automorphisms, and coverings of algebras, with applications to their representation theory. We analyze classical construction as pull-up and push-down functors, graded categories, and smash products, and relate such constructions to the representation type of wide classes of algebras. We start Chap. 6 with useful numerical properties of Dynkin and Euclidean graphs known as Vinberg's characterizations and apply them to study the structure of Auslander-Reiten components of algebras. We also study the Weyl group associated with graphs and derive some interesting characterizations of wild behavior in these groups. In Chap. 7, we study fundamental groups and a family of algebras known as (strongly) simply connected. We introduce their basic properties and propose a constructive characterization of such algebras, expanding on the so-called weakly separating families of coils. In Chap. 8, we turn our attention to classical ideas on the geometry of algebras and their module varieties, mainly concerning degenerations of algebras and how their representation type is affected with through topological constructions. Some final comments and historical remarks are collected at the end in Chap. 9.

It is our intention to bring forward some important aspects of representation theory that are usually not covered in classical introductions to the topic, such as the books of Auslander, Reiten, and Smalø [4], of Assem, Simson, and Skowroński [5], and of Gabriel and Roiter [6] (see also recent introductions by Barot [7] and Assem and Coelho [8]), or even in the homological or model theoretical treatments by Zimmermann [9] or Jensen and Lenzing [10], respectively. In this sense, this work is parallel to, for instance, the books of Ringel [11], of Simson and Skowroński [12], and of Erdmann and Holm [13]. The text is suitable for advanced undergraduate students wishing to deepen in the representation theory of associative algebras (for example, as a second course), and for young researchers trying to find a path among the vast literature in the area published in the last few decades.

México, Mexico José-Antonio de la Peña

References

1. Roiter, A.V. *The unboundeness of the dimension of the indecomposable representations of algebras that have an infinite number of indecomposable representations*, Izv. Acad. Nauk SSSR, Ser. Mat., **32** (1968), 1275–82 (in Russian).
2. Nazarova, L.A. and Roiter, A.V. *Kategorielle Matrizen-Probleme und die Brauer-Thrall-Vermutung*, Mitt. Math. Sem. Giessen **115** (1975), 1–153.
3. Barot, M., Jiménez González, J.A. and de la Peña, J.A. *Quadratic Forms: Combinatorics and Numerical Results*, Algebra and Applications, Vol. **25** Springer Nature Switzerland AG 2018
4. Auslander, M. and Reiten, I. and Smalø, S. *Representation theory of Artin algebras*, Cambridge University Press **36** Cambridge Studies in Advanced Mathematics (1995)
5. Assem, I. and Simson, D. and Skowronski, A., *Elements of the representation theory of associative algebras*, Cambridge University Press (2006), London Mathematical Society Student Texts **65**
6. Gabriel, P. and Roiter, A., *Representations of finite-dimensional algebras*, London Mathematical Society, LNS **362**. Springer-Verlag Berlin Heidelberg. (1997)
7. Barot, M., *Introduction to the Representation Theory of Algebras*, Springer (2015)
8. Assem,I. and Coelho, F., *Basic Representation Theory of Algebras*. Graduate Texts in Mathematics, 283 (2020) Springer.
9. Zimmermann, A., *Representation theory: a homological algebra point of view*, Heidelberg, Springer, 2014
10. Jensen, C.U. and Lenzing, H., *Model theory and representation theory*. Lecture Notes in Mathematics 832. Springer-Verlag, 1980

11. Ringel, C.M. *Tame algebras and integral quadratic forms*, Springer LNM, **1099** (1984)
12. D. Simson and A. Skowroński, *Elements of the Representation Theory of Associative Algebras 2: Tubes and Concealed Algebras of Euclidean Type*. London Mathematical Society Student Texts, Vol. **71**, Cambridge University Press, Cambridge, 2007.
13. Erdmann, K. and Holm, T., *Algebras and Representation Theory*. Springer Undergraduate Mathematics Series, 2018

Contents

Chapter 1
Introduction and First Examples

1.1 General Algebraic Problems

1.1.1 Throughout the text K will denote an algebraically closed field. Let A be a K-algebra (meaning a finite dimensional K-vector space with an associative product and having a unit element, unless otherwise stated). By A-mod we denote the category of left finite dimensional A-modules.

The main purpose of the representation theory of algebras is the study of the category A-mod. Partial solutions to this problem received a decisive impulse at the beginning of the 70s of last century in the works of Auslander, Gabriel, Roiter and others. We shall build on their works and that of others.

One of the problems that played a central role in the theory is the determination of the representation type of an algebra. We shall consider the following classical problem and solve it by different points of view to illustrate the different approaches of the historic schools:

The algebra A is said to be **representation-finite** if there are only finitely many isoclasses of indecomposable A-modules (see details in Sect. 2.1 below). We say that A is **representation-bounded** if there is a number N which serves as upper bound for the dimension of indecomposable A-modules. An early success was the proof by Auslander [3, 1973] (see also [4]) and, independently, Tachikawa [28, 1974] that representation-bounded algebras are just the representation-finite ones. This result is known as **first Brauer-Thrall conjecture** (see also Ringel's work [24, 26]).

1.1.2 Theorem *If A is a representation-bounded algebra then A is representation-finite.*

The concepts needed to complete the, rather elementary, proof will be developed in Chap. 2. Nevertheless, the proof is a nice (maybe, brilliant) example of the gain obtained just by reinterpreting modules as a category of functors, that we briefly

J.-A. de la Peña, *Representations of Algebras*, Algebra and Applications 30,
https://doi.org/10.1007/978-3-031-12288-0_1

summarize as follows. Consider Bass's Theorem [9] characterizing left perfect rings (cf. [21] or [22]).

Theorem (Bass's Theorem) *The following are equivalent for a ring R:*

(a) Every left R module (not necessarily finitely generated) has a projective cover.
(b) R is semilocal, and for any $a_1, a_2, \ldots$ *in the Jacobson radical* $\mathrm{rad}(R)$ *of R, there is some m with* $a_1a_2\cdots a_m = 0$.
(c) R satisfies the descending chain condition on principal right ideals.

A proof of Theorem 1.1.2 will be given in Chap. 2.3.2. The argument, that we sketch here, depends on a categorical version of Bass's Theorem (see 2.2.6.1). Considering a small additive category **A** as a generalization of a ring (*a ring with several objects*), one applies Harada-Sai's result 2.2.2.2 to verify that condition (b) is satisfied when **A** is the category of representatives of isoclasses of indecomposable A-modules and A is a representation bounded algebra. Hence, an appropriate version of claim (c) implies that the representable functors $\mathrm{Hom}_A(-, X)$ have finite length for any module X in A-mod. In particular, this holds for X a simple A-module S, which can be shown to imply that the support of $\mathrm{Hom}_A(-, S)$ is finite. Since every indecomposable A-module is in the support of some functor $\mathrm{Hom}_A(-, S)$ with S a simple module (and the number of isoclasses of simple A-modules is finite), then A is representation finite.

1.1.3 Also the representation-infinite algebras have been considered in further detail. The first precise definitions of tameness and wildness are due to Freislich and Donovan in [15, 1973]. They have been modified by many people since then, but consensus seems to have settled on something like the following. We say that the algebra A is tame if for every number n, almost every indecomposable A-module of dimension n is isomorphic to a module belonging to a finite number of 1-parameter families of modules.

Formally, an algebra A is **tame** if for every $n \in \mathbb{N}$ there is a finite family of $A - K[t]$-bimodules $M_1, \ldots, M_{\ell(n)}$ with the following properties:

(i) M_i is finitely generated free as a right $K[t]$-module;
(ii) almost every indecomposable left A-module X with $\dim_K X = n$ is isomorphic to a module of the form $M_i \otimes_{K[t]} S_\lambda$ for some $\lambda \in K$, where S_λ is the simple $K[t]$-module of dimension one and $t \cdot 1_{S_\lambda} = \lambda 1_{S_\lambda}$.

The book of Ringel [25, 1984] is an excellent example of the depth of the understanding on tame algebras at the time, in particular when considering associated quadratic forms and their corresponding root systems.

1.1.4 On the other hand, the algebra A is called wild if the classification of the indecomposable A-modules implies the classification of the indecomposable modules over the associative algebra $K\langle x, y\rangle$ in two indeterminates.

More formally, A is **wild** if there exists a functor $F : K\langle x, y\rangle\text{-mod} \to A\text{-mod}$ which *insets indecomposable modules*, that is, such that

(i) F preserves indecomposability of modules;
(ii) If $F(X)$ and $F(Y)$ are isomorphic, then X and Y are isomorphic.

Other notions of wildness were originally explored by Brenner in the early seventies [11, 12], after work of Corner [13] (cf. [20]). Equivalently (this is not obvious), A has wild type if for every finite dimensional K-algebra B, there is a representation embedding of B-mod into A-mod. This means an exact functor from B-mod to A-mod which preserves non-isomorphy and indecomposability.

1.1.5 Proposition *Let B be any finitely generated K-algebra, then there exists a fully faithful functor $F : B\text{-mod} \to K\langle x, y\rangle\text{-mod}$.*

Proof Let $b_1, \ldots, b_s$ be a system of generators of B. Define the $K\langle x, y\rangle - B$-bimodule M as $M_B = B^{\oplus(s+2)}$ (meaning the direct sum of $s+2$ copies of B, sometimes denotes simply by B^{s+2}), and the structure of left $K\langle x, y\rangle$-module given by the $(s+2) \times (s+2)$-matrices

$$
{}_xM = \begin{bmatrix} 0 & 1 & & & 0 \\ & 0 & \ddots & & \\ & & \ddots & \ddots & \\ & & & 0 & 1 \\ 0 & & & & 0 \end{bmatrix}
\qquad
{}_yM = \begin{bmatrix} 0 & & & & & 0 \\ 1 & 0 & & & & \\ b_1 & 1 & & & & \\ & \ddots & \ddots & & & \\ & & \ddots & 1 & 0 & \\ 0 & & & b_s & 1 & 0 \end{bmatrix}
$$

We set $F = M \otimes_B : B\text{-mod} \to K\langle x, y\rangle\text{-mod}$, and check that F is full and faithful. □

1.1.6 Proposition *Let p be a prime number ≥ 3, and assume that K has characteristic p. Then the group algebra $A = K[\mathbb{Z}_p \times \mathbb{Z}_p]$ is wild.*

Proof Let $\varphi : K[u, v] \to A$, $u \mapsto g - 1$, $v \mapsto h - 1$, where $\mathbb{Z}_p \times \mathbb{Z}_p = \langle g\rangle \times \langle h\rangle$. Then $A \cong K[u, v]/\operatorname{Ker}\varphi = K[u, v]/(u^p, v^p)$. Moreover $K[u, v]/(u^p, v^p) \twoheadrightarrow K[u, v]/(u, v)^3 = K[u, v]/(u^3, v^3, uv^2, vu^2) =: B$. It is enough to show that B is wild.

Consider the $B - K\langle x, y\rangle$-bimodule M defined as $M_{K\langle x, y\rangle} = K\langle x, y\rangle^4$ and the structure as B-module defined by the matrices

$$
{}_uM = \begin{bmatrix} 0&0&0&0 \\ 0&0&0&0 \\ 1&0&0&0 \\ 0&x&y&0 \end{bmatrix}
\qquad
{}_vM = \begin{bmatrix} 0&0&0&0 \\ 1&0&0&0 \\ 0&0&0&0 \\ 0&1&x&0 \end{bmatrix}
$$

One checks that $_BM$ is well-defined, and that

$$M \otimes_{K\langle x,y\rangle} -: K\langle x, y\rangle\text{-mod} \to B\text{-mod},$$

insets indecomposable modules. □

1.1.7 Theorem (Drozd [16], Crawley–Boevey [14]) *Let A be a finite dimensional algebra over an algebraically closed field A. Then A*-mod *has either tame type or wild type, and not both.*

For certain families of algebras there are good criteria to determine the representation type. For instance, if $A = KQ/I$ with Q a tree quiver (that is, Q has no cycles, oriented or not) and I and admissible ideal, then A is of tame type if and only if the Tits form q_A is weakly non-negative (that is, $q_A(v) \geq 0$ for any vector v with non-negative coordinates). Similar characterizations in the general situation remain open.

We present the "not both" proof in Chap. 8. The "either" proof is much involved and required (historically) the development of *BOCS theory* by the Ukrainian school. We will not do this and just give references. Nevertheless, there are arguments that could be used for a proof if we assume extra-hypothesis on the ground algebra A.

1.2 Some Terminology on Quadratic Forms

1.2.1 Let A be a **basic** finite dimensional K-algebra (that is, if $\mathbf{e}_1, \dots, \mathbf{e}_n$ is a complete set of primitive orthogonal idempotents of A, then $A\mathbf{e}_i$ and $A\mathbf{e}_j$ are non-isomorphic A-modules for $i \neq j$). This definition is clearly independent of the choice of $\mathbf{e}_1, \dots, \mathbf{e}_n$, see [2, I.6] for details. Column vectors of $\mathbb{Z}^n$ will be denoted by $x = (x_1, \dots, x_n)$ with $x_i \in \mathbb{Z}$. Let us consider the following quadratic forms (that is, homogeneous polynomials of degree two).

Choosing $\{S_i\}_{i=1,\dots,n}$ a set of representatives of isoclasses of simple A-modules and considering the extension functors $\mathrm{Ext}_A^k(-,-)$ for $k \geq 0$, take

$$\chi_A(x) = \sum_{i,j\in Q_0} \sum_{k=0}^{\infty} (-1)^k \dim_K \mathrm{Ext}_A^k(S_i, S_j) x_i x_j,$$

and

$$q_A(x) = \sum_{i\in Q_0} x_i^2 - \sum_{i\to j} \dim_K \mathrm{Ext}_A^1(S_i, S_j) x_i x_j$$
$$+ \sum_{i,j\in Q_0} \dim_K \mathrm{Ext}_A^2(S_i, S_j) x_i x_j.$$

The quadratic form q_A is called **Tits form** of A. If A has **finite global dimension** (that is, the supremum of the set of projective dimensions of all left A-modules is finite, cf. [2, Appendix A.4]), then χ_A is well-defined and is called **Euler (quadratic) form** of A. In some cases q_A is a truncation of χ_A. Special emphasis is made in the characterization of the representation type of algebras via arithmetical properties of their Tits forms. Also, we consider related results about the structure of an algebra A or of its category of modules.

1.2.2 Let $q : \mathbb{Z}^n \to \mathbb{Z}$ be a quadratic form, that is, a homogeneous polynomial of degree two with integer coefficients. We denote by

$$q(-,-) : \mathbb{Z}^n \times \mathbb{Z}^n \to \mathbb{Z}$$

the corresponding symmetric bilinear form, that is $q(x, y) = q(x + y) - q(x) - q(y)$. The study of certain quadratic forms, in particular the question of whether a given integer can be the value of a quadratic form over the integers, dates back many centuries. Among other famous representability problems we shall mention the following great theorems:

- Lagrange, 1772: any non-negative integer m can be written as a sum of four integer squares.
- Legendre, 1798: a non-negative integer m can be written as a sum of three integer squares if and only if m is not of the form $4^a(8k + 7)$ for some $a, k \in \mathbb{Z}$.
- Fermat, 1640: a prime number $p > 2$ can be written as a sum of two integer squares if and only if $p \equiv 1(mod\ 4)$.

While we can ask about which integers are represented by a given quadratic form q, it is also interesting to ask in how many ways it is represented by q. For particular cases, there are also some classic answers to this problem. Let $r_q(m)$ count the number of ways of representing m by q (that is, the cardinality of the set of vectors x in $\mathbb{Z}^n$ such that $q(x) = m$). We recall:

- Jacobi, 1828: if $q(x) = x_1^2 + x_2^2 + x_3^2 + x_4^2$ and m is a positive integer, then

$$r_q(m) = 8 \sum_{\substack{0<d|m \\ d\neq 4k}} d.$$

1.2.3 It is important to notice that two quadratic forms q_1 and q_2 look the same (so we will call them **equivalent**) if they differ from each other by some invertible (linear) change of variables with integer coefficients. We write this equivalence as $q_1 \sim q_2$. Notice that if q_1 and q_2 are equivalent, then $r_{q_1}(m) = r_{q_2}(m)$ for all integers m.

It will also be important to keep track of how many ways a form q is equivalent to itself. We call these self-equivalences the **automorphisms** of q, and collectively refer to them as **group of automorphisms** $\mathrm{Aut}(q)$ of q. To a quadratic form q one usually associates the integral symmetric **Gram matrix** G_q of $q(-,-)$, determined by symmetry and the equation

$$q(x) = \frac{1}{2}x^t G_q x, \quad \text{for all } x \in \mathbb{Z}^n.$$

Notice that two quadratic forms q_1 and q_2 are equivalent precisely when $G_{q_1} = T^t G_{q_2} T$ for some $\mathbb{Z}$-invertible matrix T. Observe that, in particular, in this situation we have $\det(G_{q_1}) = \det(G_{q_2})$.

The quadratic forms we shall consider in this book are **unitary** or **unit forms**, that is, a quadratic form of the shape

$$q(x_1, \ldots, x_n) = x_1^2 + \ldots + x_n^2 + \sum_{i<j} q_{ij} x_i x_j,$$

or equivalently, the diagonal entries of G_q are all equal to 2. These matrices are known as **symmetric quasi-Cartan matrices** (cf. [7]).

1.2.4 Our interest in quadratic forms and their arithmetic properties comes from the study of finite dimensional algebras. For instance, if A is the algebra of upper triangular $n \times n$-matrices with entries in $\mathbb{C}$, then obviously A is a finite dimensional $\mathbb{C}$ algebra. Any A-module is automatically a $\mathbb{C}$-vector space, and so are the sets of A-module homomorphisms $\mathrm{Hom}_A(M, N)$ between two fixed modules. The property of finitely generated modules translates into finite dimensionality of vector spaces over $\mathbb{C}$.

The left module A decomposes into a sum of projective modules, and in our example we have $A = \bigoplus_{i=1}^n P_i$ where $P_i = AE_{ii}$ and E_{ii} is the matrix with a unique non zero entry, located in the i-th row and i-th column. For any left A-module M we define $\mathbf{dim}(M)$ to be the vector whose i-th entry is $\dim_{\mathbb{C}} \mathrm{Hom}_A(P_i, M)$. There we have the vectors which will serve as arguments for our quadratic form. Take $C_{ij} = \dim_{\mathbb{C}} \mathrm{Hom}_A(P_j, P_i)$.

Consider the quadratic form $q(v) = v^t C^{-1} v$ for $C = (C_{ij})_{i,j=1}^n$. It turns out that in the example above we have $q(\mathbf{dim}(M)) = 1$ if and only if M is indecomposable, that is, it cannot be properly decomposed into a direct sum of A-submodules. In our example

$$\begin{aligned} q(x) &= x_1^2 - x_1x_2 + x_2^2 - x_2x_3 + x_3^2 + \cdots + x_{n-1}x_n + x_n^2 \\ &= \frac{1}{2}\left[x_1^2 + (x_1 - x_2)^2 + (x_2 - x_3)^2 + \cdots + (x_{n-1} - x_n)^2 + x_n^2\right] \end{aligned}$$

and thus we see that if $x \neq 0$ then $q(x) > 0$. As we shall see later, this implies that there are only finitely many vectors x with $q(x) = 1$. In conclusion: there are only finitely many possibilities for $[\dim_{\mathbb{C}} \mathrm{Hom}_A(P_i, M)]_{i=1}^n$ if M is indecomposable.

Before going further, we recommend to read the section *Motivation and Problems* of the book *Quadratic Forms. Combinatorics and numerical results* by M.

Barot, J. A. Jiménez G. and J.A. de la Peña [8], where the authors study abstract integral quadratic forms and some arithmetic properties.

1.2.5 Given a non-empty subset $I = \{i_1, ..., i_m\}$ of $\{1, ..., n\}$ we denote by q^I the quadratic form restricted to the variables indexed by the set I, that is, if i_I is the linear map taking a canonical k-th vector in $\mathbb{Z}^I$ to the corresponding canonical vector in $\mathbb{Z}^n$, then $q^I = qi_I$. If $I = \{l, ..., n\} \setminus \{i\}$, we will denote q^I by $q^{(i)}$.

1.2.6 The subset of $\mathbb{Z}^n$ given by

$$\mathrm{rad}(q) = \{y \in \mathbb{Z}^n \mid q(x + y) = q(x) + q(y) \text{ for all } x \text{ in } \mathbb{Z}^n\}$$

is called **radical** of q. Clearly $y \in \mathrm{rad}(q)$ if and only if $y^t G_q = 0$, thus $\mathrm{rad}(q)$ is a subgroup of $\mathbb{Z}^n$. Alternatively, observe that

$$\mathrm{rad}(q) = \{y \in \mathbb{Z}^n \mid q(x) = q(x + y), \text{ for all } x \in \mathbb{Z}^n\}.$$

We say that q is **regular** if $\mathrm{rad}(q) = 0$. If the symmetric matrix G_q of $q : \mathbb{Z}^n \to \mathbb{Z}$ has diagonal block form,

$$G_q = G_1 \oplus G_2 = \begin{bmatrix} G_1 & 0 \\ 0 & G_2 \end{bmatrix},$$

then we say that q **decomposes** as $q = q_1 \oplus q_2$ where $q_1 : \mathbb{Z}^{n_1} \to \mathbb{Z}$ is the form given by $q_1(x_1) = \frac{1}{2}x_1^t G_1 x_1$ (for $x_1 \in \mathbb{Z}^{n_1}$, $0 \leq n_1 \leq n$) and $q_2 : \mathbb{Z}^{n-n_1} \to \mathbb{Z}$ is given by $q_2(x_2) = \frac{1}{2}x_2^t G_2 x_2$ for $x_2 \in \mathbb{Z}^{n-n_1}$. For convenience, we set $q_{ji} = q_{ij}$ for $i < j$. A form $q : \mathbb{Z}^I \to \mathbb{Z}$ is called **disconnected** if there exists a (non-trivial) partition $I = I_1 \cup I_2$ such that $q_{ij} = 0$ for $i \in I_1$ and $j \in I_2$. Otherwise q is called **connected**.

1.2.7 A **bigraph** B we mean a graph (B_0, B_1) admitting loops and multiple-edges where B_1 admits both solid and dotted edges, and such that between any two vertices there are no solid and dotted edges simultaneously (see for instance [27]). We label the elements of B_0, called **vertices**, with a subset of natural numbers (usually $\{1, 2, \ldots, n, \cdots\}$). Define the $n \times n$ (upper triangular) **adjacency matrix** T_B of the bigraph B as follows. Let a_{ij} be the number of edges between vertices i and j.

$$(T_B)_{ij} = \begin{cases} a_{ij}, & \text{if } i \leq j \text{ and any edge between } i \text{ and } j \text{ is solid;} \\ -a_{ij}, & \text{if } i \leq j \text{ and any edge between } i \text{ and } j \text{ is dotted;} \\ 0, & \text{if } i > j. \end{cases}$$

Define the integral quadratic form $q_B : \mathbb{Z}^n \to \mathbb{Z}$ associated to a bigraph B with n vertices as

$$q_B(x) = x^t(\mathbf{Id}_n - T_B)x, \quad \text{for } x \in \mathbb{Z}^n.$$

1.2.8 Now, related to an integral quadratic form $q = \sum_{1 \leq i \leq j \leq n} q_{ij}x_i x_j$ we define a bigraph B_q associated to q as follows. Let $(B_q)_0$ be the set $\{1, \dots, n\}$. For each pair of different vertices i, j, the bigraph B_q contains $|q_{ij}|$ edges connecting i and j. These edges are solid if $q_{ij} \leq 0$ or dotted if $q_{ij} \geq 0$. In addition, B_q has $|1 - q_{ii}|$ solid loops attached to the vertex i if $1 - q_{ii} \geq 0$, otherwise there are $|1 - q_{ii}|$ dotted loops on it. By construction we have $q_{B_q} = q$ and $B_{q_B} = B$, since B has no solid and dotted edges simultaneously between any pair of vertices.

1.2.9 The symmetric Gram matrix G_q of q and the adjacency matrix T_{B_q} of B_q satisfy $T_{B_q} + T^t_{B_q} = 2\mathbf{Id}_n - G_q$. Notice that q is a unit form if and only if the bigraph B_q has no loops.

Observe that if q is an integral quadratic form given by a bigraph without loops and multiple edges, then

$$q(x, e_i) = 2x_i + \sum_{j \neq i} q_{ij}x_j = 2x_i - \sum_{\substack{\text{there is} \\ \text{an edge} \\ i-j}} x_j,$$

where $e_1, \dots, e_n$ is the canonical basis of $\mathbb{Z}^n$. For a vertex i in a bigraph B, denote by $B^{(i)}$ the bigraph obtained from B by removing vertex i and all edges containing it. Observe that $B_q^{(i)} = B_{q^{(i)}}$.

1.3 Geometrical and Linear Aspects of Unit Forms

A nice quotation from P. Etingof, O. Golberg, S. Hensel, T. Liu, A. Schwendner, E. Udovina and D. Vaintrob: Introduction to representation theory. *arXiv:0901.0827*:

> If we needed to make contact with an alien civilization and show them how sophisticated our civilization is, perhaps showing them Dynkin diagrams would be the best choice!

1.3.1 Let $q(x_1, \dots, x_n) : R^n \to R$ be a quadratic form over a ring R of characteristic zero. Consider the canonical basis for R^n, $e_1, \dots, e_n$. We say that $0 \neq x \in R^n$ is a **reflection vector** if $\frac{q(x,e_i)}{q(x)} \in R$ (that is, if there exists $a_i \in R$ such that $q(x)a_i = q(x, e_i)$) for all $i = 1, \dots, n$. For a reflection vector x define the **reflection morphism** $\sigma_x : R^n \to R^n$ at x, by

$$\sigma_x(y) = y - \frac{q(x, y)}{q(x)}x.$$

If q is a unit quadratic form, then all canonical vectors $e_1, \dots, e_n$ are reflection vectors. The reflection σ_{e_i} is called **simple reflection** at i and is denoted by σ_i. Also, if x is a **root** of q (that is, if $q(x) = 1$), then x is a reflection vector. Consider for example the integral quadratic form $q(x_1, x_2) = x_1^2 + x_2^2 - x_1x_2$. Then $v = e_1 - e_2 \in \mathbb{Z}^2$ is a reflection vector with $q(v) = 3$. Observe that $\sigma_v : \mathbb{Z}^2 \to \mathbb{Z}^2$ has matrix $\left[\begin{smallmatrix} 0 & 1 \\ 1 & 0 \end{smallmatrix}\right]$ which cannot be obtained as a product of simple reflections σ_i $(1 \le i \le n)$.

1.3.2 In the following lemma we summarize some basic facts about reflections.

Lemma *Let $x \in R^n$ be a reflection vector.*

(a) The reflection σ_x maps x to $-x$ and acts as the identity in $x^\perp = \{y \in R^n \mid q(x, y) = 0\}$.
(b) $\sigma_x^2 = \mathbf{Id}_{R^n}$.
(c) $q(\sigma_x(y)) = q(y)$ for all $y \in R^n$.
(d) If y is a reflection vector, then $\sigma_x(y)$ is a reflection vector.
(e) Let y be a reflection vector, $x \neq y$, then

$$\sigma_x\sigma_y = \sigma_y\sigma_x \quad \textit{if and only if} \quad q(x, y) = 0.$$

(f) For indices $i \neq j$, we have $\sigma_i\sigma_j = \sigma_j\sigma_i$ if and only if $q_{ij} = 0$.
(g) For indices $i \neq j$, we have $\sigma_i\sigma_j\sigma_i = \sigma_j\sigma_i\sigma_j$ if and only if $q_{ij} \in \{1, -1\}$.

Proof Points (a) and (b) are clear. For (c), observe that

$$q(\sigma_x(y)) = q(y) + \left(\frac{q(x, y)}{q(x)}\right)^2 q(x) - \frac{q(x, y)}{q(x)} q(x, y) = q(y).$$

For (d) and (e) let y be a reflection vector and take $z \in \mathbb{R}^n$. Then

$$q(z, \sigma_x(y)) = q(z, y) - \frac{q(x, y)}{q(x)} q(z, y),$$

where $\frac{q(x,z)}{q(x)} z \in R$ since x is a reflection vector and $q(z, y)$ is divisible by $q(y) = q(\sigma_x(y))$ since y is a reflection vector. This shows (d), and (e) follows from the observation

$$\sigma_x(\sigma_y(z)) - \sigma_y(\sigma_x(z)) = \frac{q(x, y)[q(y, z)x - q(x, z)y]}{q(x)q(y)}.$$

Thus if $x \neq y$, then σ_x and σ_y commute if and only if $q(x, y) = 0$. Point (f) follows from (e), and (g) is left as an exercise. □

Point (c) tells us that reflections are isometries of the quadratic form q. Notice that an isometry σ of q has $\det(\sigma)^2 = 1$.

1.3.3 Let B be a graph (that is, a bigraph with only solid edges) with n vertices. Throughout this section we assume that B is connected and has no loops. In this section q_B will be considered as a function from $V = \mathbb{R}^n$ to $\mathbb{R}$.

A closed subset C of V is a **cone** if it satisfies

(i) $C + C \subset C$, and
(ii) $\lambda C \subset C$, for every $\lambda \geq 0$.

Important examples are the following:

(a) The positive cone $V^+ = \{v \in V \mid v_i \geq 0, \text{ for } i = 1, \dots, n\}$. We write $v \geq 0$ for the elements in V^+, and denote the **boundary** of V^+ by δV^+. If $v \in V^+$ but $v \notin \delta V^+$, we write $v \gg 0$ and $-v \ll 0$.
(b) If $G : V \to V$ is a linear transformation and C is a cone in V, then the inverse image $G^{-1}(C)$ is also a cone in V.
(c) Let C be a cone in V. Consider the dual space $V^* = \mathrm{Hom}_{\mathbb{R}}(V, \mathbb{R})$. The **orthogonal cone** $C^\perp$ in V^* is defined as

$$C^\perp = \{f \in V^* : f(v) \geq 0, \text{ for every } v \in C\}.$$

Let $\phi : V \to V^{**}$ be the natural isomorphism defined by $\phi(v)(g) = g(v)$, for $v \in V$ and $g \in V^*$. One can easily verify that $C^\perp$ is a cone in V^*, and $C^{\perp\perp} = \phi(C)$.

1.3.4 Set $G = G_B$ as a linear transformation $V \to V$, with $G(v) = Gv$.

Lemma *If B is a connected graph, then $G^{-1}(V^+) \cap \delta V^+ = \{0\}$.*

Proof Assume $0 \neq y \in G^{-1}(V^+) \cap \delta V^+$. By connectivity of B there is an edge $i - j$ such that $y_i > 0$ and $y_j = 0$. Then

$$0 \leq G(y)_j = \sum_k a_{jk} y_k = a_{jj} y_j + a_{ji} y_i + \sum_{k \neq i,j} a_{jk} y_k \leq a_{ji} y_i < 0,$$

where $G = (a_{ij})$. □

1.3.5 The next result presents a first division of connected graphs into classes depending on the behavior of the Gram matrix on the positive cone.

Proposition *The Gram matrix G of B satisfies one and only one of the following properties:*

(a) $G^{-1}(V^+) \subset V^+$
(b) $G^{-1}(V^+) = \mathbb{R}u$, *for some vector* $u \gg 0$. *In this case,* $G(u) = 0$.
(c) $G^{-1}(V^+) \cap V^+ = \{0\}$.

Proof Clearly G satisfies one of the conditions (a), (b') and (c), where

(b') $G^{-1}(V^+) \not\subset V^+$ and $G^{-1}(V^+) \cap V^+ \neq \{0\}$.

Moreover, G cannot satisfy simultaneously (a) and (b') or (b') and (c). If both (a) and (c) hold, then $G^{-1}(V^+) = G^{-1}(V^+) \cap V^+ = \{0\}$. In particular, $\mathrm{Ker}(G) \subset G^{-1}(V^+) = \{0\}$, that is, G is an isomorphism satisfying $G^{-1}(V^+) = \{0\}$, which is absurd. Thus, it is enough to show that G satisfies (b) if and only if it satisfies (b').

Assume that (b') holds. Let $v \in G^{-1}(V^+)$ be such that $v \notin V^+$ and let $0 \neq u \in G^{-1}(V^+) \cap V^+$. By Lemma 1.3.4, $u \gg 0$. We shall prove that $G^{-1}(V^+) = \mathbb{R}u$. Let $z \in G^{-1}(V^+)$. If $z \notin V^+$, we find $\lambda > 0$ such that $z + \lambda u \in \delta V^+$. Hence $z + \lambda u \in \delta V^+ \cap G^{-1}(V^+) = \{0\}$ and $z \in \mathbb{R}u$. (In particular, $-v \in \mathbb{R}^+ u$, where $\mathbb{R}^+$ is the set of positive real numbers). If $0 \neq z \in V^+$, a similar argument yields $v \in \mathbb{R}z$. Then $z \in \mathbb{R}u$. Moreover, $\mathbb{R}u = \mathbb{R}v \subset G^{-1}(V^+)$. Since $G(u) \geq 0$ and $G(v) \geq 0$ with $-v \in \mathbb{R}^+ u$, it follows that $G(u) = 0$. This shows that (b') implies (b), and the converse is clear. □

1.3.6 We say that a connected loopless graph B is **elliptic** if G satisfies the condition (a) in Proposition 1.3.5, B is **parabolic** if G satisfies (b), and finally, it is **hyperbolic** if G satisfies (c).

Lemma *Let B be a connected loopless graph with G the Gram matrix of q_B.*

(i) B is elliptic if and only if there exists $u \gg 0$ with $G(u) \gg 0$.
(ii) B is parabolic if and only if there exists $u \gg 0$ with $\mathrm{Ker}(G) = \mathbb{R}u$.
(iii) B is hyperbolic if and only if there exists $u \gg 0$ with $G(u) \ll 0$.

Proof

(i) Assume that B is elliptic. Then $\mathrm{Ker}(G) \subset G^{-1}(V^+) \subset V^+$, but since $\mathrm{Ker}(G)$ is a vector space, $\mathrm{Ker}(G) = 0$. In particular, for any $v \gg 0$, there exists $u \in V$ with $G(u) = v$. Then $u \in V^+$, and by Lemma 1.3.4, $u \gg 0$.

Assume that $u \gg 0$ satisfies $G(u) \gg 0$. Then B does not satisfy (b) or (c) in 1.3.5, thus B is elliptic.

(ii) Suppose that B is parabolic. Then $\mathrm{Ker}(G) \subset G^{-1}(V^+) = \mathbb{R}u$ and $G(u) = 0$ with $u \gg 0$. Therefore, $\mathrm{Ker}(G) = \mathbb{R}u$. If $\mathrm{Ker}(G) = \mathbb{R}u$ with $u \gg 0$, then G does not satisfy (a) or (c), and B is parabolic.

(iii) Assume that B is hyperbolic, that is, $G^{-1}(V^+) \cap V^+ = 0$. Consider the $2n \times n$ matrix

$$M = \begin{pmatrix} \mathbf{Id}_n \\ G \end{pmatrix},$$

with rows $M_1, \ldots, M_{2n}$. Here $\mathbf{Id}_n$ denotes the $n \times n$ identity matrix.

Let C be the cone generated by the column vectors $M_1^t, \ldots, M_{2n}^t$. We shall see that $C = V$. Indeed, if $w \in C^\perp$ (which may be seen as a row vector

$w = (w_1, \ldots, w_n)$), then

$$0 \leq w(M_i^t) = w_i, \quad \text{for } 1 \leq i \leq n.$$

Moreover, by the symmetry of G,

$$G(w^t) = \begin{pmatrix} M_{n+1}w^t \\ \vdots \\ M_{2n}w^t \end{pmatrix} = \begin{pmatrix} w(M_{n+1}^t) \\ \vdots \\ w(M_{2n}^t) \end{pmatrix} \geq 0.$$

Therefore, $w^t \in G^{-1}(V^+) \cap V^+$ and $w = 0$. That is, $C^\perp = 0$, and by 1.3.3 we have $C = V$.

Take any vector $v \ll 0$. There are numbers $\lambda_1, \ldots, \lambda_{2n} \geq 0$ with $v = \sum_{i=1}^{2n} \lambda_i M_i^t$. Let $u' = (\lambda_{n+1}, \ldots, \lambda_{2n})$. Then $v = (\lambda_1, \ldots, \lambda_n) + G(u')$, and therefore $G(u') \ll 0$. By continuity, there exists $u \gg 0$ (close enough to u') with $G(u) \ll 0$. The converse follows from 1.3.5. □

1.3.7 We establish the main result of the section.

Theorem *Let B be a connected loop-less graph.*

(i) B is elliptic if and only if q_B is positive.
(ii) B is parabolic if and only if q_B is non negative and not positive.
(iii) B is hyperbolic if and only if q_B is indefinite.

For the proof of the theorem we need some further considerations. Let $1 \leq i \leq n$. By $G^{(i)}$ we denote the **principal submatrix** of G which is obtained from G by deleting the i-th row and i-th column. Clearly $G^{(i)} = G_{B^{(i)}}$, where $B^{(i)}$ is the full subgraph of B obtained by deleting the vertex i. In general, if B' is a full subgraph of B, we notice that G' is a submatrix of $G = G_B$ (obtained by deleting the corresponding rows and columns). The determinant $\det(G^{(i)})$ is called a **principal minor** of G.

Lemma *If B is an elliptic or parabolic graph, then every proper connected full subgraph of B is elliptic.*

Proof Follows easily from 1.3.5 and 1.3.6 and the comments above. □

1.3.8 Proof of Theorem 1.3.7 (i) We show that B elliptic implies $q = q_B$ is positive by induction on the number n of vertices. If $n = 1$ the result is clear, thus we may suppose that $n > 1$.

By Lemma 1.3.7 and induction hypothesis, the quadratic form $q^{(i)}$ is positive for any index i. In particular, all the proper principal minors of G are positive, and therefore it is enough to show that $\det(G) > 0$ (see for instance [8, Proposition 1.32]).

Assume that $\det(G) \leq 0$. As is well known, if $\mathrm{adj}(G)$ is the adjoint (or adjunction) matrix of G, we have (cf. [19, p. 89]),

$$G[\mathrm{adj}(G)] = (\det G)\mathbf{Id}_n.$$

Let M_i be the i-th column of $\mathrm{adj}(G)$. Then $G(M_i) \leq 0$, that is, $-M_i \in G^{-1}(V^+)$ which is contained in V^+. But the i-th entry of M_i is $\det(G^{(i)}) > 0$, a contradiction.

Conversely, it is clear that the quadratic form q_B associated with a parabolic or hyperbolic graph is not positive. Therefore, if q_B is positive then Q is elliptic.

(ii) Assume that B is parabolic and let $u \geq 0$ be such that $G(u) = 0$. Then $\det(G) = 0$. By the lemma above $\det(B) > 0$ for every proper principal submatrix B of G. Then q_B is non-negative(see [8, Proposition 1.33]).

The rest of the proof is clear.

1.3.9 It is easy to construct the graphs corresponding to each type of the above classification 1.3.6.

(a) Let B be a parabolic graph, that is, there is a vector $u \gg 0$ with $G(u) = 0$. These vectors satisfy the linear equations

$$uGe_x{}^t = q(u, e_x) = 0, \text{ for all vertices } x,$$

which provides a simple method to construct the vector u and the actual graph B (see [25]). As shown in [5, 25] (see also [8]), B is one of the following **extended Dynkin diagrams** (also known as **Euclidean diagrams**). We label the vertices with the entries of such minimal u with integral coefficients:

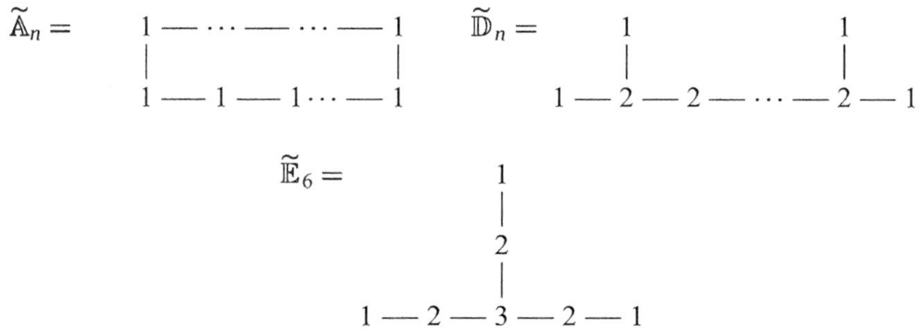

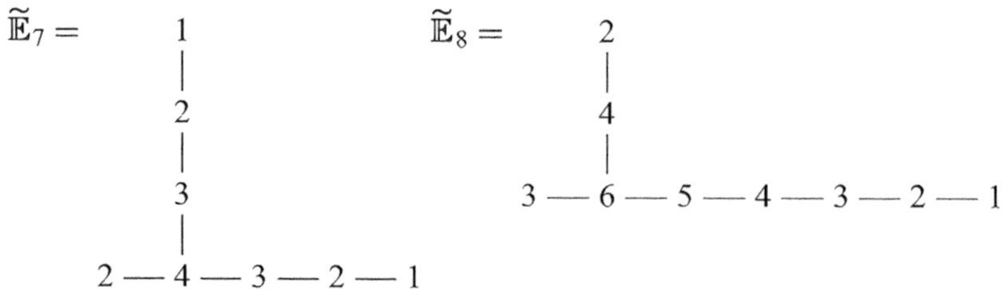

(b) Let B be an elliptic graph. Then it has no parabolic subgraphs (by Lemma 1.3.7), and it is easy to check that B is one of the following **Dynkin diagrams**. We label the vertices with the entries of a maximal root u of q_B:

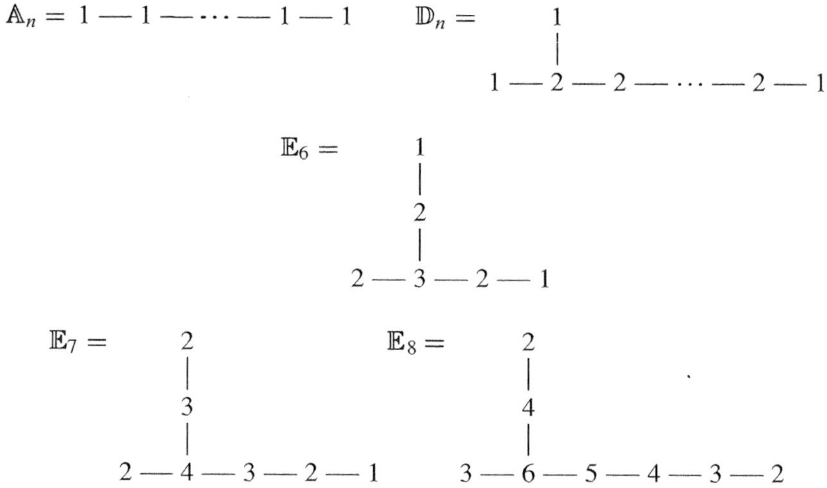

(c) If B is not a Dynkin nor a Euclidean diagram, then it is a hyperbolic graph.

1.4 Fundamental Examples

1.4.1 Much of linear algebra may be formulated for modules over a division ring D instead of vector spaces over a field. Doing so it must be specified whether one is considering right or left modules, and some care is needed in properly distinguishing left and right in formulas. The dual of a right module is a left module, and vice versa. The transpose of a matrix must be viewed as a matrix over the opposite division ring D^{op} in order for the rule $(AB)^t = B^t A^t$ to remain valid.

Every module over a division ring is free (that is, has a basis), and all bases of a module have the same number of elements. Linear maps between finite dimensional modules over a division ring can be described by matrices, and the Gaussian elimination algorithm remains applicable (even though we do not meet applications in this book). The column rank of a matrix is the dimension of the right module generated by the columns, and the row rank is the dimension of the left module generated by the rows; the same proof as for the vector space case can be used to show that these ranks are the same, and define the rank of a matrix. The following is important (see [23]):

1.4.2 Theorem (Frobenius) *The only finite dimensional associative division algebras over $\mathbb{R}$ are $\mathbb{R}$ itself, the complex numbers $\mathbb{C}$ and the quaternions $\mathbb{H}$.*

Many of the algebras considered in the book are triangular. By definition, a finite dimensional algebra is called **triangular** if it has triangular matrix shape

$$A = \begin{bmatrix} M_{1,1} & M_{1,2} & \cdots & M_{1,n} \\ 0 & M_{2,2} & \cdots & M_{2,n} \\ & & \ddots & \vdots \\ 0 & 0 & \cdots & M_{n,n} \end{bmatrix},$$

where the diagonal entries $D_i := M_{i,i}$ are division rings and the off-diagonal entries $M_{i,j}$, $j > i$, are $D_i - D_j$-bimodules. Each triangular algebra has finite global dimension.

If M is an $R-S$-bimodule and L is an $T-S$-bimodule, then the set $\mathrm{Hom}_S(M, L)$ of all S-module homomorphisms from M to L becomes a $T - R$-bimodule in a natural fashion. These statements extend to the derived functors Ext^k and Tor_k.

1.4.3 The Artin–Wedderburn theorem states that a semi-simple algebra that is finite dimensional over a field K, is isomorphic to a finite product $\prod M_{n_i}(D_i)$ where the n_i are natural numbers, the D_i are finite dimensional division algebras over K (possibly finite extension fields of K), and $M_{n_i}(D_i)$ is the algebra of $n_i \times n_i$-matrices over D_i. Again, this product is unique up to permutation of the factors.

If D is a finite dimensional division algebra over an algebraically closed field K, then $K = D$. Indeed, if $x \in D$, then consider the inverse closed subring of D generated by K and x. This is a finite, hence algebraic, extension of K, hence it must be equal to K.

The dimension-matrix $C_A = (m_{i,j})$ of a triangular algebra $A = [M_{i,j}]$ as above, is the integral matrix with entries $m_{i,j} = \dim_K M_{i,j}$ for all i and j. That is, C_A is an upper triangular matrix, with diagonal $m_{i,i} = 1$ if K is algebraically closed, called the **Cartan matrix** of A.

1.4.4 The algebra of polynomials in one variable, $A = K[t]$, will be of great importance in our discussion. This is an infinite dimensional K-algebra with unity. An A-module is a K-vector space V with action of A determined (by linearity) by the action of the variable t, which corresponds to an endomorphism f of V. Therefore, modules correspond to pairs (V, f) with $f : V \to V$ linear, called representation of A.

A morphism from a pair (V, f) to another pair (W, g) is a linear map $u : V \to W$ such that $uf = gu$. In particular, the endomorphisms of the pair (V, f) are exactly the endomorphisms of V that commute with f.

Given a representation (V, f) having a prescribed finite dimension n, we may choose a basis $(v_1, ..., v_n)$ of V, and hence identify f with an $n \times n$ matrix F. Choosing another basis amounts to replacing F with a conjugate BFB^{-1}, where B is an invertible matrix.

1.4.5 It follows that the isomorphism classes of n-dimensional representations of $A = K[t]$ correspond bijectively to the conjugacy classes of $n \times n$ matrices. These are classified in terms of the Jordan canonical form. Consider the matrix

$$M = \begin{pmatrix} 1 & 0 & 0 \\ 1 & 2 & 0 \\ 1 & 0 & 2 \end{pmatrix}.$$

We find the eigenvalues of M as the solutions of the equation $\det(M - \lambda I) = 0$, hence:

$$\lambda_1 = 1, \quad \lambda_2 = 2, \quad \lambda_3 = 2.$$

That is, we have a single root and a double root eigenvalue (we shall see that the latter has algebraic multiplicity one). To find the eigenvectors, we set up and solve

$$[M - \lambda_i I]v_i = 0.$$

We get: $v_1 = (1, 0, 0)$, $v_2 = (1, 0, 1)$ and $v_3 = (1, 1, 0)$. Finally, if we write P using the eigenvectors as columns, we get the **Jordan Normal Form** J, using the corresponding eigenvalues:

$$J = P^{-1}MP.$$

The Jordan Normal Form is an arrangement of blocks (cf. [19]), where $J(\lambda_i)$ is the $n \times n$ block

$$J(\lambda) = \begin{pmatrix} \lambda & 1 & 0 & & 0 \\ & \lambda_1 & 1 & & \\ & & \lambda_1 & & \\ & & & \ddots & \\ 0 & & & & \lambda_k \end{pmatrix}.$$

See Brechenmacher [10] for an interesting historical account of the controversy between C. Jordan and L. Kronecker, in relation to their corresponding techniques to classify linear transformations up to similarity (the former using canonical reduction, the latter computing Weierstrass's elementary divisors).

1.4.6 Theorem (Jordan) *Let $f : V \to V$ be a linear transformation with V finite dimensional. Assume that K is algebraically closed. Then there exists a basis of V in which f has the form:*

$$J_{m_1}(\lambda_1) \oplus \cdots \oplus J_{m_k}(\lambda_k),$$

this is a Jordan block decomposition where $(t-\lambda_1)^{m_1} \cdots (t-\lambda_k)^{m_k}$ is the **minimal polynomial** *of f.*

In particular, there are infinitely many isomorphism classes of representations of the polynomial algebra having a prescribed dimension.

More generally, for any integer $r \geq 1$ consider the algebra of polynomials $A = K\langle t_1, \ldots, t_r\rangle$. The representations of A consist of a vector space V equipped with r endomorphisms f_1, ..., f_r. Thus, the isomorphism classes of representations of A having a prescribed dimension correspond bijectively to the r-tuples of $n \times n$ matrices up to simultaneous conjugation.

1.4.7 Let $S(r)$ denote the group of permutations of r elements (thus, the order of $S(r)$ is $r!$). A representation M of $S(r)$ consists of $r+1$ vector spaces V_1, ..., V_r, W together with r linear maps $f_i : V_i \to W$. By associating with M the images of the f_i, one obtains a bijection between the isomorphism classes of representations with dimension vector $(m_1, ..., m_r, n)$, and the orbits of the general linear group $G\ell(n)$ acting on r-tuples $(E_1, ..., E_r)$ of subspaces of K^n such that $\dim(E_i) = m_i$, for all i, via the action

$$\dot{g}(E_1, ..., E_r) = (g(E_1), ..., g(E_r)).$$

In other words, classifying representations of $S(r)$ is equivalent to classifying r-tuples of subspaces of a fixed vector space.

When $r = 1$, one recovers the classification of representations of K restricted to $S(1)$. When $r = 2$, one easily checks that the pairs of subspaces (E_1, E_2) of K^n are classified by the triples

$$(\dim(E_1), \dim(E_2), \dim(E_1 \cap E_2)),$$

that is, by those triples $(a, b, c) \in \mathbb{Z}^3$ such that $0 \leq c \leq \min(a, b)$. In particular, there are only finitely many isomorphism classes of representations having a prescribed dimension vector. This finiteness property may still be proved in the case where $r = 3$, but fails whenever $r = 4$. Indeed, consider the representations with dimension vector $(1, 1, ..., 1, 2)$, such that the maps f_1, ..., f_r are all non-zero. The isomorphism classes of these representations are in bijection with the orbits of the projective linear group $PGL(2)$ acting on the product $\mathbb{P}_1(K) \times \ldots \times \mathbb{P}_1(K)$ of r

copies of the projective line. Since $r = 4$, there are infinitely many orbits; an explicit infinite family is provided by the representations

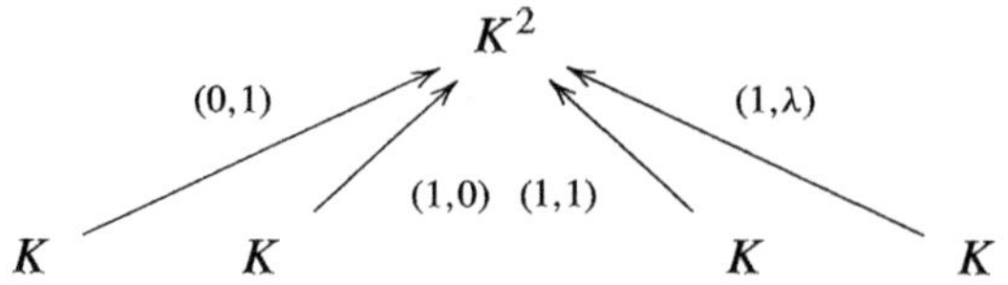

which are indecomposable and pairwise non-isomorphic.

1.4.8 The r-th Kronecker algebra $\mathbf{K}_r$ is the triangular algebra of the form

$$\mathbf{K}_r = \begin{pmatrix} K & K^r \\ 0 & K \end{pmatrix}.$$

The representations of $\mathbf{K}_r$ consist of two vector spaces V, W together with r linear maps $f_1, \ldots, f_r : V \to W$. The dimension vectors are pairs of non-negative integers.

As in the preceding example, the isomorphism classes of representations with dimension vector (m, n) correspond bijectively to the r-tuples of $n \times m$ matrices, up to simultaneous multiplication by invertible $n \times n$ matrices on the left, and by invertible $m \times m$ matrices on the right.

When $r = 1$, these representations are classified by the rank of a unique $n \times m$ matrix. In particular, they form only finitely many isomorphism classes. In the case $r = 2$, the classification of representations of $\mathbf{K}_2$ contains that of the algebra of polynomials $A = K[t]$.

1.4.9 By **quiver** we mean a directed multi-graph (admitting loops and multiple edges). For convenience, we consider a quiver Q as a tuple $Q = (Q_0, Q_1, \mathbb{s}, \mathbb{t})$ where Q_0 and Q_1 are the sets of **vertices** and **arrows** of Q, and for an arrow a in Q, $\mathbb{s}(a)$ and $\mathbb{t}(a)$ are respectively the **source vertex** and **target vertex** of a (also referred to as starting point and final point of a). We use the notation $a : \mathbb{s}(a) \to \mathbb{t}(a)$. For example, the quiver with vertices $\{1, 2, 3\}$ and arrows $a : 1 \to 2$, $b : 2 \to 1$, $c : 2 \to 3$ and $d : 3 \to 2$, is depicted as follows:

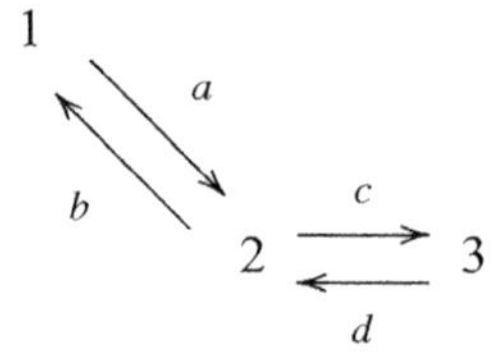

A non-trivial path in Q is a sequence of arrows $\alpha = a_1 a_2 \cdots a_m$ satisfying $\mathbb{s}(a_i) = \mathbb{t}(a_{i+1})$ for all $i = 1, \ldots, m-1$. We also call $\mathbb{s}(\alpha) := \mathbb{s}(a_m)$ and $\mathbb{t}(\alpha) = \mathbb{t}(a_1)$ the source and target of α respectively. We say that such a path α has length m. By $\mathbf{e}_i$ we will denote the trivial path which starts and ends at i. Note that if α and β are walks in Q with $\mathbb{t}(\alpha) = \mathbb{s}(\beta)$, then the concatenation $\beta\alpha$ is again a walk in Q.

The **path algebra** $A = KQ$ of a quiver Q is the K-algebra with basis the paths in Q, and with the product of two paths α and β given by the concatenation $\beta\alpha$ if $\mathbb{t}(\alpha) = \mathbb{s}(\beta)$, and zero otherwise. Notice that for a vertex i of Q, the trivial path e_i is an idempotent element of Q, and if Q has finitely many vertices, the identity element of KQ is given by $1 = \sum_{i \in Q_0} \mathbf{e}_i$ (we say that $\mathbf{e}_1, \ldots, \mathbf{e}_n$ is a **complete system of primitive orthogonal idempotents** of KQ). The left A-module $A\mathbf{e}_i$ has basis those paths starting at i. It is called the **projective module** at i. Clearly, KQ is finite dimensional if and only if Q has no oriented cycle. More generally, for a finite dimensional algebra A with fixed complete system of orthogonal idempotents $\mathbf{e}_1, \ldots, \mathbf{e}_n$, we have $A = \bigoplus_{i=1}^n A\mathbf{e}_i$, and $A_1, \ldots, A_n$ are projective A modules. In that case, the A-modules $A\mathbf{e}_1, \ldots, A\mathbf{e}_n$ are a complete set of representatives of projective A-modules.

For instance, the algebra of polynomials $K[t]$ is isomorphic to the path algebra KL where L is the **loop quiver** (having exactly one vertex and one arrow, called a loop of L), and the algebra $K\langle t_1, \ldots, t_r\rangle$ is isomorphic to KL_r where L_r is the quiver with one vertex and r loops. The r-th Kronecker algebra $\mathbf{K}_r$ is isomorphic to the path algebra $K\mathbb{K}_r$, where $\mathbb{K}_r$ is the connected quiver with two vertices and r arrows in the same direction (the **Kronecker quiver**).

1.4.10 The indecomposable modules over the quiver algebra $A = K\mathbb{K}_2$:

$$\mathbb{K}_2 = \quad \bullet \rightrightarrows \bullet$$

were classified by Weierstrass and Kronecker in the following families:

$$K^{n+1} \underset{\begin{bmatrix} 0 & \\ \vdots & \mathbf{Id}_n \\ 0 & \end{bmatrix}}{\overset{\begin{bmatrix} & 0 \\ \mathbf{Id}_n & \vdots \\ & 0 \end{bmatrix}}{\rightrightarrows}} K^n \qquad\qquad K^n \underset{\begin{bmatrix} 0 \cdots 0 \\ \mathbf{Id}_n \end{bmatrix}}{\overset{\begin{bmatrix} \mathbf{Id}_n \\ 0 \cdots 0 \end{bmatrix}}{\rightrightarrows}} K^{n+1}$$

(preinjective representation) (post-projective representation)

$$R_n(\lambda): K^n \underset{J_n(\lambda)}{\overset{\mathbf{Id}_n}{\rightrightarrows}} K^n \qquad\qquad R_n(\infty): K^n \underset{\mathbf{Id}_n}{\overset{J_n(0)}{\rightrightarrows}} K^n$$

(regular representations)

with $\lambda \in K$, where $J_n(\lambda)$ is the Jordan block of size n and eigenvalue λ. It is simple to verify that these modules are indecomposable.

1.5 Some Categorical Notions

We remind the reader some basic concepts involving categories and functors, fixing some notation along the section.

1.5.1 Recall that a category where Hom-spaces are abelian groups and the composition is bilinear, is called a **preadditive category**. A functor $F : C_1 \to C_2$ between preadditive categories is **additive** if F is linear in the set of morphisms (all functors between preadditive categories will be assumed to be additive). The functor F is **fully faithful** if F is an isomorphism of abelian groups on every set of morphisms, and F is **dense** (or **essentially surjective**) if for every object $X \in C_2$, there is some $Y \in C_1$ such that X and $F(Y)$ are isomorphic.

1.5.2 Definition Let C_1 and C_2 be two categories, and let $F, G : C_1 \to C_2$ be two functors. A (functorial) morphism $g : F \to G$ is a natural transformation, that is, a collection of maps

$$g_X : F(X) \to G(X),$$

for each $X \in C_1$ such that for any morphism $f : X \to Y$ with ($Y \in C_1$), the following diagram commutes:

$$\begin{array}{ccc} F(X) & \xrightarrow{g_X} & G(X) \\ {\scriptstyle F(f)}\downarrow & & \downarrow{\scriptstyle G(f)} \\ F(Y) & \xrightarrow[g_Y]{} & G(Y) \end{array}$$

If each map g_X is an isomorphism, then g is also called a (functorial) isomorphism (written $G \cong G$). By **equivalence of categories** we mean a (covariant) functor $F : C_1 \to C_2$ such that there is a functor $G : C_2 \to C_1$ satisfying $\mathbf{Id}_{C_1} \cong GF$ and $\mathbf{Id}_{C_2} \cong FG$. Recall that a fully faithful and dense contravariant functor is also called a **duality**.

The most common way to establish an equivalence of categories is provided by the following result (cf. [2, Theorem 2.5, Appendix A] and [5, Theorem 2.5, Ch. II]),

Lemma *Let C_1 and C_2 be two preadditive categories, and let $F : C_1 \to C_2$ be a fully faithful, dense functor. Then F is an equivalence of categories.*

1.5.3 Let Ab be the category of abelian groups. For any preadditive category C, the (additive) covariant functors from C to Ab form a category $\mathrm{Add}(C, \mathrm{Ab})$. Observe that any object $X \in C$, gives rise to the functor $\mathrm{Hom}_C(X, -) : C \to \mathrm{Ab}$. It is straightforward to check that this assignment determines a natural contravariant functor $C \to \mathbf{Fun}(C, \mathrm{Ab})$.

1.5.4 A category C is **additive** if:

(A1) For every pair of objects X and Y, the morphism set $\mathrm{Hom}_C(X, Y)$ is an abelian group such that the composition maps

$$\mathrm{Hom}_C(Y, Z) \times \mathrm{Hom}_C(X, Y) \to \mathrm{Hom}_C(X, Z),$$

are bilinear (that is, C is preadditive).

(A2) There is a zero object 0 in C (such that the identity $\mathbf{Id}_0$ is the zero in $\mathrm{End}_C(0)$).

(A3) Every pair of objects X, Y admits a biproduct $X \times Y$.

Recall that in a preadditive category C, a **biproduct** of objects $X_1, \ldots, X_r$ is an object

$$X = X_1 \times \cdots \times X_r,$$

together with morphisms $\iota_i : X_i \to X$ and $\pi_i : X \to X_i$ for all $1 \leq i \leq r$, satisfying

$$\sum_{i=1}^{r} \iota_i \pi_i = \mathbf{Id}_X, \quad \text{and} \quad \pi_i \iota_j = \delta_{ij}(\mathbf{Id}_{X_i}), \quad \text{for all } 1 \leq i, j \leq r,$$

where δ_{ij} denotes the Kronecker delta.

Lemma *In an additive category C, the above morphisms ι_i and π_i induce isomorphisms:*

$$X_1 \times \ldots \times X_r \cong \bigoplus_{i=1}^{r} X_i \cong \prod_{i=1}^{r} X_i,$$

where $\bigoplus$ and $\prod$ denote the categorical coproduct and product.

Proof A morphism $X \to Y$ in C is an isomorphism if it induces for each object $A \in C$ an isomorphism $\mathrm{Hom}_C(A, X) \to \mathrm{Hom}_C(A, Y)$ of abelian groups. The covariant functor $\mathrm{Hom}_C(A, -)$ sends the direct sum $\bigoplus_{i=1}^r X_i$ in C to a direct sum $\bigoplus_{i=1}^r \mathrm{Hom}_C(A, X_i)$ of abelian groups. It is a standard fact that finite direct sums and products of abelian groups are isomorphic. Thus the following composite is in fact an isomorphism.

$$\bigoplus_{i=1}^{r} \mathrm{Hom}_C(A, X_i) \cong \mathrm{Hom}_C(A, \bigoplus_{i=1}^{r} X_i) \cong \mathrm{Hom}_C(A, \prod_{i=1}^{r} X_i) \cong \prod_{i=1}^{r} \mathrm{Hom}_C(A, X_i).$$

This establishes the first isomorphism; the other isomorphism follows by symmetry.

□

1.5.5 In particular, the product $X_1 \times \ldots \times X_r$ solves a universal problem, and is therefore unique up to a unique isomorphism.

Let $X = \bigoplus_{i=1}^r X_i$ and $Y = \bigoplus_{i=1}^s Y_i$ be two direct sums. Then we get

$$\mathrm{Hom}_C(X, Y) \cong \bigoplus_{\substack{i=1,\ldots,r \\ j=1,\ldots,s}} \mathrm{Hom}_C(X_i, Y_j),$$

and therefore each morphism $f : X \to Y$ can be written uniquely as a $r \times s$ matrix $f = (f_{i,j})$ with entries $f_{i,j} = \pi_j f \iota_i$ in $\mathrm{Hom}_C(X_i, Y_j)$ for all pairs i, j. An object in an additive category is called **indecomposable** if it is not isomorphic to a direct sum of non-zero objects.

Given an object X in an additive category, we denote by add(X) the full subcategory consisting of all finite direct sums of copies of X and their direct summands. This is the smallest **additive subcategory** which contains X and is closed under direct summands.

1.5.6 An additive category $\mathscr{A}$ is **abelian**, if every morphism $f : X \to Y$ in $\mathscr{A}$ has kernel and cokernel, and the canonical factorization of f,

$$\begin{array}{ccccccccc}
0 \longrightarrow \mathrm{Ker}(f) & \xrightarrow{\mathrm{ker}(f)} & X & & \xrightarrow{\quad f \quad} & & Y & \xrightarrow{\mathrm{coker}(f)} & \mathrm{CoKer}(f) \longrightarrow 0\ , \\
 & & {\scriptstyle \mathrm{coker}(\mathrm{ker}(f))}\downarrow & & & & \uparrow{\scriptstyle \mathrm{ker}(\mathrm{coker}(f))} & & \\
 & & \mathrm{CoKer}(\mathrm{ker}(f)) & & \xrightarrow[\theta]{} & & \mathrm{Ker}(\mathrm{coker}(f)) & &
\end{array}$$

induces an isomorphism θ.

An object P of an abelian category $\mathscr{A}$ is said to be **projective** if the functor $\mathrm{Hom}_{\mathscr{A}}(P, -) : \mathscr{A} \to \mathrm{Ab}$ is exact. We say that a subcategory of $\mathscr{A}$ is **fully exact** if it is full and closed under the formation of isomorphic objects, kernels, cokernels, direct sums and direct products.

1.5.7 Let C be an additive category with arbitrary direct sums. For any set S, and an object G of C we denote by G^S the direct sum of copies of G indexed by the set S. An object G of C is called a **generator** if $\mathrm{Hom}_C(G, A)$ is non-zero for every nonzero object $A \in C$.

Proposition *Let A be a K-algebra. Consider a module V in A-Mod and $\overline{V}$ the intersection of all fully exact subcategories of A-Mod containing V. Let W be the intersection of all submodules of V^V in $\overline{V}$ containing the element $w = (w_v)_{v \in V}$, where $w_v = v$. Then W is a generator of $\overline{V}$.*

Proof Given a full subcategory F of A-Mod, we denote by SF (resp. HF) the full subcategory of $\overline{V}$ formed by all submodules (resp. homomorphic images) of modules in F. Let P be the full subcategory of $\overline{V}$ formed by all powers V^I. Obviously HSP is fully exact. So $\overline{V} = HSP$.

Let $G \in \overline{V}$ and $g \in G$. By the above, there is a morphism $q : T \to G$ in $\overline{V}$, where $T \subset V^I$ for some I. Let $t = (t_i)_{i \in I}$ be an inverse image of g in T and $f : V^V \to V^I$ the morphism whose i-th component is the t_i-th projection $V^V \to V$. Clearly, $f(w) = t$, hence $W \subset f^{-1}(T)$. So each $g \in G$ is the image of w under some $W \to G$. □

The above result makes direct use of techniques from **universal algebra**, which is a little surprise. Let C be an additive category with arbitrary direct sums. An object X of C is called **compact** if, for an arbitrary set of objects $\{M_i\}$ of C and a morphism $f : X \to \bigoplus_{i \in I} M_i$, there exists a finite subset $J \subset I$ such that the image $\mathrm{Im}(f)$ is contained in $\bigoplus_{i \in J} M_i$.

1.5.8 An easy consequence of the definition of compactness is the following

Lemma *An object X in an abelian category $\mathscr{A}$ is compact if and only if the functor $\mathrm{Hom}_{\mathscr{A}}(X, -)$ commutes with arbitrary direct sums.*

1.5.9 Lemma *Let A be a K-algebra and M an A-module. Then:*

(i) If M is a finitely generated A-module, then M is a compact object of A-mod.
(ii) If M is projective and a compact object of A-mod, then M is finitely generated.

1.5.10 The following result provides a useful criterion for an abelian category to be equivalent to the category of left modules over a ring.

Proposition *Let $\mathscr{A}$ be an abelian subcategory of A-Mod. Then there exists a compact projective generator P of $\mathscr{A}$ and for $B = \mathrm{End}_{\mathscr{A}}(P)^{op}$, the functor $\mathrm{Hom}_{\mathscr{A}}(P, -)$ yields an equivalence of categories between $\mathscr{A}$ and B-mod.*

Proof The existence of the projective generator P is given by 1.5.7. We will show that the functor $F = \mathrm{Hom}_{\mathscr{A}}(P, -)$ is fully faithful and then apply Lemma 1.5.2. First, we show that F is faithful: suppose $f : X \to Y$ is such that $F(f) = 0$. Since P is a generator, we find a $0 \neq g \in \mathrm{Hom}_{\mathscr{A}}(P, X)$. Consider the submodule

$L = \mathrm{Im}(g)$ of X, if it is not all of M then X/L is nonzero, hence there is some nonzero map $P \to X/L$. But this map lifts to a map to X, and the image of this lift must include points not in L, which is a contradiction. So, every A-module X can be written as a quotient of a direct sum P^S for some set S and an onto map $P^S \to X$, but then $F(f) \neq 0$. Now we show that F is full: let X be an A-module and $P^S \to X$ an onto map. Since P is compact, then

$$\mathrm{Hom}_{\mathscr{A}}(P, P)^S \to \mathrm{Hom}_{\mathscr{A}}(P, X) \to 0.$$

is exact. Let $L = \mathrm{Im}(f)$ which has length smaller than X. We can assume that $F(Y) = L$ such that $0 \longrightarrow L \longrightarrow X \xrightarrow{p} X/L \longrightarrow 0$ is exact and X/L is quotient of P^T for some set T. Lifting p to $\bar{p} : P^T \to X$ we get the following exact and commutative diagram

$$\begin{array}{ccccccccc}
0 & \longrightarrow & K & \xrightarrow{i} & P^T & \longrightarrow & X & \longrightarrow & 0\,. \\
 & & \downarrow & & \downarrow{\scriptstyle \bar{p}} & & \downarrow{\scriptstyle p} & & \\
0 & \longrightarrow & L & \longrightarrow & X & \xrightarrow{p} & X/L & \longrightarrow & 0
\end{array}$$

Since $\kappa : K \to L$ and p are surjective, then the **5-lemma** asserts that $\bar{p}$ is surjective, that is, F is full. □

Define the **center** $\mathbb{Z}(C) = \mathrm{End}(\mathbf{Id}_C)$ of an additive category C where $\mathbf{Id}_C : C \to C$ is the identity functor.

Lemma *Let A be a K-algebra. Then the center of* A-mod *is the center of A,* $\mathbb{Z}(A\text{-mod}) = \mathbb{Z}(A)$.

Proof Choose an element $z \in \mathbb{Z}(A)$ and take $C = A$-mod. Define an endomorphism of $\mathbf{Id}_C : C \to C$ by setting g_X to be the action of z on the module X. Since z is central, this is a homomorphism and it commutes with all module maps, that is, the diagram commutes,

$$\begin{array}{ccc}
X & \xrightarrow{g_X} & X \\
{\scriptstyle f}\downarrow & & \downarrow{\scriptstyle f} \\
Y & \xrightarrow[g_Y]{} & Y.
\end{array}$$

Conversely, suppose an endomorphism $g : \mathbf{Id}_C \to \mathbf{Id}_C$ is given. Set $z = g_A(1_A)$. We only need to check that this element is indeed central. □

1.5.11 Recall that if Q is a $B - A$-bimodule for two K-algebras A and B, then the functor $Q \otimes_A (-) : A$-mod $\to B$-mod is an additive right exact covariant functor (see for instance [1]). The converse also holds, as can be easily shown.

Theorem *Let A and B be two K-algebras, and $F : A\text{-mod} \to B\text{-mod}$ an additive right exact (covariant) functor. Then there exists a $B-A$-bimodule Q, unique up to isomorphism, such that F is isomorphic to the functor from A-mod to B-mod, sending $M \in A$-mod to $Q \otimes_A M$.*

Proof Take the left B-module $Q = F({}_AA)$ with structure of right A-module as follows. For a fixed $a \in A$ and $m \in Q$, considering the morphism $f_a : {}_AA \to {}_AA$ given by $x \mapsto xa$, take

$$m \cdot a = F(f_a)(m).$$

By additivity and covariance of F, this is a well-defined action of A on Q. Moreover, ${}_BQ_A$ is indeed a bimodule (that is, both left and right actions are compatible), since $F(f_a)$ is a morphism of left B-modules.

For any module M in A-mod, there is an exact sequence of the form

$$A^m \longrightarrow A^n \longrightarrow M \longrightarrow 0,$$

that is, M is **finitely presented**. Again by additivity and right exactness of F, and since $Q \otimes_A -$ is also a right exact functor, we have exact sequences

$$\begin{array}{ccccccc}
Q^m & \longrightarrow & Q^n & \longrightarrow & F(M) & \longrightarrow & 0 \\
\downarrow \cong & & \downarrow \cong & & \downarrow \psi_M & & \\
Q \otimes_A A^m & \longrightarrow & Q \otimes_A A^n & \longrightarrow & Q \otimes_A M & \longrightarrow & 0.
\end{array}$$

The exactness of the rows, and the two natural isomorphisms of the columns, determine a family of isomorphisms $(\psi_M)_M$, and hence an isomorphism between the functors $F(-)$ and $Q \otimes_A (-)$. For uniqueness, if Q' is any other $B-A$-bimodule with $F(-)$ isomorphic to $Q' \otimes_A (-)$, then

$$Q = F({}_AA) \cong Q' \otimes_A A \cong Q',$$

as B-modules, and clearly such an isomorphism is also of $B-A$-bimodules. □

1.5.12 Here we consider **locally finite quivers** (that is, quivers with possible infinite vertices, such that each vertex has a finite number of arrows incident to it). The **path category** of a quiver Q over a field K, also denoted by KQ, is the preadditive category having as objects the vertices of Q, and for vertices x and y the set of morphisms $KQ(x, y)$ is the free K-module generated by the paths from x to y, with composition determined by concatenation of paths.

A **representation** of a locally finite quiver Q is an additive functor $V : KQ \to K$-Mod from the path category KQ to the vector space category K-Mod (we denote the category of such functors by REP(Q)). If V can be restricted to a functor $KQ \to K$-mod we say that V is **locally finite dimensional**, and if in addition the support of V is finite, then V is **finite dimensional**. The corresponding full subcategories of REP(Q) will be denoted by Rep(Q) and rep(Q) respectively.

Notice that a morphism $f : V \to W$ between representations of the quiver Q is a collection of linear maps $f_x : V(x) \to W(x)$ (one for each vertex x of Q) such that for every arrow a in Q from x to y we have

$$W(a)f_x = f_y V(a).$$

A morphism f is an **isomorphism**, if f_x is invertible for all vertices x in the quiver.

Given two morphisms $f : V \to W$ and $g : W \to Z$ we have $gf : V \to Z$ defined by the pointwise composition

$$(gf)_{x\in Q_0} = (g_x)(f_x)_{x\in Q_0}.$$

1.5.13 The following claim is straightforward:

Proposition *If Q is a finite quiver, then the category* Rep(Q) *is equivalent to* KQ-Mod, *and this equivalence restricts to an equivalence between* rep(Q) *and* KQ-mod.

Consider the ideal $K^m Q$ of a path algebra KQ generated by all the paths of length m. Let I be an ideal in the path algebra KQ. We say that an ideal $I \subset KQ$ is **admissible** if it is contained in the ideal $K^2 Q$, and there is an $m \geq 2$ such that $K^m Q$ is contained in I. In that case, the quotient KQ/I is a basic finite dimensional K-algebra. Most of the algebras we shall consider are of the form $A = KQ/I$ with I and admissible ideal. In that case, for any **cycle** of arrows in Q (that is, a path $\alpha = \beta_1 ... \beta_s$ with $\mathbb{s}(\alpha) = \mathbb{t}(\alpha)$, the corresponding class $\overline{\alpha}$ in the quotient KQ/I are such that the power $(\overline{\alpha})^k$ is zero. For the following well know result see [2, 6]. Notice that an algebra $A = KQ/I$ as above, with I an admissible ideal, is finite dimensional over K, and therefore satisfies both chain conditions.

1.5.14 The reasons to consider algebras KQ/I are not only due to their simplicity but to the following fundamental result. We recall here that two K-algebras A and B are **Morita equivalent** if there is an additive covariant functor which is an equivalence $F : A\text{-mod} \to B\text{-mod}$ of categories. In that case, the following holds. Let A and B be Morita equivalent K-algebras via the additive equivalences $F : A\text{-mod} \to B\text{-mod}$ and $G : B\text{-mod} \to A\text{-mod}$. Set $P = F(A)$ and $Q = G(B)$. Then P and Q are natural bimodules ${}_BP_A$ and ${}_AQ_B$ such that there are natural equivalences as follows (cf. 1.5.11):

$$F = \mathrm{Hom}_A(Q, -) \text{ and } G = \mathrm{Hom}_B(P, -), \quad \text{and} \quad F = P \otimes_A - \text{ and } G = Q \otimes_B -.$$

Theorem **([17])** *Let A be a finite dimensional K-algebra with K an algebraically closed field. Then there exists a quiver Q and an admissible ideal I of KQ such that the quotient KQ/I is Morita equivalent to A.*

Sketch of Proof We actually show that if A is a basic algebra, then A is isomorphic to KQ/I for some quiver Q and some admissible ideal I of KQ (which is enough for Morita equivalence, see for instance [6, Proposition 3.16]).

Fix a complete set S of orthogonal, primitive idempotents of A, and consider the category $\mathbf{A}$ with objects S and morphisms $\mathrm{Hom}_{\mathbf{A}}(x, y) = yAx$ for two idempotents x and y in S, composition given by multiplication. In the basic case, any two different objects in $\mathbf{A}$ are non-isomorphic (the category $\mathbf{A}$ is usually called *spectroid* of A, see [18] or [6]). The quiver Q associated to a basic algebra A (also known as **Gabriel's quiver** of A), has as set of vertices S, the objects of $\mathbf{A}$, and the number of arrows from vertex x to vertex y is the dimension of the K-vector space $\mathrm{rad}(yAx)/\mathrm{rad}^2(yAx)$. Fixing a set of elements in $\mathrm{rad}(yAx)$ whose classes modulo $\mathrm{rad}^2(yAx)$ form a basis of the quotient above, then we have a *display functor*

$$\psi_A : KQ \longrightarrow \mathbf{A},$$

which is a bijection on objects. Since K is algebraically closed, then

$$\mathrm{End}_{\mathbf{A}}(x)/\mathrm{rad}(\mathrm{End}_{\mathbf{A}}(x)) \cong K,$$

for any idempotent x in S. This implies that ψ_A is a full and dense functor. Using that A is finite dimensional, and therefore its radical $\mathrm{rad}(A)$ is nilpotent, then it can be shown that the kernel I of ψ_A is an admissible ideal of KQ. Moreover, the quiver Q is independent of the choice of set of primitive idempotents S (see details in [2, 6, 18]). Then the categories KQ/I and $\mathbf{A}$ are equivalent (by Lemma 1.5.2), and this equivalence induces an isomorphism of K-algebras $KQ/I \cong A$. □

1.5.15 We need some additional notation: A representation V of a quiver Q is said to be trivial if $V(x) = 0$ for all vertices x in Q. If V and W are representations of a quiver Q, then the direct sum of these representations, $V \oplus W$, is defined by $(V \oplus W)(x) = V(x) \oplus W(x)$ for all vertices x in Q and $(V \oplus W)(a)$ is the direct sum of the linear mappings $V(a)$ and $W(a)$.

A representation is said to be **decomposable** if it is isomorphic to a direct sum of non-zero representations, and is called **indecomposable** otherwise. The **dimension vector** of a finite dimensional KQ-module V is the vector $\mathbf{dim}V \in \mathbb{N}^{Q_0}$, with $(\mathbf{dim}V)_x = \dim(V(x))$.

1.5.16 Proposition *Let A be a finite dimensional K-algebra with Gabriel quiver Q, and let* proj-A *(resp.* inj-A *) be the full subcategory of* A-mod *composed of all projective (resp. injective) modules. Then there exists a natural equivalence $\mathcal{N}$:* proj-$A \to$ inj-A *defined by $\mathcal{N}(P_x) = I_x$ for each vertex $x \in Q_0$.*

Proof Indeed, observe the following equivalences:

$$\mathrm{Hom}_A(P_x, P_y) = P_y(x) = e_x A e_y = DD(e_x A e_y) = DI_x(y) = \mathrm{Hom}_A(I_x, I_y).$$

Note that the map in the middle is the natural isomorphism between a vector space and its double dual. Moreover, note that all maps are bijections on objects and functorial. □

The functor $\mathcal{N}$ is called **Nakayama functor**.

1.6 A Quick Overview on Algebraic Geometry

1.6.1 An **affine algebraic set** is the set of solutions in an algebraically closed field K of a system of polynomial equations with coefficients in K. More precisely, if $f_1, \ldots, f_m$ are polynomials with coefficients in K, they define an affine algebraic set

$$V(f_1, \ldots, f_m) = \left\{(a_1, \ldots, a_n) \in K^n \mid f_1(a_1, \ldots, a_n) = \ldots = f_m(a_1, \ldots, a_n) = 0\right\}.$$

An affine (algebraic) variety is an affine algebraic set which is not the union of two proper affine algebraic subsets. Such an affine algebraic set is said to be **irreducible**.

If X is an affine algebraic set defined by an ideal I, then the quotient ring $R = K[x_1, \ldots, x_n]/I$ is called the coordinate ring of X. If X is an affine variety, then I is prime, so the coordinate ring is an integral domain. The elements of the coordinate ring R are also called the regular functions or the polynomial functions on the variety. They form the ring of regular functions on the variety, or, simply, the ring of the variety.

1.6.2 Examples The complement of a hypersurface in an affine variety X (that is $X - f = 0$ for some polynomial f) is affine. Its defining equations are obtained by saturating by f the defining ideal of X. The coordinate ring is thus the localization $K[X][f^{-1}]$. In particular, $\mathbb{C} - 0$ (the affine line with the origin removed) is affine. On the other hand $\mathbb{C}^2 - 0$ (the affine plane with the origin removed) is not an affine variety.

The subvarieties of codimension one in the affine space K^n are exactly the hypersurfaces, that is the varieties defined by a single polynomial. The normalization of an irreducible affine variety is affine; the coordinate ring of the normalization is the integral closure of the coordinate ring of the variety.

For an affine variety $V \subseteq K^n$ over an algebraically closed field K, and a subfield k of K, a k-rational point of V is a point $p \in V \cap k^n$. That is, a point of V whose coordinates are elements of k. The collection of k-rational points of an affine variety V is denoted $V(k)$.

1.6.3 Let V be an affine variety defined by the polynomials $f_1, \dots, f_r \in k[x_1, \dots, x_n]$, and $a = (a_1, \dots, a_n)$ be a point of V. The **Jacobian matrix** $JV(a)$ of V at a is the matrix of the partial derivatives

$$\frac{\partial f_j}{\partial x_i}(a_1, \dots, a_n).$$

The point a is regular if the rank of $JV(a)$ equals the dimension of V, and singular otherwise. If a is regular, the tangent space to V at a is the affine subspace of k^n defined by the linear equations

$$\sum_{i=1}^{n} \frac{\partial f_j}{\partial x_i}(a_1, \dots, a_n)(x_i - a_i) = 0, \quad j = 1, \dots, r.$$

1.6.4 The affine algebraic sets of K^n form the closed sets of a topology on K^n, called the **Zariski topology**. This follows from the fact that

$$V(0) = K[x_1, \dots, x_n], \quad V(1) = \emptyset,$$

$$V(S) \cup V(T) = V(ST), \quad \text{and} \quad V(S) \cap V(T) = V(S + T),$$

(in fact, a countable intersection of affine algebraic sets is an affine algebraic set).

The Zariski topology can also be described by way of basic open sets, where Zariski-open sets are countable unions of sets of the form

$$U_f = \{p \in K^n : f(p) \neq 0\} \quad \text{for} \quad f \in K[x_1, \dots, x_n]$$

1.6.5 The geometric structure of an affine variety is linked in a deep way to the algebraic structure of its coordinate ring. Let I and J be ideals of $K[V]$, the coordinate ring of an affine variety V. Let $I(V)$ be the set of all polynomials in $K[x_1, \dots, x_n]$, which vanish on V, and let $\sqrt{I}$ denote the radical of the ideal I, the set of polynomials f for which some power of f is in I. The reason that the base field is required to be algebraically closed is that affine varieties automatically satisfy **Hilbert's Nullstellensatz**: for an ideal J in $K[x_1, \dots, x_n]$, where K is an algebraically closed field, $I(V(J)) = \sqrt{J}$.

The function taking an affine algebraic set W and returning $I(W)$, the set of all functions which also vanish on all points of W, is the inverse of the function assigning an algebraic set to a radical ideal, by the Nullstellensatz. Moreover, observe that the correspondence between affine algebraic sets and radical ideals is a bijection. The coordinate ring of an affine algebraic set is reduced (nilpotent-free), as an ideal I in a ring R is radical if and only if the quotient ring R/I is reduced.

Under this correspondence, prime ideals of the coordinate ring correspond to affine subvarieties. Maximal ideals of $K[V]$ correspond to points of V.

References

1. Anderson, F. W. and Fuller, K. R., *Rings and categories of modules*. Graduate texts in mathematics **13**, Springer-Verlag, New York 1974
2. Assem, I. and Simson, D. and Skowronski, A., *Elements of the representation theory of associative algebras*, Cambridge University Press (2006), London Mathematical Society Student Texts **65**
3. M. Auslander, *The representation dimension of artin algebras,* Queen Mary College Mathematics Notes (1971)
4. Auslander M. (1982) *A functorial approach to representation theory*. In: Auslander M., Lluis E. (eds) Representations of Algebras. Lecture Notes in Mathematics, vol 944. Springer, Berlin, Heidelberg. https://doi.org/10.1007/BFb0094058
5. Auslander, M. and Reiten, I. and Smalø, S. *Representation theory of Artin algebras*, Cambridge University Press **36** Cambridge Studies in Advanced Mathematics (1995)
6. Barot, M., *Introduction to the Representation Theory of Algebras*, Springer (2015)
7. Barot, M., Geiss, C. and Zelevinsky A., *Cluster algebras of finite type and positive symmetrizable matrices*. J. London Math. Soc., **73** (2006) 545–564
8. Barot, M., Jiménez González, J.A. and de la Peña, J.A. *Quadratic Forms: Combinatorics and Numerical Results*, Algebra and Applications, Vol. **25** Springer Nature Switzerland AG 2018
9. Bass, H.: *Finitistic dimension and a homological generalization of semiprimary rings*, Trans. Amer. Math. Soc. **95** (1960), 466–488
10. Brechenmacher, F., *Algebraic generality vs arithmetic generality in the controversy between C. Jordan and L. Kronecker (1874)*. (2008) ffhal-00340071v1
11. Brenner, S.: *Modular representations of p groups*, J. Algebra **15** (1970) 89–102
12. Brenner, S.: *Decomposition properties of some small diagrams of modules*, Symposia Math. XIII, Ac. Press (1974) 127–141.
13. Corner, A. L. S.: *Endomorphism algebras of large modules with distinguished submodules*, J. Algebra **11** (1969) 155–185
14. W. Crawley-Boevey, *On tame algebras and bocses*, Proc. London Math. Soc. **56** (1988), 451–483.
15. Donovan, P. and Freislich, M.-R.: *Some evidence for an extension of the Brauer-Thrall conjecture*. Sonderforschungsbereich Theor. Math. **40**, Bonn, 1973.
16. Y. A. Drozd, *Tame and wild matrix problems*, In: Representation Theory II, Lecture Notes in Mathematics, Vol. **832**, Springer-Verlag, Berlin-Heidelberg, 1980, pp. 242–258.
17. Gabriel, P.: *Indecomposable representations II*, Symposia Math. Ist. Naz. di Alta Mat., Vol. XI (1973)
18. Gabriel, P. and Roiter, A., *Representations of finite-dimensional algebras*, London Mathematical Society, LNS **362**. Springer-Verlag Berlin Heidelberg. (1997)
19. Gantmacher, F.R., *The Theory of Matrices*, Vols. **1** and **2**, Chelsea Publishing Company, New York, N.Y. (1960)
20. Han, Y. *Controlled wild algebras*. Proceedings of the London Mathematical Society, Volume **83** , Issue 2 , September 2001 , pp. 279–298
21. Lam, T.Y.: *Bass's Work in Ring Theory and Projective Modules*, Contemporary Mathematic.
22. Lambek, J.: *Lectures on Rings and Modules*, Blaisdell, Waltham, Mass. (1966).
23. McLaughlin, T.G.: *C. S. Peirce's Proof of Frobenius' Theorem on Finite-Dimensional Real Associative Division Algebras* Transactions of the Charles S. Peirce Society Vol. **40**, No. 4 (Fall, 2004), pp. 701–710 (Published By: Indiana University Press)
24. Ringel C.M. (1980) *Report on the Brauer-Thrall conjectures: Rojter's theorem and the theorem of Nazarova and Rojter (on algorithms for solving vectorspace problems. I)*. In: Dlab V., Gabriel P. (eds) Representation Theory I. Lecture Notes in Mathematics, vol **831**. Springer, Berlin, Heidelberg. https://doi.org/10.1007/BFb0089780
25. Ringel, C.M. *Tame algebras and integral quadratic forms*, Springer LNM, **1099** (1984)

26. Ringel, C. M., *The first Brauer–Thrall conjecture.* In: Models, Modules and Abelian Groups, pp. 371–376 (de Gruyter, 2008)
27. Simson, D. *A computational technique in Coxeter spectral study of symmetrizable integer Cartan matrices*. Linear Algebra Appl., **586**:190–238 (2020)
28. Tachikawa, H.: *$QF - 3$ Rings and Categories of Projective Modules*. J. of Algebra **28**, (1974) 408–413

Chapter 2
A Categorical Approach

In this chapter we introduce Krull-Schmidt categories and discuss their basic properties, making a connection to the existence of projective covers by passing from objects to their endomorphism rings. The crucial concept is that of a **semi-perfect ring** which is due to H. Bass (cf. [13]). We also discuss Harada-Sai sequences, and basic concepts of the Auslander-Reiten theory of K-algebras. As before, K denotes an algebraically closed field.

2.1 Morphisms Between Indecomposable Modules

Since by Krull-Schmidt theorem (cf. 2.1.4 below), every module admits, in an essentially unique manner, a decomposition as a direct sum of indecomposable modules, we may concentrate our study to the morphisms between indecomposable modules of finite dimension. Here we collect some of the special properties of these maps, starting by the following fundamental observation. Recall that a module is said to be **indecomposable** if it is nonzero and whenever $M \cong N \oplus N'$, then either $N = 0$ or $N' = 0$.

2.1.1 Lemma (Fitting's Lemma) *If X is an indecomposable A-module of finite dimension, then* $\mathrm{End}_A(X)$ *is a local ring.*

Proof Let $f : X \to X$ be an endomorphism of X, and consider an integer k such that $\mathrm{Ker}(f^k) = \mathrm{Ker}(f^{k'})$ and $\mathrm{Im}(f^k) = \mathrm{Im}(f^{k'})$ for any $k' \geq k$ (this is possible by the finite dimension of X). A straightforward calculation shows that $X = \mathrm{Ker}(f^k) \oplus \mathrm{Im}(f^k)$, and by indecomposability of X, f is either an isomorphism or a nilpotent morphism. □

2.1.2 Recall that the **radical** $\mathrm{rad}(M)$ of an A-module is the intersection of all maximal submodules of M.

J.-A. de la Peña, *Representations of Algebras*, Algebra and Applications 30,
https://doi.org/10.1007/978-3-031-12288-0_2

Theorem *Let A be a K-algebra and M an A-module of finite dimension. Then the following holds.*

(a) The radical of M is the intersection $\bigcap_{g:M\to S}\operatorname{Ker}(g)$ over all simple modules S.
(b) The quotient $M/\operatorname{rad}(M)$ is a semi-simple A-module.
(c) The module M is indecomposable if and only if $E = \operatorname{End}_A(M)$ is a local ring, that is, E has any one of the following equivalent properties:

(i) E has a unique maximal left ideal.
(ii) E has a unique maximal right ideal.
(iii) $\mathbf{Id}_M \neq 0$ and the sum of any two non-units in E is a non-unit.
(iv) $\mathbf{Id}_M \neq 0$ and if e is any element of E, then e or $\mathbf{Id}_M - e$ is a unit.
(v) $\mathbf{Id}_M \neq 0$ and every member of E is either a unit or a nilpotent element.

If these properties hold, then the unique maximal left ideal coincides with the unique maximal right ideal and with the radical $\operatorname{rad}(E)$.

Proof A discussion on local rings and equivalent definitions may be found, for instance, in [1]. □

2.1.3 An additive category is called a **Krull-Schmidt category** if every object decomposes into a finite direct sum of objects having local endomorphism rings. A ring is **semi-perfect** if it satisfies the equivalent conditions in the following result.

Proposition *For a ring R the following are equivalent.*

(1) The category of finitely generated projective R-modules is a Krull-Schmidt category.
(2) The module R_R admits a decomposition $R = P_1 \oplus P_2 \oplus \ldots \oplus P_s$ such that each P_i has a local endomorphism ring.
(3) Every simple R-module admits a projective cover.
(4) Every finitely generated R-module admits a projective cover.

Recall that a R-module M admits a **projective cover** P if there is an epimorphism $p : P \to M$ with P projective, such that $\operatorname{Ker}(p)$ is superfluous. We say that X is a **superfluous submodule** of Y if whenever Z is a submodule of Y such that $Z + X = Y$, then $Z = Y$.

Proof That (1) implies (2) is clear.

To show that (2) implies (3), let S be a simple R-module. Then we have a non-zero morphism $R \to S$ and therefore a non-zero morphism $f : P \to S$ for some indecomposable direct summand P of R. The module P is the wanted projective cover. Indeed, since S is simple, then f is onto and for any proper submodule N of P, if $f(N) = S$ there is a morphism $\rho : P \to P$ satisfying $f\rho = f$. Since $\operatorname{End}_R(P)$ is local then ρ is an isomorphism. Thus $\operatorname{Ker}(\rho) = 0$, a contradiction. □

As consequence, notice that semi-perfect is a left-right symmetric property. The uniqueness of direct sum decompositions in Krull-Schmidt categories can be derived from the existence and uniqueness of projective covers over semi-perfect rings, as follows. If X is an object in an additive category C, denote by add(X) the full subcategory of C determined by direct sums of direct summands of X.

2.1.4 Theorem (Krull-Schmidt-Remak) *Let X be an object of an additive category and suppose we have two decompositions*

$$X_1 \oplus \cdots \oplus X_r = X = Y_1 \oplus \cdots \oplus Y_s,$$

into objects with local endomorphism rings. Then $r = s$ and there exists a permutation σ such that $X_i \cong Y_{\sigma(i)}$, for all $1 \leq i \leq r$.

Proof Let $C = \text{add}(X)$ and identify C via $\text{Hom}_C(X, -)$ with a full subcategory of the category of finitely generated projective modules over $\text{End}_C(X)$. Thus we may assume that X is a finitely generated projective module over a semi-perfect ring. It follows from Proposition 2.1.3 that for every index i the radical $\text{rad}(X_i)$ is a maximal submodule of X_i and that the canonical morphism $X_i \to X_i/\text{rad}X_i$ is a projective cover. Therefore, $X_i \cong Y_j$ if and only if $X_i/\text{rad}X_i \cong X_j/\text{rad}X_j$, and the result follows. □

Some examples and comments are in order. In the category of A modules, if M is already a projective module, then the identity map from M to M is a superfluous epimorphism. Hence, projective modules always have projective covers. The **injective envelope** of a module always exists, however over certain rings modules may not have projective covers. For example, the module $\mathbb{Z}_2$ has no projective cover. The class of rings which provides all of its right modules with projective covers is the class of right perfect rings. Any R-module M has a flat cover, which is equal to the projective cover if R has a projective cover.

2.1.5 Corollary *Let X be an object of a Krull-Schmidt category and suppose there are two decompositions*

$$X_1 \oplus \cdots \oplus X_n = X = X' \oplus X'',$$

such that each X_i is indecomposable. Then there exists an integer $t \leq n$ such that $X = X_1 \oplus \cdots \oplus X_t \oplus X'$ after reindexing the X_i.

Proof Let $X' = Y_1 \oplus \cdots \oplus Y_s$ and $X'' = Z_1 \oplus \cdots \oplus Z_t$ be decompositions into indecomposable objects. It follows from the uniqueness of these decompositions that $n = s + t$ and that $X'' \cong X_1 \oplus \cdots \oplus X_t$ after some reindexing of the X_i. Composing the decomposition $X = X' \oplus X''$ with that isomorphism yields the assertion. □

2.1.6 Corollary *An additive category is a Krull-Schmidt category if and only if it has split idempotents and the endomorphism ring of every object is semi-perfect.*

Proof It is enough to show that a Krull-Schmidt category has split idempotents. But this is clear since there is an equivalence

$$\mathrm{add}(X) \cong \mathrm{proj}\text{-}G,$$

for $G = \mathrm{End}(X)$, due to Proposition 2.1.3 and Corollary 2.1.5. □

2.2 Harada-Sai Sequences

2.2.1 Let A be a K-algebra. We will use throughout that for a nonzero morphism $f : M \to N$ between indecomposable modules the following are equivalent: f is a section, f is a retraction, f is an isomorphism. Moreover, we care about indecomposable modules in light of Theorem 2.1.4. We shall call a sequence of homomorphisms of A-modules

$$e: M_1 \xrightarrow{f_1} M_2 \xrightarrow{f_2} \quad \xrightarrow{f_{s-1}} M_s\ ,$$

a **Harada-Sai sequence** if each M_i is an indecomposable A-module of finite dimension $d(M_i)$, each morphism f_i is a non-isomorphism, and the composite map $f_{s-1}\cdots f_1 : M_1 \to M_s$ is nonzero.

2.2.2 Lemma (Harada-Sai, [10]) *Let A be a K-algebra. Suppose that $M_1, \ldots, M_{2^n}$ are indecomposable A-modules all of dimension at most b, and that we are given morphisms $f_i : M_i \to M_{i+1}$, for $i = 1, \ldots, 2^n - 1$ that are not isomorphisms. Then the image of the composite $f_{2^n-1}\cdots f_1$ has dimension at most $b - n$. Thus the composite is zero if $n = b$.*

Proof By induction on n. If $n = 1$, there is a single morphism $f_1 : M_1 \to M_2$ that is not an isomorphism. If the image of f_1 has dimension at least b, then M_2 must have dimension exactly b, and in particular f_1 is an epimorphism. This means that M_1 has dimension at least that of M_2, so it must have dimension b, too. This means f_1 is an epimorphism between modules of equal dimension, and this implies that f_1 is an isomorphism, which cannot be (a result which goes by the name of Schur's lemma). Hence $\dim_K \mathrm{Im}(f_1) \leqslant b - 1$. Assume the case n is already proven and consider $n + 1$. We split the $2^{n+1} - 1$ arrows into three collections:

$$\{f_1, \ldots, f_{2^n-1}\}, \{f_{2^n}\}, \{f_{2^n+1}, \ldots, f_{2^{n+1}-1}\},$$

and note that the non-trivial collections have both $2^n - 1$ arrows. Call the composite of the first collection f, let $f_{2^n} = g$ and let h denote the composite of the arrows in the third collection. The inductive hypothesis entails that $\dim_K \mathrm{Im}(f) \leq b - n$ and $\dim_K \mathrm{Im}(h) \leq b - n$, and we would like to conclude that $\dim_K \mathrm{Im}(hgf) \leqslant b - n - 1$.

We note that if any of the two inequalities are strict then we may conclude, for $\dim_k \mathrm{Im}(hgf)$ is at most both these dimensions. We thus assume that $\dim_K \mathrm{Im}(f) = \dim_K \mathrm{Im}(h) = b - n$. We will show that $\dim_K \mathrm{Im}(hgf) = b - n$ implies g is an isomorphism, which is contrary to our assumption, thus proving the theorem.

Consider $\mathrm{Im}(f) \xrightarrow{\iota} M_{2^n} \xrightarrow{g} M_{2^n+1} \xrightarrow{h} \mathrm{Im}(h)$. We have inclusions $\mathrm{Im}(hgf) \subseteq \mathrm{Im}(hg) \subseteq \mathrm{Im}(h)$, and since the endpoints have equal dimensions, we have equalities throughout. Now $\mathrm{Im}(hg)\iota = \mathrm{Im}(hgf) = \mathrm{Im}(h)$ so that $hg\iota$ is onto and hence an isomorphism, since the endpoints have equal dimension. It follows that the inclusion ι is a section whence f must be an epimorphism, so that $gh = gh\iota$ is an isomorphism. But this means g is a retraction so it is an isomorphism. □

2.2.3 In the following paragraphs we sharpen the Harada-Sai Lemma, following [9], by showing exactly which dimension sequences $(d(M_1), ..., d(M_s))$ are possible, and what the dimension of the image of the composite map $M_1 \to M_s$, which we call the **composite rank** of e, can be.

First some simple examples. There is no Harada-Sai sequence with dimension sequence $(2, 3, 3, 3, 4, 4)$. There are Harada-Sai sequences with dimension sequence $(3, 2, 3, 3, 3)$, but none with the permuted sequences $(3, 3, 2, 3, 3)$ or $(2, 3, 3, 3, 3)$.

Observe that if $d(M_i) = d(M_{i+1})$ in the Harada-Sai sequence e, then $\mathrm{Im}(f_i)$ is a proper submodule of M_{i+1} with indecomposable decomposition $\mathrm{Im}(f_i) = \bigoplus_{j=1}^m N_i$ for some $m \geq 1$.The restriction of f_i to N_j in the codomain will be denoted by $f_{i,j}$, and the inclusion $N_j \to M_{i+1}$ by g_j. Then we have $f_i = \sum_{j=1}^m g_j f_{i,j}$, and

$$0 \neq f_{s-1} \cdots f_1 = \sum_{j=1}^{m} f_{s-1} \cdots g_j f_{i,j} \cdots f_1.$$

Therefore there is j with $0 \neq f_{s-1} \cdots g_j f_{i,j} \cdots f_1$, and since none of the factors of the composition is an isomorphism, we obtain a Harada-Sai sequence of length $s+1$ of the form

$$e' : M_1 \xrightarrow{f_1} M_2 \xrightarrow{f_2} \cdots \xrightarrow{f_{i-1}} M_i \xrightarrow{f_{i,j}} N_j \xrightarrow{g_j} M_i \xrightarrow{f_{i+1}} \cdots \xrightarrow{f_{s-1}} M_s$$

Moreover, any composition $g := f_j f_{j-1} \cdots f_i : M_i \to M_j$ of consecutive maps in a Harada-Sai sequence is a non-isomorphism. For if g were an isomorphism then $g^{-1} f_j f_{j-1} \cdots f_{i+1}$ is a retraction of f_i, and therefore f_i is an isomorphism, contradicting the definition.

2.2.4 We introduce the **universal sequences of dimensions** $d^{(b)}$. Let $d^{(1)} = (1)$, and define the sequences of length $2^b - 1$ as:

$$d^{(2)} = (2, 1, 2),$$
$$d^{(3)} = (3, 2, 3, 1, 3, 2, 3),$$
$$d^{(4)} = (4, 3, 4, 2, 4, 3, 4, 1, 4, 3, 4, 2, 4, 3, 4).$$

That is, if $d^{(b)}$ is already defined, we define the sequence $d^{(b+1)}$ as

$$d^{(b+1)}(j) = \begin{cases} d^{(b)}(j'), & \text{if } 2j' = j \text{ is even}, \\ b+1, & \text{if } j \text{ is odd}. \end{cases}$$

We say that a sequence λ is **embeddable** in μ if there is a strictly increasing function $a : \{1, \ldots, m\} \to \{1, \ldots, n\}$ with $\lambda(i) = \mu(a(i))$ for all i (we write $a : \lambda \to \mu$).

2.2.5 Theorem (Embedding) *There is a Harada-Sai sequence with dimension sequence λ if and only if there is an embedding $a : \lambda \to d^{(b)}$ for some b.*

The classic Harada-Sai Lemma follows from the "only if" statement, since the length of the sequence $d^{(b)}$ is less or equal than $2^b - 1$.

Proof We begin by showing that the dimension sequence $d = (d_1, \ldots, d_s)$ of a Harada-Sai sequence e as above is embeddable in $d^{(b)}$, where $b = \max_i\{d_i\}$. We do this by induction on b, the case $b = 1$ being trivial.

By 2.2.3, we may assume that no two consecutive d_i are equal to b. Let $d' = (d_{i_1}, \ldots, d_{i_t})$ be the subsequence of d consisting of those terms that are $< b$. By induction there is an embedding of d' to $d^{(b-1)}$. Since every second element of $d^{(b)}$ is b, and the remaining elements of $d^{(b)}$ make up $d^{(b-1)}$, this embedding can be extended to the desired embedding of d into $d^{(b)}$, completing the proof of embeddability. The other implication of the claim follows from the existence of a maximal Harada-Sai sequence, constructed below. □

We define the **rank** of an embedding $a : d \to d^{(b)}$ to be $\mathrm{rk}(a) = \min_j\{d_j^{(b)} \mid a(1) \leq j \leq a(s)\}$. The following more precise version of Theorem 2.2.5 is the main result of the section.

2.2.6 Theorem (Rank) *There is a Harada-Sai sequence with dimension sequence d and composite rank r if and only if there is an embedding $a : d \to d^{(b)}$, such that $r = \mathrm{rk}(a)$.*

A proof of the theorem above and of the following claims may be found in [9].

Corollary *If e as above is a Harada-Sai sequence of length* $s = 2^b - 1$ *with* $d(M_i) \leq b$ *for all* i*, then the composite rank of any subsequence*

$$M_{i_1} \to M_{i_2} \to \cdots \to M_{i_t},$$

is precisely $\min_j \{d_j^{(b)} \mid i_1 \leq j \leq i_t\}$.

From the embedding theorem we also get a sharper Harada-Sai bound, taking into account the structure of the ring:

Corollary *Let* R *be an Artinian ring and let* $t = t(R)$ *be the maximum, over all simple left* R*-modules* N*, of the minimum of the lengths of the projective cover and the injective hull of* N*. The length* s *of any Harada-Sai sequence* e *of left modules with lengths* $\leq b$*, is bounded by* $s \leq 2^{b-t+1}(2^{t-1} - 1) + 1$.

2.3 The Remak-Krull-Schmidt-Azumaya Decomposition

2.3.1 We begin this section with a more general Krull-Schmidt decomposition for groups and rings. The category of finitely generated torsion-free abelian groups admits unique decompositions into indecomposable objects. However, the unique indecomposable object $\mathbb{Z}$ does not have a local endomorphism ring. We say that a group G satisfies the **ascending chain condition** (ACC) on subgroups if every sequence of subgroups of G:

$$1 = G_0 \leq G_1 \leq G_2 \leq \ldots$$

is eventually constant, i.e., there exists N such that $G_N = G_{N+i}$ for all $i \geq 1$. We say that G satisfies the ACC on normal subgroups if every such sequence of normal subgroups of G eventually becomes constant. Consider similarly the **descending chain condition** (DCC) on subgroups.

2.3.2 Proposition *If* G *is a group that satisfies ACC and DCC on normal subgroups, then there is a unique way of writing* G *as a direct product*

$$G = G_1 \times G_2 \times \cdots \times G_k,$$

of finitely many indecomposable subgroups of G*. Here, uniqueness is to be understood as direct decompositions into indecomposable subgroups have the exchange property. That is: suppose* $G = H_1 \times H_2 \times \cdots \times H_s$ *is another expression of* G *as a product of indecomposable subgroups. Then* $k = s$ *and there is a reindexing of the*

H_i's satisfying G_i and H_i are isomorphic for each i, and

$$G = G_1 \times \cdots \times G_r \times H_{r+1} \times \cdots \times H_s,$$

for each r.

Proof Proving existence is easy: let S be the set of all normal subgroups of G that cannot be written as a product of indecomposable subgroups. Moreover, any indecomposable subgroup is (trivially) the one-term direct product of itself, hence decomposable. If Krull-Schmidt fails, then $G \in S$; so we may iteratively construct a descending series of direct factors; this contradicts the DCC. One can then invert the construction to show that all direct factors of G appear in this way.

The proof of uniqueness, on the other hand, is quite long and requires a sequence of technical lemmas. For a complete exposition, see [8]. □

2.3.3 Proposition *Let R be a ring. If $E \neq 0$ is a R-module satisfying the ACC and DCC on submodules (that is, it is both noetherian and artinian), then E is a direct sum of indecomposable modules. Up to a permutation, the indecomposable direct summands in such a direct sum are uniquely determined up to isomorphism.*

In general, the Proposition fails if we only assume that the module is noetherian or artinian.

Let R be a noetherian ring. Then there might exist infinite descending chains of ideals in R. As an intermediate between noetherian and artinian rings, we will allow the descending chain condition to creep into the noetherian ring in a restricted manner. Consider all the ideals which may be written as the intersection of such an infinite descending chain. If R is non-artinian, this is a non-empty set. Let J be a maximal element of this set. Then R/J has a curious property: any infinite descending chain of ideals in this ring has zero intersection.

A ring R is said to be a **zero-minimum ring** if the intersection of an infinite decreasing chain of ideals in R is zero.

2.3.4 Proposition *Let R be a domain. Then the following are equivalent:*

(1) R is noetherian and 1-dimensional.
(2) R is a zero-minimum ring.
(3) Every finitely generated R-module with a non-zero annihilator is of finite length.

Proof Suppose that R is noetherian and 1-dimensional. Take an infinite descending chain of ideals in R and consider their intersection, say J. If J is not 0, R/J is a noetherian ring having all its prime ideals maximal (the minimal prime ideal 0 has been eliminated by taking R/J). Hence R/J is an artinian ring and we have a decreasing infinite chain in R/J. This is clearly a contradiction.

We assume now that R is a zero-minimum ring. Then each R/J is artinian, for non-zero J. Thus, each R/J is noetherian for non-zero J, and by the same argument as in the last result, this shows that R is noetherian and that all non-zero prime ideals in R are maximal. Since R is a domain, we have one more prime ideal, namely 0, which makes the ring 1-dimensional.

Let M be a R-module of non-zero annihilator I. Then M may be considered as a R/I-module. Since R/I is a 0-dimensional noetherian ring, M is of finite length. Conversely, we consider the case $M = R/J$, where J is a non-zero ideal, as a R-module. We see that it has non-zero annihilator, i.e. J, and hence must be of finite length. Thus R/J is artinian, i.e. noetherian and 0-dimensional. Hence, R is noetherian. Also, since R is a domain, (0) is a prime ideal. Thus R is 1-dimensional. □

2.3.5 Proposition *Let R be a noetherian ring (and non-artinian) and let J be maximal among the ideals that can be written as the intersection of an infinitely long, decreasing chain of ideals in R. Then J is prime.*

Proof As a quotient of R, R/J is noetherian but not artinian (there exists an infinite chain intersecting to 0). But R/J is a zero intersection ring. From the previous result, if R/J is not a domain, it must be artinian. Hence R/J is a domain. Thus, J is a prime ideal. □

2.3.6 Proposition *If R is not a domain, and R is a zero minimum ring implies R is artinian (and thus there is no infinite decreasing chain at all).*

Proof We see that R is a noetherian ring with all prime ideals maximal. Hence, R is artinian.

Let R be a ring with ACC on two sided ideals. Let $I \leq R$ be an ideal of R. We show that there exist prime ideals $P_1, \ldots, P_n$ such that $P_1 \cdots P_n \leq I$. Indeed, let $\mathscr{F} = \{L \leq R : L$ does not contain a product of primes $\}$. Then each chain has a maximal element by the ACC condition and so from Zorn's Lemma we have that this family has a maximal element. Denote it by M.

Now as M is not prime (else it would not be in $\mathscr{F}$) and we get two ideals $A, B \leq R$ such that $AB \leq M$ but A and B are not contained in M. This then gives that $M \leq (A + M)$ and $M \leq (B + M)$ so that both $(A + M)$ and $(B + M)$ are not in $\mathscr{F}$, which means that they contain a product of primes.

However from the above condition we have that $(A + M)(B + M) = AB + AM + MB + M^2 \leq M$ and as from above we have that $(A + M)(B + M)$ contains a product of primes so does M hence the family $\mathscr{F}$ is empty and we are done. □

2.3.7 Theorem *If R is a zero-minimum ring with non-zero Jacobson radical, then R has only finitely many maximal ideals.*

Proof Suppose that R has infinitely many maximal ideals. Choose countably many of those, say $m_1, m_2, \cdots$. Consider the following chain of ideals:

$$m_1 \geq m_1 \cap m_2 \geq m_1 \cap m_2 \cap m_3 \geq \cdots$$

The inclusions are strict because: if there is equality at the k-th stage, then $m_k = m_1 \cap m_2 \cap \cdots \cap m_{k-1} \leq m_{k-1}$. This will imply that m_k contains some other maximal ideal, which is absurd. Thus, the chain of maximal ideals in R grows properly. Since R is a zero minimum ring, the intersection of this chain is (0). But, the intersection contains the Jacobson radical J. Thus, $J = 0$ which is a contradiction. □

A remarkable fact about an Artin ring is that it can be written as a finite direct product of Artin local rings. The *proof* proceeds along the lines of the Chinese Remainder Theorem: An Artin ring R has only finitely many maximal ideals, say $m_1, \ldots, m_n$ and the Jacobson radical J is nilpotent, say $J^k = 0$. Consider the product $\Pi_i m_i^k$. Since the m_i are mutually co-prime, this product equals $\bigcap m_i^k$. But $\bigcap m_i^k \subset (\bigcap m_i)^k = (0)$. Hence we have the natural isomorphism from $R = R/\bigcap m_i^k$ which is isomorphic to the direct product of the artinian local rings R/m_i^k.

2.4 First Elements of Auslander-Reiten Theory

Let us start with a short presentation from [12] on Auslander's defect formula.

2.4.1 Let A be a K-algebra and consider an exact sequence of A-modules $0 \longrightarrow L \xrightarrow{f} M \xrightarrow{g} N \longrightarrow 0$. The **covariant defect** d_* and the **contravariant defect** d^* are the functors determined by the exactness of the following sequences:

$$0 \to \mathrm{Hom}_A(N, -) \to \mathrm{Hom}_A(M, -) \to \mathrm{Hom}_A(L, -) \to d_* \to 0,$$

$$0 \to \mathrm{Hom}_A(-, L) \to \mathrm{Hom}_A(-, M) \to \mathrm{Hom}_A(-, N) \to d^* \to 0.$$

For a finite dimensional A-module X having a projective presentation

$$P_1 \to P_0 \to X \to 0,$$

and taking $P^* = \mathrm{Hom}_A(P, A_A)$, the **transpose** $\mathrm{Tr}(X)$ of X is defined by the exactness of the following sequence of A^{op}-modules

$$P_0^* \to P_1^* \to \mathrm{Tr}(X) \to 0.$$

2.4.2 Theorem (Auslander's Defect Formula) *Let $d : 0 \to L \to M \to N \to 0$ be an exact sequence of A-modules and X be a finite dimensional A-module. Then* *there is an isomorphism*

$$Dd^*(X) \cong d_*(D\operatorname{Tr}(X)),$$

which is functorial on d and X.

Proof For A-modules M and N, denote $\operatorname{Hom}_A(M,N)$ by (M,N). A projective presentation $P_1 \to P_0 \to X \to 0$ of X induces the following commutative diagram with exact rows

$$\begin{array}{ccccccccc} & & L\otimes_A P_0^* & \longrightarrow & L\otimes_A P_1^* & \longrightarrow & L\otimes_A \operatorname{Tr}(X) & \longrightarrow & 0 \\ & & \downarrow\cong & & \downarrow\cong & & & & \\ 0 \longrightarrow (X,L) & \longrightarrow & (P_0,L) & \longrightarrow & (P_1,L). & & & & \end{array}$$

Therefore d induces the commutative diagram with exact columns and rows,

$$\begin{array}{ccccccccccc} & & 0 & & 0 & & 0 & & & & \\ & & \downarrow & & \downarrow & & \downarrow & & & & \\ 0 & \to & (X,L) & \to & (P_0,L) & \to & (P_1,L) & \to & L\otimes_A \operatorname{Tr}(X) & \to & 0 \\ & & \downarrow & & \downarrow & & \downarrow & & \downarrow & & \\ 0 & \to & (X,M) & \to & (P_0,M) & \to & (P_1,M) & \to & M\otimes_A \operatorname{Tr}(X) & \to & 0 \\ & & \downarrow & & \downarrow & & \downarrow & & \downarrow & & \\ 0 & \to & (X,N) & \to & (P_0,N) & \to & (P_1,N) & \to & N\otimes_A \operatorname{Tr}(X) & \to & 0 \\ & & & & \downarrow & & \downarrow & & \downarrow & & \\ & & & & 0 & & 0 & & 0 & & \end{array}$$

Using the snake lemma and the adjunction isomorphism $D(M \otimes_A Y) \cong \operatorname{Hom}_A(M, DY)$ for a left A module Y, we obtain

$$\begin{aligned} Dd^*(X) &= D\operatorname{CoKer}\operatorname{Hom}_R(X,g) \cong D\operatorname{Ker}(f\otimes_R \operatorname{Tr}(X)) \\ &\cong \operatorname{CoKer}D(f\otimes_R \operatorname{Tr}(X)) \cong \operatorname{CoKer}\operatorname{Hom}_R(f, D\operatorname{Tr}(X)) = d_*(D\operatorname{Tr}(X)). \end{aligned}$$

Then $Dd^*(X) \cong d_*(D\operatorname{Tr}(X))$. □

2.4.3 Consider the subgroups

$$P(M,N) = \{f \in \operatorname{Hom}_A(M,N) \mid f \text{ factors through a projective module}\}.$$

We set $\underline{\operatorname{Hom}}_A(M,N) = \operatorname{Hom}_A(M,N)/P(M,N)$, and let $\underline{\text{mod}}\text{-}A$ be the category with the same objects as mod-A and morphisms $\underline{\operatorname{Hom}}_A(M,N)$. It is called the

stable category of mod-A **modulo projectives**. We leave the proof of the following statements as exercise.

Proposition *The following holds:*

(1) There is a group isomorphism $\mathrm{Hom}_A(M, N) \to \mathrm{Hom}_A(\mathrm{Tr}\, N, \mathrm{Tr}\, M)$*;*
(2) $\mathrm{End}_A(M)$ *is local if and only if* $\mathrm{End}_A(\mathrm{Tr}\, M)$ *is local.*
(3) Tr *induces a duality* A-$\underline{\mathrm{mod}} \to \underline{\mathrm{mod}}$-$A$.

Corollary (Auslander-Reiten Formula) *Let C be an A-module and X be a finite dimensional A-module. Then there is an isomorphism*

$$D\underline{\mathrm{Hom}}_A(X, C) \cong \mathrm{Ext}^1_A(C, D\,\mathrm{Tr}(X)),$$

which is functorial on C and X.

The defect formula 2.4.2 appears, without a proof, in the book [8] of Auslander, Reiten, and Smalø. Originally, it is Theorem III.4.1 in Auslander's Philadelphia Notes [5]. About the same time, the Auslander-Reiten formula appeared in the classical paper [7].

2.4.4 Definition Let A be a K-algebra.

(1) A homomorphism $g : M \to N$ in A-Mod is called **right minimal** if any endomorphism $h : M \to M$ with $gh = g$ is an isomorphism.
(2) A homomorphism $g : M \to N$ is called **right almost split** in A-Mod if:
    (a) g is not a split epimorphism, and
    (b) if $h : X \to N$ is a morphism in A-Mod that is not a split epimorphism, then h factors through g.
(3) A homomorphism $g : M \to N$ in A-Mod is called **minimal right almost split** in A-Mod if it is right minimal and right almost split in A-Mod.

Observe that if $g : M \to N$ is right almost split, then $\mathrm{End}_A(N)$ is a local ring, and if N is not projective, then g is an epimorphism.

2.4.5 Proposition *The following statements are equivalent for an exact sequence in* A-Mod,

$$0 \longrightarrow L \xrightarrow{\ f\ } M \xrightarrow{\ g\ } N \longrightarrow 0\,.$$

(a) f is left almost split and g is right almost split in A-Mod.
(b) $\mathrm{End}_A(N)$ *is local and f is left almost split in* A-Mod.
(c) $\mathrm{End}_A(L)$ *is local and g is right almost split in* A-Mod.

(d) f is minimal left almost split in A-Mod.
(e) g is minimal right almost split in A-Mod.

Definition An exact sequence $0 \longrightarrow L \xrightarrow{f} M \xrightarrow{g} N \longrightarrow 0$ in A-Mod is called **almost split (or Auslander-Reiten) sequence** if it satisfies one of the equivalent conditions above.

Remark Almost split sequences starting (or ending) at a given module are uniquely determined up to isomorphism (cf. [8, Theorem 1.16]). More precisely, if

$$0 \to L \to M \to N \to 0 \quad \text{and} \quad 0 \to L_0 \to M_0 \to N_0 \to 0,$$

are almost split sequences, then

(i) $L \cong L_0$ if and only if $N \cong N_0$.
(ii) if and only if there are isomorphisms a, b, c making the following diagram commute,

$$\begin{array}{ccccccccc} 0 & \longrightarrow & L & \longrightarrow & M & \longrightarrow & N & \longrightarrow & 0 \\ & & \downarrow{\scriptstyle a} & & \downarrow{\scriptstyle b} & & \downarrow{\scriptstyle c} & & \\ 0 & \longrightarrow & L_0 & \longrightarrow & M_0 & \longrightarrow & N_0 & \longrightarrow & 0 \end{array}$$

2.4.6 Theorem ([4]) *Let N be a non-projective indecomposable A-module of finite dimension. Then there is an almost split sequence*

$$0 \to L \to M \to N \to 0,$$

in A-mod, *with* $L = D(\operatorname{Tr} N)$.

This result was originally proved in [5]. Auslander-Reiten sequences, also known as almost split exact sequences, are one of the central tools in the representation theory of Artin algebras. The precise definition can be found in many places; the book [8] gives an excellent presentation. In our context, we give a short proof below.

2.5 The Category of Additive Functors

2.5.1 Ring operations mimic the additivity and composition properties of functor categories, and this in a quite good way! Indeed, every ring R may be viewed as an additive category $[R]$ having just one object $\underline{R}$ and morphism group

$$\operatorname{Hom}_{[R]}(\underline{R}, \underline{R}) = R,$$

composition given by the multiplication. Thus, an additive functor $M : [R] \to \text{Ab}$ is just an abelian group $M([R])$ equipped with a linear R-action. In this way, the category R-Mod is the functor category Add($[R]$, Ab).

Similarly, if A is a K-algebra and S is a complete system of indecomposable idempotents of A, let A_S be the category with objects the set S and morphism $\text{Hom}_{A_S}(x, y) = yAx$, composition given by multiplication. Then the category of additive functors from A_S to Ab is the category of representations of A, and therefore equivalent to the category of A-modules (see 1.1.5.14).

Conversely, a small additive category can be viewed as a generalization of a ring, called **ring with several objects**, correspondingly the functor category Add(**A**, Ab), with **A** a small additive category, is a generalization of the module category A-Mod. In this way, we can adapt arguments from the well-developed theory of rings to the realm of functors (this "philosophy" is explained in [2]). An example of the possibilities of this transfer is illustrated by the simple and powerful:

2.5.2 Proposition (Yoneda's Lemma) *For any additive functor $F : \mathbf{A} \to \mathbf{Ab}$, the mapping*

$$\text{Hom}(\mathbf{A}(A, -), F) \longrightarrow F(A),$$

$$w \longmapsto w_A(1_A)$$

is an isomorphism of abelian groups with inverse

$$e \mapsto (u_X)_{X \in \mathbf{A}}, \quad \textit{where} \quad u_X(f) = F(f)(e).$$

We need the following definitions that generalize to the category Add(A^{op}-mod, Ab) known concepts in the module category. Recall that a sequence of additive functors $F_1 \to F_2 \to F_3$ is exact, if the evaluation $F_1(M) \to F_2(M) \to F_3(M)$ is exact for every M in A-mod. Notice that by Yoneda's lemma, this happens exactly when the sequence

$$\text{Hom}(\text{Hom}_A(-, M), F_1) \to \text{Hom}(\text{Hom}_A(-, M), F_2) \to \text{Hom}(\text{Hom}_A(-, M), F_3)$$

is exact.

2.5.3 Let F be a functor in Add(A^{op}-mod, Ab).

- F is called **projective** if for any exact sequence $F_1 \to F_2 \to F_3 \to 0$ of additive functors, the following sequence is exact,

$$\text{Hom}(F, F_1) \to \text{Hom}(F, F_2) \to \text{Hom}(F, F_3) \to 0.$$

In particular, the Hom functors form a generating system of projective objects in Add(A^{op}-mod, Ab).

- F is called **finitely generated** if there is an epimorphism $\mathrm{Hom}_A(-, M) \to F$ with M in A-mod.
- F is called **finitely presented** if there is an exact sequence

$$\mathrm{Hom}_A(-, N) \to \mathrm{Hom}_A(-, M) \to F \to 0,$$

  with M and N in A-mod.
- A morphism $f : \mathrm{Hom}_A(-, M) \to F$ is called a **projective cover** of F, if for any endomorphism $h : \mathrm{Hom}_A(-, M) \to \mathrm{Hom}_A(-, M)$, the identity $f = fh$ implies that h is an isomorphism.
- F is called **simple** if $F \neq 0$ and any non-zero morphism $F' \to F$ is an epimorphism.
- The **radical** of F is the intersection $\mathrm{rad}(F)$ of all maximal subfunctors of F.
- F is said to have **finite length**, if there is a chain of subfunctors $F_0 \subset F_1 \subset \ldots \subset F_n$ with $F_0 = 0$, $F_n = F$ and F_{i+1}/F_i a simple functor for $i = 0, \ldots, n-1$.
- The **support** of a functor F is the full subcategory $\mathrm{supp}(F)$ of A-mod determined by indecomposable modules M with $F(M) \neq 0$. We say that F has **finite support** if there are only finitely many isoclasses of indecomposable A-modules in $\mathrm{supp}(F)$.

The following useful characterization of simple functors implies that there is a bijection between (isomorphism classes of) simple functors and indecomposable projective functors in $\mathrm{Add}(A^{op}\text{-mod}, \mathrm{Ab})$.

Proposition *A functor S in* $\mathrm{Add}(A^{op}\text{-mod}, \mathrm{Ab})$ *is simple if and only if there is an indecomposable A-module M, unique up to isomorphism, such that*

$$S \cong \mathrm{Hom}_A(-, M)/\mathrm{radHom}_A(-, M).$$

In that case, the natural morphism $\mathrm{Hom}_A(-, M) \to S$ *is a projective cover. Moreover, if $S(N) \neq 0$ there is an A-module isomorphism $N \cong M$.*

Lemma *Let F be a functor in* $\mathrm{Add}(A^{op}\text{-mod}, \mathrm{Ab})$ *of finite length. Then F has finite support.*

Proof By hypothesis, there is a chain $0 = F_0 \subset F_1 \subset \ldots \subset F_n = F$ of subfunctors such that F_{i+1}/F_i is simple for all $i = 0, \ldots, n-1$. Notice that $\mathrm{supp}(F) = \bigcup_{i=0}^{n-1} \mathrm{supp}(F_{i+1}/F_i)$. But if S is a simple functor and M, N are indecomposable modules such that $S(M) \neq 0$ and $S(N) \neq 0$, then by the proposition above $M \cong N$. Hence $\mathrm{supp}(F_{i+1}/F_i) < \infty$ for all $i = 0, \ldots, n-1$ and so $\mathrm{supp}(F) < \infty$. □

2.5.4 The functor category $\mathrm{Add}(A^{op}\text{-mod}, \mathrm{Ab})$ is an abelian category with exact direct limits, and the representable functors $\mathrm{Hom}_A(M, -)$ with M in A-mod form a system of finitely generated projective functors. Recall that an abelian category is said to have **exact direct limits** if each directed system (η_α) of exact sequences has an exact limit sequence $\lim_{\to} \eta_\alpha$. Moreover, a category which has exact direct limits and having all the direct sums is said to be a **Grothendieck category**. In particular, the functor category Add(A^{op}-mod, mod-K) is Grothendieck.

2.5.5 Proposition *Any finitely generated subfunctor of a finitely presented functor F in* Add(A^{op}-mod, mod-K) *is itself finitely presented.*

We refer the interested reader to [6] for a proof. Recall that a ring is called (left) *coherent* if any of its finitely generated (left) ideals is finitely presented, or equivalently, if every finitely generated submodule of a finitely presented (left) module is itself finitely presented. By the proposition above, one says that the abelian category Add(A^{op}-mod, mod-K) is coherent (cf. [3] or [11]).

Corollary *A functor F in* Add(A^{op}-mod, mod-K) *is finitely presented if and only if DF is finitely presented.*

Proof By adjunction we have,

$$D\mathrm{Hom}_A(A, -) \cong \mathrm{Hom}_K(-, K) \cong \mathrm{Hom}_K((-) \otimes_A A, K) \cong \mathrm{Hom}_A(-, DA),$$

which implies that $D\mathrm{Hom}_A(A, -)$ is finitely presented.

To show the claim it is enough to verify that DF is finitely presented if F is finitely presented in Add(A^{op}-mod, mod-K). In this case, there are modules M and N in A-mod and an exact sequence $\mathrm{Hom}_A(M, -) \to \mathrm{Hom}_A(N, -) \to F \to 0$, and therefore $0 \to DF \to D\mathrm{Hom}_A(N, -) \to D\mathrm{Hom}_A(M, -)$. By the proposition above, it is enough to show that $D\mathrm{Hom}_A(N, -)$ is finitely presented.

We have an exact sequence $A^{\oplus m} \to A^{\oplus n} \to N \to X$, which induces an exact sequence in Add(A^{op}-mod, mod-K),

$$D\mathrm{Hom}_A(A^{\oplus m}, -) \to D\mathrm{Hom}_A(A^{\oplus n}, -) \to D\mathrm{Hom}_A(N, -) \to 0.$$

By the first claim of the proof, the first two terms of the sequence are finitely presented, hence so is $D\mathrm{Hom}_A(N, -)$. □

The aim of this section is to prove the following important result.

2.5.6 Theorem *Every simple functor in* Add(A^{op}-mod, Ab) *is finitely presented.*

Proof First observe that for a finitely generated functor F in Add(A^{op}-mod, mod-K), if DF is also finitely generated then F is finitely presented. Indeed, we have an epimorphism $\mathrm{Hom}_A(-, M) \to F$ with M in A-mod, and hence an exact sequence

$$0 \to DF \to D\mathrm{Hom}_A(-, M).$$

Trivially, the functor $\mathrm{Hom}_A(-, M)$ is finitely presented, hence by Corollary 2.5.5 the functor $D\mathrm{Hom}_A(-, M)$ is also finitely presented. Since DF is finitely generated, by Proposition 2.5.5 the functor DF is finitely presented, and so is F again by Corollary 2.5.5, since $F \cong DDF$.

Now, if S is a simple functor in $\text{Add}(A^{op}\text{-mod}, \text{Ab})$, by Proposition 2.5.3 it is a finitely generated functor in $\text{Add}(A^{op}\text{-mod}, \text{mod-}K)$. Notice that DS is also a simple functor, and therefore finitely generated. By the above discussion, S is finitely presented, completing the proof. □

2.5.7 We are now ready to sketch a proof of Theorem 2.4.6.

Let N be an indecomposable module in A-mod, and consider the family of non-isomorphisms $\{f_i : M_i \to N\}_{i\in I}$ with M_i indecomposable. Fix the functors $F = \bigoplus_{i\in I} \text{Hom}_A(-, M_i)$ and $G = \text{Hom}_A(-, N)$, and the morphism $\eta = \bigoplus_{i\in I} \text{Hom}_A(-, f_i)$ from F to G in $\text{Add}(A^{op}\text{-mod}, \text{Ab})$.

Step 1 *The cokernel S of η is a simple functor.* First, if $\eta : F \to G$ is an epimorphism, then $\eta_N : F(N) \to \text{End}_A(N)$ is an epimorphism of $\text{End}_A(N)^{op}$ modules. Since $\text{End}_A(N)^{op}$ is a local ring, there is $i \in I$ such that $\text{Im}(\text{Hom}_A(N, f_i))$ is not contained in the maximal ideal of $\text{End}_A(N)^{op}$, thus there is $g_i : N \to M_i$ such that $f_i g_i = \mathbf{Id}_N$. This is impossible since M_i is indecomposable and f_i is not an isomorphism. Then η is not an epimorphism, and $S \neq 0$.

Let now G' be the image of η, and assume that H is a proper subfunctor of G. We want to show that H is a subfunctor of G'. With that purpose, let X be an indecomposable module in A-mod, and take $x \in H(X)$. Notice that there is a morphism $g : X \to N$ such that the inclusion $\text{Hom}_A(-, X) \hookrightarrow \text{Hom}_A(-, N)$ has the shape $\text{Hom}_A(-, g)$. Since H is a proper subfunctor of G, then g is not an isomorphism and there is $i \in I$ such that $g = f_i$. In particular $x \in G'(X)$, which implies that $H(X) \subset G'(X)$. Since X is arbitrary, then H is a subfunctor of the image G' of η. Then G' is the unique maximal subfunctor of G, and by Proposition 2.5.3, the functor S is simple.

Step 2 Considering an exact sequence of additive functors

$$F \xrightarrow{\ \eta\ } G \xrightarrow{\ \theta\ } S \longrightarrow 0\ ,$$

let $\text{Ker}(\theta)$ denote the kernel of θ. By Theorem 2.5.6, since S is simple there is a module M in A-mod and an epimorphism $\zeta : \text{Hom}_A(-, M) \to \text{Ker}(\theta)$. Therefore, by the universal property of kernels, and by the projectivity of F and $\text{Hom}_A(-, M)$ (since ξ is an epimorphism), there are morphisms ξ, χ and χ' such that the following diagram is commutative,

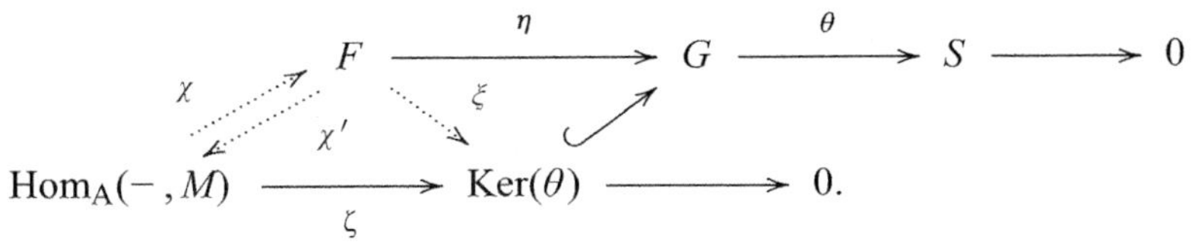

Moreover, since $\mathrm{Ker}(\theta) \subset G$, there is a morphism $f : M \to N$ such that the following composition coincides with $\mathrm{Hom}_A(-, f)$.

$$\mathrm{Hom}_A(-, M) \xrightarrow{\zeta} \mathrm{Ker}(\theta) \hookrightarrow G = \mathrm{Hom}_A(-, N) \ .$$

Step 3 *There is a finite subset $J \subset I$ and a commutative diagram as below right, and for each $i \in I$ there is a commutative diagram as below left,*

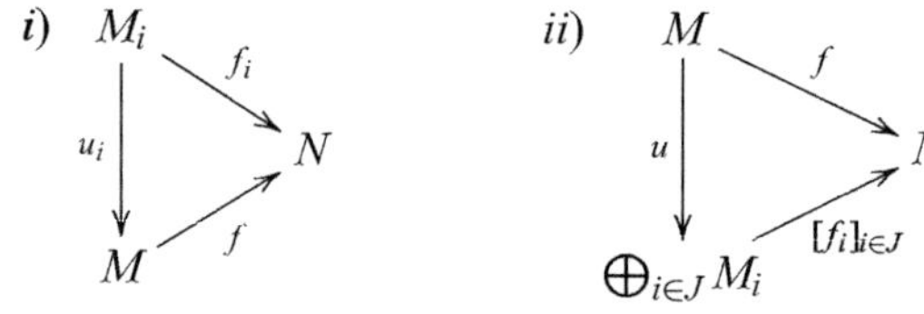

Indeed, diagram (i) is consequence of the identity $\eta = \mathrm{Hom}_A(-, f)\chi'$, and (ii) follows from $\mathrm{Hom}_A(-, f) = \eta\chi$, both identities coming from the commutative diagram above and the definition of $\mathrm{Hom}_A(-, f)$.

Step 4 *The morphism $f : M \to N$ is right almost split.* From Definition 2.4.4, we have to show: (a) that f is not a split epimorphism, and (b) that any non-split epimorphism $h : X \to N$ factors through f. For the first claim, using the diagram (ii) above, if ℓ is such that $\mathbf{Id}_N = f\ell = [f_i]_i(u\ell)$, then at least one of the morphism f_i is an isomorphism, which is impossible. For the second claim, since h is not a split epimorphism, we may write $X = \bigoplus_{i\in J'} M_i$ and $h = [f_i]_{i\in J'}$ for some finite subset J' of I. Using the diagram (i) we have

$$h = [f_i]_{i\in J'} = [fu_i]_{i\in J'} = f[u_i]_{i\in J'},$$

hence the claim.

Step 5 Observe now that if $\zeta : \mathrm{Hom}_A(-, M) \to \mathrm{Ker}(\theta)$ is a projective cover, then f is a minimal right almost split morphism. In case N is a projective module, it can be shown that f is isomorphic to the inclusion of the radical of N into N, that is, there is a commutative diagram

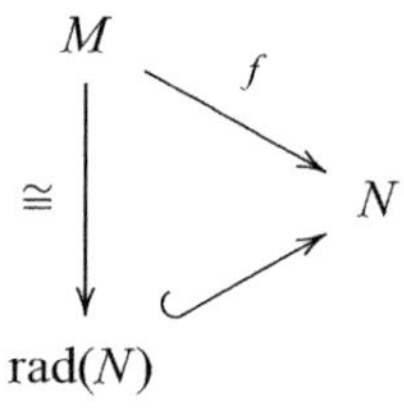

Assume that N is not projective. Then $f : M \to N$ is an epimorphism, for otherwise the projective cover of N which is not a split epimorphism, could not be factored through f. By Proposition 2.4.5, the exact sequence

$$0 \longrightarrow L \xrightarrow{g} M \xrightarrow{f} N \longrightarrow 0 ,$$

is almost split, and L is indecomposable.

Step 6 *If N is not projective, there is an isomorphism $L \cong D(\mathrm{Tr}\, N)$.*

2.6 First Brauer-Thrall Conjecture

Let $\mathbf{A}$ be a set representing isomorphism classes of finite dimensional A-modules, for a K-algebra A. Observe that by restriction, the categories

$$\mathrm{Add}(\text{mod-}A^{op}, \mathrm{Ab}) \quad \text{and} \quad \mathrm{Add}(\mathbf{A}^{op}, \mathrm{Ab}),$$

are equivalent. Moreover, the radical $\mathrm{rad}(\mathbf{A})$ is generated by the non-isomorphisms between indecomposable modules in $\mathbf{A}$, since those $M \in \mathbf{A}$ have a local endomorphism ring.

2.6.1 We include next a characterization of functor categories $\mathrm{Add}(\mathbf{A}, \mathrm{Ab})$ for $\mathbf{A}$ small and additive.

Theorem (H. Bass) *For a small additive category $\mathbf{A}$ the following are equivalent:*

(a) Each functor $F : \mathbf{A} \to \mathbf{Ab}$ has a projective cover $\pi : P \to F$.
(b) The following conditions hold:

(i) For every sequence

$$\cdots M_{n+1} \xrightarrow{r_n} M_n \xrightarrow{r_{n-1}} M_{n-1} \xrightarrow{r_{n-2}} \cdots \xrightarrow{r_1} M_1 \xrightarrow{r_0} M_0,$$

of morphisms r_i in $\mathrm{rad}(\mathbf{A})$, there is an integer m such that $r_0 r_1 \cdots r_m = 0$;
(ii) The quotient $\mathbf{A}/\mathrm{rad}(\mathbf{A})$ is semi-simple, that is, it decomposes as direct sum of simple functors.

(c) For each $M \in \mathbf{A}$, the hom-functor $\mathrm{Hom}_{\mathbf{A}}(-, M)$ satisfies the descending chain condition on finitely generated subfunctors.

2.6.2 We proceed to present a proof of the first Brauer-Thrall Conjecture. For this purpose, we say that a K-algebra A is of **bounded type** if there is an integer $n \in \mathbb{N}$ such that every indecomposable finite dimensional A-module has length

$\leq n$. We say that A is of **finite representation type** if the number of isoclasses of indecomposable A-modules is finite.

Theorem (**[14]**) *If A is a representation-bounded algebra then A is representation-finite.*

Proof That the quotient $\mathbf{A}/\mathrm{rad}\mathbf{A}$ is semi-simple follows from Proposition 2.5.3. Moreover, since A has bounded representation type, Harada-Sai's Lemma 2.2.2 implies that $\mathrm{rad}(\mathbf{A})$ has property (b) in Bass's Theorem 2.6.1.

Then Theorem 2.6.1(c) holds, and in particular any non-zero contravariant functor from $\mathbf{A}$ to Ab has a simple subfunctor, which is finitely presented by Theorem 2.5.6. We show that this implies that any finitely generated contravariant functor $M : \mathbf{A} \to \mathrm{Ab}$ has finite length. Indeed, let M' be the (directed) union of all finite length subfunctors M_γ of M. If M/M' is non-zero, then it has a simple subfunctor S, and there is a finitely generated subfunctor U of M such that the following short sequence is exact

$$0 \longrightarrow U \cap M' \longrightarrow U \xrightarrow{k'} S \longrightarrow 0\ ,$$

where k' is the restriction of the natural epimorphism $k : M \to M/M'$ to the simple subfunctor S. Since S is finitely presented, the functor $U \cap M'$ is finitely generated, and therefore contained in some M_γ. In particular, $U \cap M'$ and U have finite length, and therefore $U \subseteq M'$ and $S = 0$, a contradiction.

Now, let $S_1, \ldots, S_n$ be a complete set of representatives of simple A-modules, and consider the contravariant representable functors $\mathrm{Hom}_A(-, S_i)$ for $i = 1, \ldots, n$. Since each $\mathrm{Hom}_A(-, S_i)$ is finitely generated, by the above, $\mathrm{Hom}_A(-, S_i)$ has finite length. For any indecomposable A-module M, the quotient $M/\mathrm{rad}(M)$ is semi-simple, and thus there is an index i with $\mathrm{Hom}_A(M, S_i) \neq 0$. This means that any indecomposable A-module M is in the union

$$\bigcup_{i=1}^{n-1} \mathrm{supp}(\mathrm{Hom}_A(-, S_i)),$$

which is finite by Lemma 2.5.3. □

References

1. Anderson, F. W. and Fuller, K. R., *Rings and categories of modules*. Graduate texts in mathematics **13**, Springer-Verlag, New York 1974
2. Assem, I. and Simson, D. and Skowronski, A., *Elements of the representation theory of associative algebras*, Cambridge University Press (2006), London Mathematical Society Student Texts **65**
3. Auslander M. (1966) *Coherent Functors*. In: Eilenberg S., Harrison D.K., MacLane S., Röhrl H. (eds) Proceedings of the Conference on Categorical Algebra. Springer, Berlin, Heidelberg.

4. Auslander, M.: *Existence theorems for almost split sequences*, Proc. Conf. on Ring Theory II, Oklahoma, Marcel Dekker (1977), 1–44
5. Auslander, M. *Applications of morphisms determined by objects*, in Proc. of the Philadelphia Conf., Lecture Notes in Pure and Applied Math. **37** (1978), pp. 245–327
6. Auslander M. (1982) *A functorial approach to representation theory*. In: Auslander M., Lluis E. (eds) Representations of Algebras. Lecture Notes in Mathematics, vol 944. Springer, Berlin, Heidelberg. https://doi.org/10.1007/BFb0094058
7. Auslander, M. and Reiten, I. *Representation theory of Artin algebras. III. Almost split sequences*, Communications in Algebra **3** (3) (1975): 239–294
8. Auslander, M. and Reiten, I. and Smalø, S. *Representation theory of Artin algebras*, Cambridge University Press **36** Cambridge Studies in Advanced Mathematics (1995)
9. Eisenbud, D. and de la Peña, J. A.: *Chains of maps between indecomposable modules*. J. Reine Angew. Math. **504** (1998), 29–35
10. Harada, M. and Sai, Y. *On categories of indecomposable modules I*, Osaka J. Math., **8** 1971, 309–321
11. Hartshorne, R., *Algebraic geometry*. Graduate Texts in Mathematics **52**, Springer-Verlag, New York (1977)
12. Krause, H.: *A short proof for Auslander's defect formula*. Linear Agebra and its Applications **365** (2003) 267–270
13. Lam, T.Y.: *Bass's Work in Ring Theory and Projective Modules*, Contemporary Mathematic.
14. Roiter, A.V.: *Unboundedness of the dimension of the indecomposable representations of an algebra which has infinitely many indecomposable representations*, Izv. Akad. Nauk SSSR. Ser. Mat. **32** (1968) 1275–1282

Chapter 3
Constructive Methods

3.1 The Lattice of Ideals

In this section we assume that the reader is familiar with the following concepts:

3.1.1 If $f : R \to S$ is a ring homomorphism, then the kernel $\ker(f) = f^{-1}(0_S)$ is a two-sided ideal of R. For each left ideal I of S, the pre-image $f^{-1}(I)$ is a left ideal. If I is a left ideal of R, then $f(I)$ is a left ideal of the subring $f(R)$ of S.

Ideal correspondence: Given a surjective ring homomorphism $f : R \to S$, there is a bijective order-preserving correspondence between the left (resp. right, two-sided) ideals of R containing the kernel of f and the left (resp. right, two-sided) ideals of S. Moreover, for commutative rings, this bijective correspondence restricts to prime ideals, maximal ideals, and radical ideals. If M is a left R-module, the **annihilator** of a set $S \subset R$ is $\mathrm{Ann}_R(S) = \{r \in R \mid rs = 0, \text{ for all } s \in S\}$, which is a left ideal of R.

Let $\mathfrak{a}_i$ be an ascending chain of left ideals in a ring R indexed by a set $i \in S$. Then the union $\bigcup_{i \in S} \mathfrak{a}_i$ is a left ideal of R. This fact together with Zorn's lemma proves the following: if $E \subset R$ is a possibly empty subset and $\mathfrak{a}_0 \subset R$ is a left ideal that is disjoint from E, then there is an ideal that is maximal among the ideals containing $\mathfrak{a}_0$ and disjoint from E. In particular, there exists a left ideal that is maximal among proper left ideals (called a maximal left ideal).

3.1.2 Given a possibly empty subset X of R, there is the smallest left ideal containing X, called the left ideal generated by X and denoted by RX, it is the intersection of all left ideals containing X. Equivalently, RX is the set of all the (finite) left R-linear combinations of elements of X over R. A left (resp. right, two-sided) ideal generated by a single element x is called the **principal left** (resp. right, two-sided) **ideal generated by** x, denoted by Rx (resp. xR, RxR). The principal two-sided ideal RxR is often also denoted by (x). If every ideal of R can be generated by a single element, then R is a **principal ideal ring**.

J.-A. de la Peña, *Representations of Algebras*, Algebra and Applications 30,
https://doi.org/10.1007/978-3-031-12288-0_3

A **poset** (= partially ordered set) $(S, \leq)$ is a (complete) **lattice** if for any two elements $s, t \in S$ there is a supremum and an infimum of $T = \{s, t\}$. In this case write $s + t$ for the supremum (also known as **join**) and $s \cap t$ for the infimum (also known as **meet**). The lattice of ideals $\mathbb{L}(R)$ is always **modular**, that is, it satisfies the modular law

$$a \leq b \quad \text{implies} \quad a + (x \cap b) = (a + x) \cap b,$$

where $\leq$ is the partial order defining the lattice, and $+$ and $\cap$ are the operations of the lattice.

The sum and the intersection of ideals is again an ideal; with these two operations as join and meet, the set of all ideals of a given ring forms a **modular lattice**. The lattice is not, in general, a distributive lattice (see definition 3.1.4). If $\mathfrak{a}$ and $\mathfrak{b}$ are ideals of a commutative ring R, then $\mathfrak{a} \cap \mathfrak{b} \subset \mathfrak{a}\mathfrak{b}$. More generally, the difference between a product and an intersection of ideals is measured by the Tor functor:

$$\operatorname{Tor}_1^R(R/\mathfrak{a}, R/\mathfrak{b}) = (\mathfrak{a} \cap \mathfrak{b})/\mathfrak{a}\mathfrak{b}.$$

For instance, an integral domain is called a **Dedekind domain** if for each pair of ideals $\mathfrak{a} \subset \mathfrak{b}$, there is an ideal $\mathfrak{c}$ such that $\mathfrak{a} = \mathfrak{b}\mathfrak{c}$. It can then be shown that every nonzero ideal of a Dedekind domain can be uniquely written as a product of maximal ideals, a generalization of the fundamental theorem of arithmetic.

3.1.3 The above remarks can be summarized by saying that the (left) ideals of a ring R form a **lattice** $\mathbb{L}(R)$ with the ring operations. Properties of this lattice are related to properties of the ring.

Observe that the **pentagon lattice** N_5 is not modular. In fact, we have:

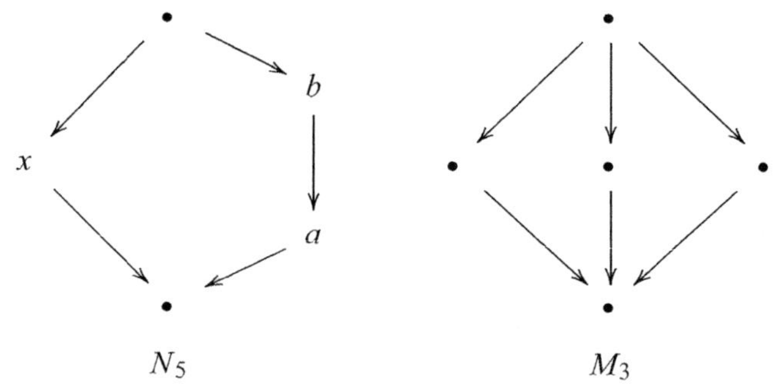

Proposition *A lattice is not modular if and only if it contains a sub lattice isomorphic to* N_5.

Proof Since N_5 is not modular, then the "if" implication follows.

Assume now that L is a non-modular lattice and let $a \leq b$ be such that

$$a + (b \cap c) \neq b \cap (a + c).$$

We first show that $a < b$, and none of the following holds: $(c = a)$, $(a = c)$, $(c = b)$ or $(b = c)$. Indeed, if $a = b$, then $a + (b \cap c) = a + (a \cap c) = a$ and $b \cap (b + c) = a \cap (a + c) = a$. Therefore $a + (b \cap c) = b \cap (a + c)$, contradicting the assumption. Hence $a \neq b$ and $a < b$. In a similar way we get $a \neq c$ and the rest of the claims.

Now consider the following elements of the lattice :

$$\begin{aligned} 1' &= a + c, \\ z &= c, \\ y &= b \cap (a + c), \\ x &= a + (b \cap c), \\ 0' &= b \cap c. \end{aligned}$$

We can prove that:

$$0' < x < y < 1',$$
$$0' < z < 1',$$

and z is not x or y. Moreover,

$$x \cap z = y \cap z = 0', \quad \text{and} \quad x + z = y + z = 1',$$

and

$$0' < x < y < 1', 0' < z < 1'.$$

Moreover, $z \cap x = 0' = z \cap y$ and $z + x = 1' = z + y$. Hence, the lattice formed by $\{0', x, y, z, 1'\}$ is embedded in L and isomorphic to N_5. □

3.1.4 A lattice is **distributive** if it satisfies the law:

$$a \cap (b + c) = (a \cap b) + (a \cap c).$$

Observe that neither the pentagon N_5 nor the diamond M_3 are distributive. In fact we have:

Proposition *For a lattice L the following are equivalent:*

(i) L is distributive.
(ii) None of the sub lattices of L is isomorphic to the pentagon N_5 or to the ***diamond*** *M_3.*
(iii) L is modular and it does not contain a diamond.

Many lattices can be defined in such a way that the elements of the lattice are represented by sets, the join operation of the lattice is represented by set union, and the meet operation of the lattice is represented by set intersection. Because set unions and intersections obey the distributive law, any lattice defined in this way is a distributive lattice. **Birkhoff's theorem** states that in fact all finite distributive lattices can be obtained this way.

For instance, consider the divisors of some composite number, such as drawing below for 120, partially ordered by divisibility. The join $a + b$ is given by the least common multiple and the meet $a \cap b$ as the unique common factor. Such a lattice is called **lattice of divisors**.

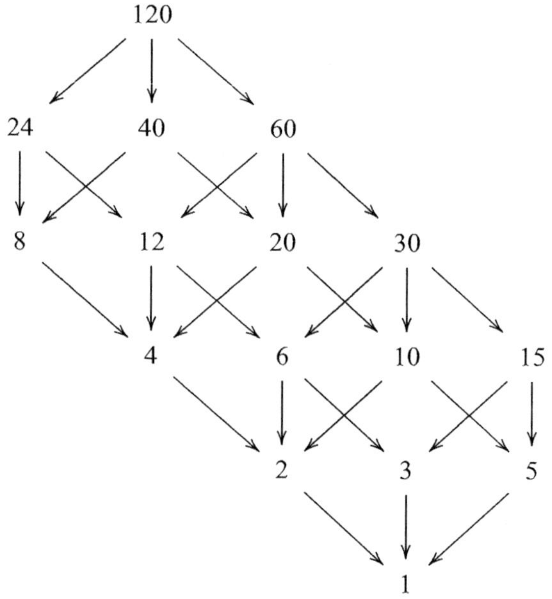

The interested reader should prove that the lattice of divisors is distributive.

3.1.5 Here we return to the main discussion of the text, that of finite dimensional associative K-algebras with unity, over an algebraically closed field K, or simply a K-algebra as in Chap. 1. Let A be K-algebra and fix a complete set $\{e_x\}$ of primitive orthogonal idempotents of A, and assume that if $e_x \neq e_y$ then Ae_x is not isomorphic to Ae_y as left A-modules (that is, A is a basic K-algebra, cf. 1.1.2.1). By Gabriel's Theorem 1.1.5.14, there is a quiver Q and an admissible ideal I of KQ such that A is isomorphic to the quotient KQ/I. In that case, we take the set of trivial paths of Q as set of primitive orthogonal idempotents of A. Denote by rad(A) the (Jacobson) radical of A, and by A_x the algebra e_xAe_x for a vertex $x \in Q_0$. For a $A_x - A_y$-bimodule J we denote by rad(J) the radical as a bimodule, and the higher radicals rad$^i(J)$ are defined by induction. We have $\text{rad}(A_x) = e_x[\text{rad}(A)]e_x$ for any x, and $\text{rad}(J) = e_x[\text{rad}(A)J + J\text{rad}(A)]e_y$ for any subbimodule J. The following straightforward claim is left as an exercise.

Lemma *Let $A = KQ/I$ be a K-algebra with vertices $x, y \in Q_0$. If J is a $A_x - A_y$-bimodule, and $\langle J \rangle$ denotes the two-sided ideal of A generated by J, then*

$$\langle J \rangle = \text{rad}(A)J + J\text{rad}(A) + J \text{ and } e_x \langle J \rangle e_y = \langle J \rangle \cap e_x A e_y = J.$$

3.1.6 Let $\mathbb{L}$ be the lattice of two-sided ideals of A. Denote by $B(x, y)$ the lattice of $A_x - A_y$-subbimodules of $e_x A e_y$, for vertices $x, y \in Q_0$. By Lemma 3.1.5, there is a bijection of lattices $e_x \mathbb{L} e_y$ and $B(x, y)$. Since the map $\mathbb{L} \to e_x \mathbb{L} e_y$ preserves intersections, sums and inclusions, the following transformation of lattices, induced by left and right multiplication with e_x and e_y respectively, is surjective

$$\mathbb{L} \longrightarrow e_x \mathbb{L} e_y \cong B(x, y).$$

Lemma *Let $A = KQ/I$ be a K-algebra with vertices $x, y \in Q_0$. The following are equivalent.*

(a) $B(x, y)$ is a distributive lattice.
(b) $\dim_K[\text{rad}^i(e_x A e_y)/\text{rad}^{i+1}(e_x A e_y)] \leq 1$, for all $i \geq 0$.
(c) $e_x A e_y$ is a uniserial bimodule, that is, it has a unique chain of subbimodules.

Proof Suppose that the vector space $V = (\text{rad}^i e_x A e_y/\text{rad}^{i+1} e_x A e_y)$ has dimension ≥ 2 for some i. Then $\text{rad}^i(e_x A e_y)$ lies in $\text{rad}(A)$. Namely, for $x \neq y$ we have $e_x A e_y \subset \text{rad}(A)$, and for $x = y$ we have $i \geq 1$. In V there is a plane containing three different lines violating the law of distributivity. Their preimages L_1, L_2, L_3 under the canonical projection are subbimodules also violating distributivity and so $B(x, y)$ is not distributive.

We have just seen that the distributivity of $B(x, y)$ implies for all i that

$$\dim_K[\text{rad}^i(e_x A e_y)/\text{rad}^{i+1}(e_x A e_y)] \leq 1.$$

It follows easily that $B(x, y)$ is uniserial whence distributive. This proves the claim. □

3.1.7 Since e_x is a primitive idempotent for any vertex $x \in Q_0$, then $A_x = e_x A e_x$ is a local algebra (cf. 2.2.1.2), therefore A_x is uniserial if and only if $\text{rad}(A_x)$ has at most one generator. Moreover,

Lemma *If A_x and A_y are uniserial, then $e_x A e_y$ is a uniserial as $A_x - A_y$-bimodule, if and only if*

$$\dim_K[\text{rad}^i(e_x A e_y)/\text{rad}^{i+1}(e_x A e_y)] \leq 1,$$

for $i = 0$ and $i = 1$. In that case, $e_x A e_y$ is uniserial as a left A_x-module and as a right A_y-module.

Proof If one of the spaces $e_x A e_x$, $e_y A e_y$ or $e_x A e_y$ has dimension ≤ 1, then the claim is obvious. Otherwise, let a be a generator of $e_x A e_y$. Then $\alpha_x a$ and $a\alpha_y$ are not linearly independent modulo $\operatorname{rad}^2(e_x A e_y)$. Up to symmetry we can assume that we have $\alpha_x a = \lambda a \alpha_y + r$ for some scalar λ and some $r \in \operatorname{rad}^2(e_x A e_y)$. Then we obtain

$$\alpha_x^p a \alpha_y^p = \lambda^p a \alpha_y^{p+q} + r(p,q).$$

with some $r(p,q) \in \operatorname{rad}^{p+q+1}(e_x A e_y)$ for all p and q by induction on p. Now the elements $\alpha_x^i a \alpha_y^{n-i}$ with $i \geq n$ generate $\operatorname{rad}^n(e_x A e_y)$ for any n and this space is zero for large n. By descending induction it follows that all $\operatorname{rad}^i(e_x A e_y)$ are generated by the $a\alpha_y^j$ with $j \geq i$. Thus $e_x A e_y$ is cyclic as a module over $e_y A e_y$, hence uniserial. □

3.1.8 A K-algebra A is called **distributive** provided its lattice of two-sided ideals $\mathbb{L}$ is distributive. Let $\mathbb{L}'$ denote the sublattice of $\mathbb{L}$ consisting of the ideals contained in $\operatorname{rad}(A)$. The following important characterization of distributive algebras, due to Jans [8] and Kupisch [9], can be found in [6].

Theorem *For a basic K-algebra A, the following are equivalent.*

(a) A is distributive, that is, $\mathbb{L}$ is a distributive lattice.
(b) $\mathbb{L}'$ is distributive.
(c) All the lattices $B(x, y)$ are distributive.

Proof If $\mathbb{L}$ is distributive so is its sublattice $\mathbb{L}'$. From this we obtain by the arguments in the proof of Lemma 3.1.7, that $\dim_K(\operatorname{rad}^i e_x A e_y / \operatorname{rad}^{i+1} e_x A e_y) \leq 1$ for all i and all vertices x, y. Thus all $B(x, y)$ are distributive by part (iii). Finally, if (c) holds, then using the relation $I = \bigoplus_{x,y} e_x I e_y$ valid for any two-sided ideal one gets that $\mathbb{L}$ is distributive. □

3.1.9 We now turn our attention to the minimal non-distributive algebras, and show that these algebras fall into three disjoint classes. Let $A = KQ/I$ be a K-algebra. A pair (a, z) of vertices is called **critical** if the bimodule $e_z A e_a$ is not uniserial. By Lemma 3.1.6, there is a smallest natural number i such that $\operatorname{rad}^i(e_z A e_a)/\operatorname{rad}^{i+1}(e_z A e_a)$ has dimension ≥ 2, referred to as the critical index $i(a, z) := i$ of such a pair (a, z). Furthermore, given a vertex x in Q, we denote by I_x the two-sided ideal generated by all paths of lengths 2 with x as the interior point. The vertex x is called a **node** if $I_x \subseteq I$. Let S be the socle of A as a bimodule.

Theorem *Let $A = kQ/I$ be an algebra which is not distributive but any proper quotient is. Then the following holds:*

(i) For any critical pair (a, z) with critical index i we have $\mathrm{rad}^{i+1} e_z A e_a = 0$ and $S(a, z) := \mathrm{rad}^i e_z A e_a$ is a bimodule of dimension 2 which is contained in S.
(ii) There is only one critical pair (a, z) and we have $S = S(a, z)$. Moreover we are in one of the following three situations:

*(a) **Type 1**: $a = z$, $i(a, z) = 1$, $e_a A e_a \cong k[X, Y]/(X, Y)^2$ and $I_a \subset I$.*
*(b) **Type 2**: $a \neq z$, $i(a, z) = 0$, a is a source, z a sink in Q and for $e = e_a + e_z$ the algebra eAe is isomorphic to the path-algebra of the Kronecker quiver $\mathbb{K}_2$ consisting of two parallel arrows.*
*(c) **Type 3**: $a \neq z$, $i(a, z) = 1$ and for $e = e_a + e_z$ the algebra eAe is isomorphic to the path algebra of the quiver with one loop α in a, one arrow β from a to z and one loop γ in z divided by the relations $\alpha^2 = \gamma^2 = \gamma\beta\alpha = 0$.*

Proof We consider the two-sided ideal J generated by $\mathrm{rad}^{i+1}(e_z A e_a)$. Then we have $e_z A e_a \cap J = \mathrm{rad}^{i+1}(e_z A e_a)$ whence the quotient A/J is still not distributive. By minimality we have $J = 0$ and a fortiori $\mathrm{rad}^{i+1}(e_z A e_a) = 0$. Similarly, if $V := [\mathrm{rad}(A)\mathrm{rad}^i(e_z A e_a) + \mathrm{rad}^i(e_z A e_a)\mathrm{rad}(A)] \neq 0$ we look at the non-zero two-sided ideal J it generates. Because of $J \cap e_z A e_a = 0$ the proper quotient A/J is again not distributive and so $J = 0$ and a fortiori $V = 0$. This means that $\mathrm{rad}^i(e_z A e_a)$ is contained in S. If the dimension of $\mathrm{rad}^i(e_z A e_a)$ is strictly greater than 2 we choose a non-zero subbimodule J of codimension 2 in $\mathrm{rad}^i(e_z A e_a)$. Then J is even a two-sided ideal and A/J is still not distributive. This contradiction shows that $\dim_K S(a, z) = 2$.

There is at least one critical pair (a, z) and we have $S = S(a, z) \oplus S'$ for some two-sided ideal S'. This ideal is zero because A/S' is still representation infinite. Thus we have $S = S(a, z)$ and there is only one critical pair. We discuss the different possibilities.

For $a = z$ we have $i = i(a, z) = 1$ and $e_a A e_a \cong k[X, Y]/(X, Y)^2$. For any path $p = \beta\alpha$ of length 2 with interior point a we consider the two-sided ideal J generated by $\overline{p}$. For any paths v, w we have that $e_a \overline{v\beta}$ and $\overline{\alpha w t} e_a$ are in $\mathrm{rad}(e_a A e_a)$ whence their product vanishes and $J \cap e_a A e_a = 0$. Thus A/J is still not distributive and we have $J = 0$ by minimality. Thus we have $I_a \subseteq I$.

For $a \neq z$ all $e_x A e_x$ are uniserial rings and we can apply the last part of Proposition 3.1.8 to see that only $i = i(a, z) = 0$ and $i = 1$ are possible. In the case $i = 0$ we have $S(a, z) = e_z A e_a$. Take an element f in some $e_a \mathrm{rad}(A) e_y$. Then the two-sided ideal J generated by f is spanned by products vfw and the intersection with $e_z A e_a$ by products $e_z v f w e_a$. This product vanishes because f annihilates the element $e_z v$ from $S(a, z)$. Thus A/J is still not distributive and we conclude $J = 0$ whence $f = 0$. It follows that x is a source. Dually z is a sink and so eAe has the wanted form.

Finally we look at the case $a \neq z$, $i = 1$ and $S = \mathrm{rad}(e_z A e_x)$. Let f be in $\mathrm{rad}^2(e_a A e_a)$ and let J be the two-sided ideal generated by f. Then the intersection of J with $e_z A e_a$ is spanned by products $e_z v f w e_a$ which are all 0. We get that $J = 0$ and $f = 0$, i.e. $\dim_K(e_a A e_a) = 2$ and dually $\dim_K(e_z A e_z) = 2$. Let f be an element in $e_a A e_z$ such that the intersection of the ideal J generated by f with $e_z A e_a$ is not 0. Then there is a product $e_z v f w e_a \neq 0$. The non-zero products $f w e_a$ and $e_z v f$ show that the quiver of eAe has no loops and so it is an oriented cycle. But then eAe is uniserial. Thus the intersection $J \cap e_z A e_a$ is zero, $J = 0$ and $f = 0$. It follows that eAe has the wanted shape. □

3.2 Other Brauer-Thrall Conjectures

3.2.1 Let A be a K-algebra. The modules to be considered are left A-modules, and not necessarily of finite dimension. A module M is said to be of **finite type**, provided M is the direct sum of countably many copies of a finite number of indecomposable modules of finite dimension, and let $M = \oplus_{j=1}^{s} M_j^{S(j)}$ be such a decomposition, that is, M_j is an indecomposable of finite dimension, and $s \in \mathbb{N}$. We have already met the following fundamental result in a weaker form:

Theorem (Krull-Schmidt-Remak-Azumaya) *If an A-module M has a direct decomposition*

$$M = \bigoplus_{i \in \mathbb{Z}} M_i,$$

where each $\mathrm{End}_A(M_i)$ *is local, then it is in fact an indecomposable decomposition and every non-zero direct summand of M has an indecomposable direct summand. In particular, this happens when all M_i are indecomposable.*

Theorem *A module M which is not of finite type contains indecomposable submodules of arbitrarily large finite length.*

3.2.2 The main result of the section was proved by Roiter for finite dimensional algebras [12] and later completed for artinian rings by Auslander [1]. Recall that A is said to be **representation finite** provided there are only finitely many isomorphism classes of indecomposable A-modules of finite dimension, otherwise A is called representation-infinite. The **first Brauer-Thrall conjecture** asserts that a representation-infinite algebra has indecomposable submodules of arbitrarily large finite dimension. The conjecture was solved by Roiter [12] in 1968 by proving the somewhat stronger Theorem above. [Indeed, assume there are infinitely many isomorphism classes of indecomposable modules M_i; take the direct sum $M = \oplus M_i$. By Theorem 3.2.1, this module M is not of finite type. Thus we obtain indecomposable modules of arbitrarily large finite length as submodules of this particular module M].

The Brauer-Thrall conjectures, for finite dimensional algebras A, first appeared in 1957 in a paper by Thrall's student J. Jans [8]. The first Brauer-Thrall conjecture characterizes the representation finite algebras as those with an upper bound for the dimensions of the indecomposables, whereas the second Brauer-Thrall conjecture says that for a representation-infinite algebra A there are infinitely many dimensions in which one finds infinitely many isomorphism classes of indecomposable A-modules. The first Brauer-Thrall conjecture was proven by Roiter in 1968 and there is the long article [10] of Nazarova and Roiter aiming at a proof of the second Brauer-Thrall conjecture, but the first complete proof was only given by Bautista in 1983 [4].

3.2.3 Jans proves the second Brauer-Thrall conjecture in his paper for algebras that are not distributive. Observe that the direct sum theorem implies the following: If an Artin algebra A is of finite type, then any A-module is of finite type. We should stress that the converse implication is an obvious consequence of the Krull-Remak-Schmidt-Azumaya theorem.

Theorem (Jans) *Assume that K is infinite. If A is not distributive, then there is an infinite family of pairwise non-isomorphic finite-dimensional indecomposable A-modules of the same length.*

3.2.4 Ringel defines recursively the notion of an **accessible module** as follows: All simple modules are accessible, and a module of length $d \geq 2$ is accessible provided it is indecomposable and it admits an accessible submodule or factor module of length $d - 1$.

Theorem (Ringel) *If A is not distributive it has an accessible module of length d for each $d \geq 1$.*

Corollary *If A is representation finite, then A is a distributive algebra.*

3.2.5 Related to the second Brauer-Thrall conjecture several efforts were invested along the years. One was to give an inductive proof of the conjecture as follows: given an Artin algebra A and a positive integer n such that there exists $\aleph \geq \aleph_0$ non-isomorphic indecomposable modules of length n (this is hypothesis $H(n)$, here $\aleph_0$ stands for the cardinality of a countable set). Then there should exist infinitely many positive integers n_t satisfying $H(n_t)$. M. Auslander used this approach to prove what he called **Brauer-Thrall one and a half**. Namely, if the number of non-isomorphic finitely generated indecomposable modules over an Artin algebra A is $\aleph \geq \aleph_0$, then for each positive integer n, $H(n)$ holds.

On the time this was shown (1978), almost-split sequences were recently defined and their properties were been deduced. The story is told by Smalø (see [15]).

Lemma *Let A be an Artin algebra and $f : M \to N$ an irreducible map between indecomposable modules, then:*

$$|\ell(M) - \ell(N)| \leq \ell(M)m^2,$$

where m is the length of A as A-module.

Proof First we observe that there is a finite chain of irreducible maps $M = M_0 \to M_1 \to \ldots \to M_s$ (for some orientation of the maps), with M_s a simple module. We state that M satisfies $H(M)$ if there exist $\aleph \geq \aleph_0$ non-isomorphic indecomposable modules of length n. This is a consequence of the above lemma and elementary properties of irreducible maps. □

3.3 The Post-Projective Components of a Triangular Algebra

In this section we characterize the existence of post-projective components in the Auslander-Reiten quiver of triangular algebras. For simplicity, we fix indecomposable representative modules of each isomorphism class, and identify such class with the chosen representative.

3.3.1 The vertices of the **Auslander-Reiten quiver** Γ_A of a K-algebra A are the isoclasses of finite dimensional indecomposable A-modules (for which we fix a representative, identified with its isomorphism class), and the number of arrows from X to Y is $\dim_K \left[\mathrm{radHom}_A(X, Y)/\mathrm{rad}^2\mathrm{Hom}_A(X, Y)\right]$.

A **path** in A-mod is a sequence $(X_0, \ldots, X_s)$ of (isomorphisms classes of) indecomposable A-modules X_i, for $0 \leq i \leq s$, such that there is a map $0 \neq f_i \in \mathrm{Hom}_A(X_i, X_{i+1})$ which is not an isomorphism, for $0 \leq i \leq s-1$. In this case we write $X_0 \leq X_s$ and we say that X_0 is a **predecessor** of X_s. If $s \geq 1$ and $X_0 = X_s$ we say that the path $(X_0, \ldots, X_s)$ is a **cycle**.

Following [13] we say that a module M is **directing** in A-mod provided there do not exist indecomposable direct summands M_1 and M_2 of M and an indecomposable non-projective module X such that $M_1 \leq \tau X$ and $X \leq M_2$. It is shown in [13] that an indecomposable module X is directing if and only if there are no cycles $(X_0, \ldots, X_s)$ with $X_0 = X = X_s$.

3.3.2 Theorem (Auslander) *Let $\mathcal{M}$ be a connected component of Γ_A. If there is an upper bound for the dimension of indecomposable modules in $\mathcal{M}$, then $\Gamma_A = \mathcal{M}$.*

Proof Assume that M and N are indecomposable modules, and that M belongs to $\mathcal{M}$. First we show that if $h : M \to N$ is non-zero, then N also belongs to $\mathcal{M}$. Indeed, if N is not in $\mathcal{M}$, then h is not a split monomorphism, and it factorizes through the right almost split morphism $g : M \to M'$. Taking $M_0 = M$, then there is an indecomposable direct summand M_1 of M' and morphisms $g_0 : M_0 \to M_1$

and $h_1 : M_1 \to N$ with non-zero composition. Clearly M_1 belongs to $\mathscr{M}$, and in this way we construct an infinite sequence

$$M_0 \xrightarrow{g_0} M_1 \xrightarrow{g_1} M_2 \xrightarrow{g_2} M_3 \xrightarrow{g_3} M_4 \dots$$

of indecomposable modules M_i in $\mathscr{M}$, with non-zero composition $g_r \dots g_1 g_0$ for any $r \geq 0$. This contradicts Harada-Sai's Lemma 2.2.2.2. Similarly it is shown that if $h : N \to M$ is non-zero, then N is an indecomposable in $\mathscr{M}$.

Now, there is an indecomposable projective A-module P with $\operatorname{Hom}_A(P, M) \neq 0$. Hence $P \in \mathscr{M}$. By connectivity, for any other indecomposable projective module P' there is a sequence of indecomposable projectives $P_0, P_1, \dots, P_t$ with $P_0 = P$ and $P_t = P'$ such that there is either a non-zero morphisms $P_i \to P_{i+1}$ or $P_{i+1} \to P_i$ for $i = 0, \dots, t-1$. By the above, all indecomposable projective modules belong to $\mathscr{M}$, which also shows that every indecomposable module belongs to $\mathscr{M}$. □

3.3.3 A connected component $\mathscr{P}$ of Γ_A is **post-projective** if $\mathscr{P}$ has no oriented cycles and each module X in $\mathscr{P}$ has only finitely many predecessors in the path order of $\mathscr{P}$. Several important classes of algebras have post-projective components: hereditary algebras [2, VIII], algebras satisfying the separation condition [4], tilted algebras [7].

Let $A = KQ/I$ be a finite dimensional K-algebra such that the quiver Q has no oriented cycles. We may consider A as a K-category with objects the set of vertices Q_0 of Q and morphisms from $x, y \in Q_0$ the space $A(x, y) = e_y A e_x$, where e_x denotes the trivial path at the vertex x. For two vertices $x, y \in Q_0$ we write $y \leq x$ if there is a path from y to x in Q.

For $x \in Q_0$, we denote by A^x the full subcategory of A whose vertices are those $y \in Q_0$ with $y \not\leq x$. Observe that the quiver Q^x of A^x is a convex (that is, path closed) subquiver of Q. The indecomposable projective A-module $P_x = Ae_x$ has radical $\operatorname{rad}(P_x)$ which is an A^x-module.

The following result will be important in our work, see [11] and [14].

3.3.4 Theorem *Let $x \in Q_0$. Then P_x is directing in A-mod if and only if $\operatorname{rad}(P_x)$ is directing in A-mod. Moreover, if x is a source, then P_x is directing in A-mod if and only if $\operatorname{rad}(P_x)$ is directing in A^x-mod.*

We state the main result of the section which provides an algorithmic criterion for the existence of post-projective components. We denote by $\operatorname{rad}(P_x) = \bigoplus_{i=1}^{n_x} R_i^x$ the indecomposable decomposition of $\operatorname{rad}(P_x)$.

3.3.5 Theorem *Let $A = KQ/I$ be a finite dimensional K-algebra such that Q has no oriented cycles. Then Γ_A has a post-projective component if and only if for each vertex $x \in Q_0$ one of the following conditions is satisfied:*

(1x) There is a post-projective component $\mathscr{P}$ of Γ_{A^x} such that $R_i^x \notin \mathscr{P}$ for every $1 \leq i \leq n_x$.

(2x) *For each* $1 \leq i \leq n_x$ *the set of predecessors* $\{Y \in \Gamma_{A^x} : Y \leq R_i^x\}$ *of* R_i^x *in* A^x-mod *is finite and formed by directing modules. Moreover, if* x *is a source, then* $\mathrm{rad}(P_x)$ *is directing in* A^x-mod.

We prove Theorem 3.3.5 after some preparation.

3.3.6 Lemma *Assume that for all* $x \in Q_0$ *the condition* $(2x)$ *is satisfied, then* Γ_A *has a post-projective component.*

Proof We claim that for every $x \in Q_0$ the following condition is satisfied:

(3x) For each $1 \leq i \leq n_x$, the set of predecessors $\{X \in \Gamma_A : X \leq R_i^x\}$ of R_i^x in A-mod is finite and formed by directing modules.

Indeed, let X be a predecessor of R_i^x in Γ_A and assume that X is not an A^x-module. We may assume that x is minimal with this property in the path order of Q. Then there exists a vertex $y \leq x$ in Q such that $X(y) \neq 0$. Therefore in A-mod we get

$$P_y \leq X \leq R_i^x \leq P_x \leq P_y.$$

Since $(2y)$ is satisfied, then by 3.3.1 y is not a source in Q. Let z be a proper predecessor of y in Q. Therefore, P_y is a non-directing predecessor of some R_j^z. By $(2y)$, P_y is not an A^z-module, contradicting the minimality of x.

We construct inductively full subquivers C_n of Γ_A satisfying (cf. [5][Theorem 2.5]):

(i) C_n is finite, connected, contains no oriented cycle and is closed under predecessors.
(ii) $\tau_A^{-1} C_n \cup C_n \subset C_{n+1}$.

Then $\bigcup_n C_n$ forms the wanted post-projective component.

Set $C_0 = \{S\}$ where S is a simple projective A-module. Assume C_n to be defined and let $M_1, \ldots, M_t$ be the modules in C_n with $\tau_A^{-1} M_i \notin C_n$. We may assume that $M_i \leq M_j$ implies $i \leq j$. If $t = 0$, set $C_{n+1} = C_n$. Otherwise we define full subquivers D_i $(0 \leq i \leq t)$ of Γ_A satisfying $D_0 = C_n$, $D_i \cup \{\tau_A^{-1} M_{i+1}\} \subset D_{i+1}$ and condition (i) imposed on D_i. Then $C_{n+1} = D_t$ will satisfy conditions (i) and (ii).

Indeed, assume D_i is well defined. Take the almost split sequence $0 \to M_{i+1} \to X \to \tau_A^{-1} M_{i+1} \to 0$ and define D_{i+1} as the full subquiver of Γ_A with vertices D_i and all predecessors of $\tau_A^{-1} M_{i+1}$. It is enough to show that for each indecomposable direct summand Y of X, the set of predecessors $\{Z \in \Gamma_A : Z \leq Y\}$ is finite and formed by directing modules. If Y is not projective, then $\tau_A Y \in C_n$ whence Y belongs to D_i and we are done. If $Y = P_y$ is projective, then $(3y)$ is satisfied. By 3.3.1, we get the result. □

3.3.7 Proof of Theorem 3.3.5.

Let $\mathscr{P}$ be a post-projective component of Γ_A. Let $x \in Q_0$. If the projective module P_x belongs to $\mathscr{P}$, then $(2x)$ is satisfied. Assume that $P_x \notin \mathscr{P}$. We show that $\mathscr{P}$ is formed by A^x-modules. Let $X \in \mathscr{P}$ and assume $X(y) \neq 0$ for some $y \leq x$ in Q. Then $P_x \leq P_y \leq X$ in A-mod, which implies $P_x \in \mathscr{P}$, a contradiction. Hence $\mathscr{P}$ is a post-projective component in Γ_{A^x} and $R_i^x \notin \mathscr{P}$ for $1 \leq i \leq n_x$, that is $(1x)$ is satisfied.

Conversely, assume that for each $x \in Q_0$, one of the conditions $(1x)$ or $(2x)$ is satisfied. If for every $x \in Q_0$, $(2x)$ is satisfied then 3.3.6 implies the result.

Assume that for $x \in Q_0$, $(2x)$ is not satisfied. Choose a minimal such x in the path order in Q. By hypothesis $(1x)$ is satisfied, that is, there is a postprojective component $\mathscr{P}$ of Γ_{A^x} such that $R_i^x \notin \mathscr{P}$ for every $1 \leq i \leq n_x$. We shall prove that $\mathscr{P}$ is a component of Γ_A. For this purpose it is enough to show that x is a source in Q.

Assume $y \leq x$ is a source in Q and $y \neq x$. The minimality of x implies that $(2y)$ is satisfied. We will show that $(2x)$ is also satisfied which yields the wanted contradiction. Indeed, let X be a predecessor of R_i^x in A-mod. Then $X \leq R_i^x \leq P_x \leq P_y$, implies that X is a predecessor of R_j^y for some $1 \leq j \leq n_y$. Moreover, since P_y is directing in A-mod, then X is an A^y-module. Thus $\{X \in \Gamma_A : X \leq R_i^x\}$ is finite and formed by directing modules. Our theorem is proved.

3.3.8 Corollary *Let $A = KQ/I$ be as in Theorem 3.3.5 and assume Q is connected. Then all indecomposable projective modules belong to a post-projective component if and only if for every $x \in Q_0$ the condition $(2x)$ is satisfied.*

Proof The "only if" direction is clear. For the converse, assume that for every $x \in Q_0$, the condition $(2x)$ is satisfied. By Theorem 3.3.5 there is a post-projective component $\mathscr{P}$ of Γ_A. Clearly we may assume that Q is connected (otherwise we take a post-projective component for each maximal connected full subcategory of A). Let x_0 be a sink in Q such that the projective $P_{x_0} \in \mathscr{P}$. Let $x \in Q_0$ and fix a walk $x_0 \overset{\alpha_1}{—} x_1 \overset{\alpha_2}{—} \cdots \overset{\alpha_s}{\cdot} x_s = x$ in Q (that is, each α_i is an arrow in Q with some orientation). By induction, we may assume that $P_{x_{s-1}} \in \mathscr{P}$. If $x_{s-1} \xrightarrow{\alpha_s} x_s$, then P_x is a predecessor of $P_{x_{s-1}}$ and $P_x \in \mathscr{P}$. Thus, assume that $x_s \xrightarrow{\alpha_s} x_{s-1}$. Then there is a morphism $f : P_{x_{s-1}} \to \mathrm{rad}(P_x)$. Since $(2x)$ is satisfied, then f is a linear combination of compositions of finitely many irreducible maps. Hence $R_i^x \in \mathscr{P}$ for some $1 \leq i \leq n_x$. Thus $P_x \in \mathscr{P}$ and we are done. □

3.3.9 Given a post-projective component $\mathscr{P}$ of Γ_A, the modules on $\mathscr{P}$ can be easily determined. Starting with the simple projective modules and using the additivity of the dimension function on Auslander-Reiten sequences, the classes $\mathbf{dim}X$ and therefore, the unique indecomposable with class $\mathbf{dim}X$. This **knitting procedure** is used at least since 1974. The purpose now is to give an algorithmic procedure to construct all post-projective components in Γ_A. Indeed, there is an algorithm which decides whether or not a given simple projective module P_i belongs to a post-projective component.

3.3.10 Lemma *Let X be an indecomposable module in a post-projective component $\mathscr{P}$ of Γ_A. Then* $\mathrm{Ext}^s_A(X, X) = 0$ *for* $s \geq 1$ *and* $\dim_K \mathrm{End}_A(X) = 1$. *Moreover, if Y is indecomposable with* $\mathbf{dim}X = \mathbf{dim}Y$*, then* $X \cong Y$.

Proof There is a quotient B of A such that X is a faithful B-module. Then $\mathrm{gl.dim}(B) \leq 2$ and $\mathrm{p.dim}(X) \leq 1$ (see for instance [11, Theorem 2.4 (7)]). Hence if Y is an indecomposable A-module with $\mathbf{dim}X = \mathbf{dim}Y$, then Y is a B-module and $1 = \langle \mathbf{dim}X, \mathbf{dim}Y\rangle_B$ and $\mathrm{Hom}_A(X, Y) \neq 0$. Similarly, $\mathrm{Hom}_A(Y, X) \neq 0$, which implies that there is a cycle passing through X and Y, which contradicts the fact that $\mathscr{P}$ is post-projective unless $X = Y$. □

3.3.11 Given $i \in Q_0$, consider the quotient A^i of A formed as the full subcategory of A with vertices j such that there is no path from j to i in Q (recall that there is always a trivial path from i to i). We shall show that if starting with the simple projective module P_i it is possible to use the knitting procedure to construct $N(\dim_K A)$ new A-modules (where $N(\dim_K A)$ is a number depending only on $\dim_K A$) and whether or not P_i lies in a post-projective component of Γ_A.

Let $\mathscr{C}$ be a component of the Auslander-Reiten quiver Γ_A of an algebra A, and let $\mathscr{S}$ be a full subquiver of $\mathscr{C}$. We say that $\mathscr{S}$ is a **section** if $\mathscr{S}$ is path-closed in $\mathscr{C}$, if $X \in \mathscr{S}$ implies $\tau_A X \notin \mathscr{S}$, and each $X \in \mathscr{S}$ is directing. Moreover, $\mathscr{S}$ is called a **m-complete section** (for $m \in \mathbb{N} \cup \{\infty\}$) if:

(a) $\mathscr{S}$ is a section.
(b) $\mathscr{S}$ admits only finitely many predecessors in Γ_A, all of them directing.
(c) If $X \to Y$ is an arrow in Γ_A, $X \in \mathscr{S}$ and $Y \notin \mathscr{S}$, then Y is non-projective and $\tau_A Y \in \mathscr{S}$.
(d) If $X \in \mathscr{S}$, $0 \leq \ell \leq m$ and Y is a predecessor of $\tau_A^{-\ell} X$ such that $Y = I_j$ or Y is a direct summand of $\mathrm{rad}(P_j)$, then Y is a proper predecessor of $\mathscr{S}$.

3.3.12 Theorem *Let $\mathscr{S}$ be a connected component of a m-complete section in a component $\mathscr{C}$ of Γ_A and assume that $\mathscr{S}$ is not of Dynkin type. Let M be the maximal of all* $\dim_K Y$*, where $Y = I_j$ or Y is a direct summand of a* $\mathrm{rad}(P_j)$*, for some $j \in Q_0$. Suppose $m + 1 > Mn^2$. Then $\mathscr{S}$ is ∞-complete section.*

Let P_x be a simple projective module. Assume that in the step s, using the knitting procedure, we have constructed $N(s)$ modules $P_x = X_0, \dots, X_{N(s)}$. We say that the procedure is successful in the step $s + 1$ if we can construct a new module $X_{N(s)+1}$ from the set $X_0, \dots, X_{N(s)}$ such that the set of new predecessors $X_{N(s)+1}, \dots, X_{N(s+1)}$ of $X_{N(s)+1}$ is finite and formed by directing modules.

3.3.13 Theorem *The simple projective P_x belongs to a post-projective component of Γ_A if and only if the procedure is successful for every step $s \geq 0$. Moreover, if the procedure fails, then it fails for some $s \leq N_0 := 2n^2 \max\{Mn^2, 16\}$.*

We consider a **maximal ∞-complete section** $\mathscr{S}$ in a post-projective component $\mathscr{P}$. Let $\mathscr{S}_1, \ldots, \mathscr{S}_r$ be the connected components of $\mathscr{S}$. We consider the full subcategory B of A in the vertices of $\bigcup_{X \in \mathscr{S}} \operatorname{supp} X$.

We get that the algebra $B = \bigoplus_{i=1}^{r} B_i$ is a coproduct of tilted algebras $B_1, \ldots, B_r$. For each i let $T^{(i)}$ be the corresponding tilting module.

Lemma *With the notation above, assume that the AR-quiver is connected. Then either all algebras $B_1, \ldots, B_r$ are representation-finite or all are representation-infinite.*

The proof (that we will omit) relies on considerations of vector space subcategories (see [11]).

3.4 A Generalization of Jacobi's Criterion

We now prove the Jacobi-Zeldych Criteria for the weak positivity or non-negativity of quadratic unit forms (see [16]), which generalize Sylvester's well known criteria for positivity and non-negativity in terms of minors (see for instance [3, Proposition 1.32] and [3, Proposition 1.33]).

3.4.1 Theorem *Let $q : \mathbb{Z}^n \to \mathbb{Z}$ be a unit form and let A be the associated symmetric matrix. The following are equivalent:*

(a) q is weakly positive.
(b) For each principal submatrix B of A either $\det(B) > 0$ or $\operatorname{adj}(B)$ is not positive (that is, it has an entry ≤ 0).

Proof Let B be a principal submatrix of A. Suppose that $\operatorname{adj}(B)$ is positive. Then there is a positive vector v and a number $\rho > 0$ such that $v\operatorname{adj}(B) = \rho v$. Then $0 < q(v) = vBv^t = \rho^{-1}\operatorname{adj}(B)Bv^t = \rho^{-1}(\det B)vv^t$. Thus $\det(B) > 0$. Conversely, let A be a $n \times n$ matrix satisfying (b). We show that q is weakly positive by induction on n.

Since property (b) is inherited to principal submatrices, we can assume that the quadratic form $q^{(i)}$ associated with each principal submatrix $A^{(i,j)}$ is weakly positive. Assume that q is not weakly positive. Therefore, we get a vector $0 \ll y \in \mathbb{N}^n$ such that $q(y) \leq 0$.

In particular, every proper principal submatrix B of a has $\det(B) > 0$. Since A is not positive, $\det(A) \leq 0$. By hypothesis, $\operatorname{adj}(A)$ is not positive. Suppose that the j-th row v of $\operatorname{adj}(A)$ has some non-positive coordinate. Therefore, there exists a number $\lambda \geq 0$ such that $0 \leq \lambda y + v$ is not omnipresent. Therefore

$$\begin{aligned} 0 < q(\lambda y + v) &= \lambda^2 q(y) + \lambda v A y^t + q(v) \leq \lambda(\det A) y(j) + (\det A) v(j) \\ &\leq (\det A)(\det A^{(j,j)}) \leq 0, \end{aligned}$$

since by the claim $q^{(j)}$ is positive. □

3.4.2 Theorem *Let q be an integral form with associated symmetric matrix A (that is, $q(x) = x^t Ax$). The following are equivalent:*

(a) The form q is weakly non-negative.
(b) For every principal minor B of A either $\det(B) \geq 0$ *or* $\operatorname{adj}(B)$ *has a negative entry.*

Proof Let B be a principal submatrix of A and assume that $\operatorname{adj}(B)$ is non-negative. By Perron-Frobenius (see for instance [3, Theorem 1.36]) there exists an eigenvector $v > 0$ of B with eigenvalue $\rho > 0$. Hence we have

$$0 \leq q(v) = v^t B v = \frac{1}{\rho} v^t B(\operatorname{adj}(B) v) = \frac{1}{\rho} \det(B) \|v\|^2,$$

and therefore $\det(B) \geq 0$.

Suppose now that q satisfies (b) but is not weakly non-negative. Since property (b) is preserved by principal minors, by induction we may assume that q is hypercritical. By [3, Corollary 6.18], every proper restriction of q is non-negative, therefore by [3, Proposition 1.33], $\det(B) \geq 0$ for each proper principal submatrix B of A.

Thus $\det(A) < 0$ since otherwise q would be non-negative. Take $\operatorname{adj}(A) = (v_{ij})$. By hypothesis there must exist i, j with $v_{ij} < 0$. Let v_j be the j-th column of $\operatorname{adj}(A)$, so that $A v_j = \det(A) e_j$ and $q(v_j) = \det(A) v_{jj}$. Further, let $w > 0$ be a sincere positive vector with $q(w) < 0$. For $\lambda = -\frac{v_{ij}}{w_i} > 0$ we have that $(v + \lambda w)_i = 0$ and that

$$\begin{aligned} 0 &\leq q(v + \lambda w) \\ &= q(v) + 2\lambda w^t A v + \lambda^2 q(w) \\ &< \det(A)[v_{jj} + 2\lambda w_j] = \\ &= \frac{\det(A)}{w_i}[v_{jj} w_i - 2 v_{ij} w_j]. \end{aligned}$$

If $v_{jj} < 0$ we take $i = j$, thus

$$0 \leq q(v + \lambda w) < \det(A)(-v_{jj}) \leq 0,$$

and if $v_{jj} \geq 0$ then $v_{jj}w_i - 2v_{ij}w_j \geq 0$ and we have

$$0 \leq q(v + \lambda w) < \frac{\det(A)}{w_i}[v_{jj}w_i - 2v_{ij}w_j] \leq 0.$$

Both cases yield a contradiction. □

Observe that Theorem 3.4.2 does not require the quadratic form to be unitary.

3.5 The Tits Quadratic Form

3.5.1 In this section we shall prove one of the important results of the book, namely a close *correspondence between the roots of the Tits form of an algebra and the indecomposable modules* (under certain restrictions for the algebra or the indecomposable modules considered).

Let $A = KQ/I$ be a K-algebra, and L a minimal set of generators of I, with $L \subset \bigcup_{x,y \in Q_0} e_x I e_y$. Fix $z \in \mathbb{N}^{Q_0}$, with entry $z(x)$ for a vertex $x \in Q_0$. Let $X \in$ mod-A be such that $\mathbf{dim} X = z$. Choosing a basis for X we may identify X with

$$(X(\alpha))_{\alpha \in Q_1} \in \prod_{\alpha: x \to y \in Q_1} K^{z(x)z(y)}.$$

The points of the affine variety $\prod_{\alpha: x \to y \in Q_1} K^{z(x)z(y)}$ satisfying the set of relations L define a closed subvariety mod-$A(z)$ called the **variety of A-modules of dimension** z.

The **affine group** $G(z) = \prod_{x \in Q_0} \mathrm{G}\ell_K(z(x))$ acts on mod-$A(z)$ by conjugation in the following way: if $X = (X(\alpha))_{\alpha \in Q_1} \in$ mod-$A(z)$ and $G = (g_x)_{x \in Q_0} \in G(z)$, then $X^g = (g_y X(\alpha) g_x^{-1})_{\alpha: x \to y \in Q_1}$. The modules $X, Y \in$ mod-$A(z)$ are isomorphic if and only if they lie on the same $G(z)$-orbit in mod-$A(z)$. Moreover, the stabilizer of X is $\mathrm{Aut}_A(X)$.

3.5.2 Lemma (Tits Argument) *If A is representation-finite, then* dim mod-$A(z) \leq \sum_{x \in Q_0} z(x)^2 - 1$.

Proof The group K^* may be embedded diagonally into $G(z)$. The action of K^* on mod-$A(z)$ is trivial, therefore $G(z)/K^*$ acts on mod-$A(z)$. If there are only finitely many orbits of $G(z)/K^*$ in mod-$A(z)$, then

$$\sum_{x \in Q_0} z(x)^2 - 1 = \dim G(z)/K^* \geq \dim \text{mod-}A(z).$$

□

Corollary *Let $A = KQ/I$ be a representation-finite algebra such that Q has no oriented cycles. Then the Tits form q_A is weakly positive.*

Proof The points in mod-$A(z)$ satisfy the $\sum_{x,y\in Q_0} r(x, y)z(x)x(y)$ equations imposed by L (where $r(x, y) = |L \cap e_y I e_x|$). Then

$$\dim \text{mod-}A(z) \geq \sum_{x\to y\in Q_1} z(x)z(y) - \sum_{x,y\in Q_0} r(x, y)z(x)z(y).$$

By Lemma 3.5.2,

$$q_A(z) \geq \sum_{x\in Q_0} z(x)^2 - \dim \text{mod-}A(z) \geq 1.$$

□

3.5.3 Lemma *A weakly positive quadratic form $q : \mathbb{Z}^n \longrightarrow \mathbb{Z}$ has only finitely many positive roots.*

Proof Consider q as a function $q : \mathbb{R}^n \longrightarrow \mathbb{R}$. By continuity $q(z) \geq 0$ in the positive cone $K = (\mathbb{R}^n)^+$. By induction on n, it can be shown that $q(z) > 0$ for any $0 \neq z \in K$. Let $0 < \gamma$ be the minimal value reached by q on $\{z \in K : \| z \| = 1\}$ (a compact set). Then a positive root z of q satisfies $\gamma \leq q\left(\frac{z}{\|z\|}\right) = \frac{1}{\|z\|^2}$, that is $\| z \| \leq \sqrt{1/\gamma}$. □

Theorem *Let $A = KQ/I$ be an algebra such that Q has no oriented cycles. Assume that Γ_A has a post-projective component. Then A is representation-finite if and only if the Tits form q_A is weakly positive. Moreover, in that case, there is a bijection $X \to$* **dim***X between the isoclasses of indecomposable A-modules and the positive roots of q_A.*

Proof The "only if" part is Corollary 3.5.2. Assume that q_A is weakly positive. Let $\mathscr{P}$ be a post-projective component of Γ_A. Let $X \in \mathscr{P}$ and let A' be a convex subalgebra of A such that X is an omnipresent A'-module. Then X belongs to a post-projective component of Γ_A. In particular, $\dim_K \text{End}_{A'}(X) = 1$ and $\text{Ext}^i_1(X, X) = 0$ for $i \geq 0$.

Since gl.dim $A' \leq 2$, then $q_A(\mathbf{dim}X) = q_{A'}(\mathbf{dim}X) = \chi_{A'}(\mathbf{dim}X) = 1$. That is, $\mathbf{dim}X$ is a root of q_A. Moreover, the map $X \to \mathbf{dim}X$, for $X \in \mathscr{P}$, is injective by Lemma 3.3.10. It follows from Lemma 3.5.3 that $\mathscr{P}$ is a finite component of Γ_A, and by Auslander's Theorem 3.3.2, $\mathscr{P} = \Gamma_A$.

Finally, let $z \in \mathbb{N}^{Q_0}$ be a root of q_A. As in the proof of Lemma 3.5.2, $\dim G(z)/K^* \geq \dim \text{mod-}A(z)$ and there is a module $X \in \text{mod-}A(z)$ with orbit $G(z)X$ of dimension $\dim G(z) - 1$. Since $\dim G(z)X = \dim G(z) - \dim \text{End}_A(X)$, we obtain that $\text{End}_A(X) = K$. Then X is an indecomposable module with $\mathbf{dim}\ X = z$. □

3.5.4 The statement of Theorem 3.5.3 may be false if A has no post-projective component. Consider the algebras A_i for $i = 1, 2$ given by the quiver Q with relations $\rho_1 = (\alpha_3\alpha_2\alpha_1 - \beta_2\beta_1)$ and $\rho_2 = \alpha_3\alpha_2\alpha_1$,

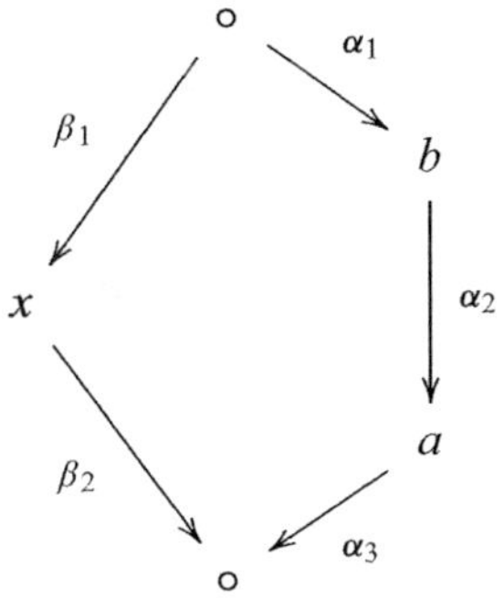

Clearly, the algebras A_1 and A_2 have the same Tits form,

$$
\begin{aligned}
q(x) &= \sum_{i=1}^{8} x_i^2 - x_1x_2 - x_2x_3 - x_3x_4 - x_1x_5 - x_4x_5 - x_5x_6 - x_6x_7 - x_7x_8 + x_1x_4 \\
&= \left(x_1 - \frac{1}{2}x_2 + \frac{1}{2}x_4 - \frac{1}{2}x_5\right)^2 + \frac{3}{4}\left(x^2 - \frac{2}{3}x_3 + \frac{1}{3}x_4 - \frac{1}{3}x_5\right)^2 + \\
&+ \frac{2}{3}\left((x_3 - \frac{1}{2}x_4 - \frac{1}{4}x_5\right)^2 + \frac{1}{2}\left(\frac{1}{2}(x_4 - \frac{1}{2}x_5\right)^2 + \frac{1}{2}(x_5 - x_6)^2 + \frac{1}{2}(x_6 - x_7)^2 + \\
&+ \frac{1}{2}(x_7 - x_8)^2 + \frac{1}{2}x_8^2.
\end{aligned}
$$

which is positive. The algebra A_1 has a post-projective component and Theorem 3.5.3 applies. However, the algebra A_2 is not representation finite: mod-A_2 contains the representations of the parabolic quiver

$\widetilde{\mathbb{E}}_7 =$

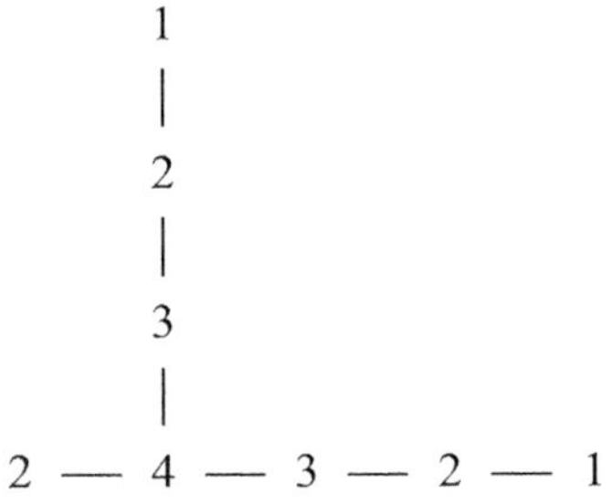

see 1.1.3.9.

References

1. M. Auslander, *The representation dimension of artin algebras,* Queen Mary College Mathematics Notes (1971)
2. Auslander, M. and Reiten, I. and Smalø, S. *Representation theory of Artin algebras*, Cambridge University Press **36** Cambridge Studies in Advanced Mathematics (1995)
3. Barot, M., Jiménez González, J.A. and de la Peña, J.A. *Quadratic Forms: Combinatorics and Numerical Results*, Algebra and Applications, Vol. **25** Springer Nature Switzerland AG 2018
4. Bautista, R. *On algebras of strongly unbounded representation type*, Comment. Math.Helv. **60** (1985), 392–399. Lecture Notes in Math., 832, Springer, Berlin, 1980.
5. Bongartz, K. *On representation-finite algebras and beyond.* Advances in Representation Theory of algebras, 65–101, EMS Ser. Congr. Rep., Eur. Math. Soc., Zurich, 2013 (arXiv:1301.4088)
6. Bongartz, K. *On minimal representation-infinite algebras*, arXiv:1705.10858 (2018)
7. Happel, D. and Ringel, C.M. *Tilted algebras*, Trans. Amer. Math. Soc. **274** (1982), 339–443.
8. Jans, J.P. *On the indecomposable representations of algebras.* Ann. of Math. (2) **66** (1957), 418–429.
9. Kupisch, H. *Symmetrische Algebren mit endlich vielen unzerlegbaren Darstellungen. I.* (German) J. Reine Angew. Math. **219** (1965) 125.
10. Nazarova, L.A. and Roiter, A.V. *Kategorielle Matrizen-Probleme und die Brauer-Thrall-Vermutung*, Mitt. Math. Sem. Giessen **115** (1975), 1–153.
11. Ringel, C.M. *Tame algebras and integral quadratic forms*, Springer LNM, **1099** (1984)
12. Roiter, A.V. *The unboundeness of the dimension of the indecomposable representations of algebras that have an infinite number of indecomposable representations*, Izv. Acad. Nauk SSSR, Ser. Mat., **32** (1968), 1275–82 (in Russian).
13. Skowroński, A and Smalø, S. *Directing modules*, J. Algebra, **147** (1992), pp. 137–146
14. Skowronski, A. and Wenderlich, M.: *Artin algebras with directing indecomposable projectives.* J. Algebra. **165** (1994), 507–530.
15. Smalø, S. *The inductive step in the second Brauer-Thrall conjecture*, Can. J. Math., vol XXXII, No. **2** (1980), pp. 342–349
16. Zeldych, M.V., *A criterion for weakly positive quadratic forms*, (Russian). In: Linear Algebra and the Theory of Representations. SSR, Kiev (1983)

Chapter 4
Spectral Methods in Representation Theory

Let A be a finite dimensional K-algebra with K an algebraically closed field. For simplicity we assume $A = K\Delta/I$ for a quiver Δ without oriented cycles, and I an admissible ideal of the path algebra $K\Delta$ (called **triangular algebras**). Let $S_1, \dots, S_n$ be a complete set of pairwise non-isomorphic simple A-modules, $P_1, \dots, P_n$ the corresponding projective covers and $I_1, \dots, I_n$ the injective envelopes. Then the **Coxeter transformation** Φ_A is the automorphism of $K_0(A)$ (the Grothendieck group of A) defined by $\Phi_A \mathbf{dim} P_i = -\mathbf{dim} I_i$, where $\mathbf{dim} X$ denotes the class of a module X in $K_0(A)$. Such transformation exists, for instance, when A has finite global dimension (for in that case the Cartan matrix C_A is $\mathbb{Z}$-invertible, and $\Phi_A = -C_A^t C_A^{-1}$).

Consider the **path algebra** $A = K\Delta$ over an algebraically closed field K. The algebra A is hereditary. Much work has been devoted in the representation theory of algebras to the study of hereditary algebras and their module categories. The Grothendieck group $K_0(A)$ is the $\mathbb{Z}$-free module $\mathbb{Z}^n$, where n is the cardinality of Δ_0. For each A-module X, the corresponding element in $K_0(A)$ is denoted by $\mathbf{dim} X$. Then the Coxeter matrix Φ_Δ of A satisfies

$$\mathbf{dim} \tau X = \Phi_\Delta(\mathbf{dim} X),$$

for any indecomposable non-projective A-module X, where τX is the Auslander-Reiten translate of X. This formula is of central importance in the theory of hereditary algebras. Properties of the matrix Φ_Δ (eigenvalues and their multiplicities, Jordan blocks, factors of the characteristic polynomial) play an important role in the behavior of the indecomposable A-modules. For instance, the spectral radius $\rho(A) = \rho(\Phi_A)$ determines the representation type of A in the following manner:

1. A is representation-finite if $1 = \rho(A)$ is not a root of the Coxeter polynomial φ_A of A.
2. A is tame if $1 = \rho(A)$ is a root of the Coxeter polynomial φ_A.

J.-A. de la Peña, *Representations of Algebras*, Algebra and Applications 30,
https://doi.org/10.1007/978-3-031-12288-0_4

3. A is wild if $1 < \rho(A)$. Moreover, if A is wild connected, Ringel [35] shows that the spectral radius $\rho(A)$ is a simple root of φ_A. Then Perron-Frobenius theory yields a vector $y^+ \in K_0(A) \otimes_{\mathbb{Z}} \mathbb{R}$ with positive coordinates such that $\Phi_A y^+ = \rho(A)y^+$. Since φ_A is self-reciprocal, there is a vector $y^- \in K_0(A) \otimes_{\mathbb{Z}} \mathbb{R}$ with positive coordinates such that $\Phi_A y^- = \rho(A)^{-1} y^-$. The vectors y^+, y^- play an important role in the representation theory of $A = K\Delta$, see [13, 33]. Namely, for an indecomposable A-module X:

   (a) X is a postprojective A-module (that is, $\tau_A^m X$ is projective for some $m \geq 0$) if and only if $\langle y^-, \mathbf{dim}X \rangle < 0$.
   (b) X is a preinjective A-module (that is, $\tau_A^{-m} X$ is injective for some $m \geq 0$) if and only if $\langle \mathbf{dim}X, y^+ \rangle < 0$.
   (c) X is regular (that is, X is not post-projective or preinjective) if and only if $\langle y^-, \mathbf{dim}X \rangle > 0$ and $\langle \mathbf{dim}X, y^+ \rangle > 0$.
   (d) If X is post-projective or regular, then $\lim_{n\to\infty} \frac{1}{\rho(A)^n} \mathbf{dim}(\tau_A^{-n} X) = \lambda_X^- y^-$, for some real number $\lambda_X^- > 0$.
   (e) If X is preinjective or regular, then $\lim_{n\to\infty} \frac{1}{\rho(A)^n} \mathbf{dim}(\tau_A^n X) = \lambda_X^+ y^+$, for some real number $\lambda_X^+ > 0$.

A **symmetry** g of Δ is a permutation of the set of vertices Δ_0 of Δ which induces an automorphism of the quiver Δ. The set of all symmetries of Δ form a group which we denote by $\mathrm{Aut}(\Delta)$. Symmetries have been studied in representation theory for the construction of skew algebras and Galois coverings. Clearly, for any symmetry $g \in \mathrm{Aut}(\Delta)$, we have $g\Phi_\Delta = \Phi_\Delta g$. In the present chapter we consider some relations between the spectrum of the Coxeter matrix Φ_Δ and the canonical representations $\gamma : G \to \mathrm{GL}(n)$ of subgroups G of $\mathrm{Aut}(\Delta)$. We get results in both directions: knowing the indecomposable rational (or complex) decomposition of γ, we get information about the set $\mathrm{Spec}(\Phi_\Delta)$ of eigenvalues of Φ_Δ. Conversely, knowing the Jordan form of Φ_Δ we get restrictions on the group $\mathrm{Aut}(\Delta)$ and decompositions of the canonical representations γ. As before, all vectors are column vectors, and the transpose of a matrix is denoted by $(-)^t$.

4.1 Hereditary Algebras and the Coxeter Transformation

4.1.1 Let Δ be a connected quiver without oriented cycles, and consider the path algebra $A = K\Delta$. Recall that for each (left) A-module X, the dimension vector $\mathbf{dim}X \in K_0(A) = \mathbb{Z}^n$ has at the i-th coordinate the number of times that the simple module S_i corresponding to vertex i appears as composition factor of X. By P_j (resp. I_j) we denote the indecomposable projective (resp. injective) A-module corresponding to the vertex j.

The indecomposable A-modules form the **Auslander-Reiten quiver** Γ_A which carries the translation τ_A defined for every non-projective vertex of Γ_A. The indecomposable A-modules are classified as postprojective, regular or preinjective according to their position in Γ_A: an indecomposable module is **postprojective** (resp. **preinjective**) if there is $m \geq 0$ such that $\tau_A^{-m} X$ is a projective module (resp. if $\tau_A^m X$ is an injective module), otherwise X is called **regular** module.

4.1.2 The **Cartan matrix** C_A of $A = K\Delta$ is the $n \times n$-matrix whose i-th column is the vector $\mathbf{dim} P_i$. It is an invertible matrix, which is in general true for algebras A with finite global dimension, cf. [2, III.3]. In the hereditary case, an inverse can be constructed easily: if m_{ij} is the number of arrows in Δ form vertex i to vertex j, and $M = M_\Delta = (m_{ij})$, since Δ has no oriented cycle then

$$C_A^t = \mathbf{Id} + M + M^2 + \dots, \quad \text{and} \quad (C_A^{-1})^t = \mathbf{Id} - M,$$

(observe that the powers M^r for $r > 0$ have as entries the number of not self-returning walks in Δ of length r). The bilinear form $\langle x, y \rangle = x^t C_A^{-t} y$ satisfies

$$\langle \mathbf{dim} X, \mathbf{dim} Y \rangle = \dim_K \mathrm{Hom}_A(X, Y) - \dim_K \mathrm{Ext}_A(X, Y),$$

(see for instance [2]). The **Coxeter matrix** $\Phi_A = -C_A^t C_A^{-1}$ is determined by the property $\Phi_A(\mathbf{dim} P_j) = -\mathbf{dim} I_j$, for $j = 1, \dots, n$. Its importance for the representation theory of hereditary algebras comes from the fact that

$$\mathbf{dim} \tau_A X = \Phi_A(\mathbf{dim} X), \quad \text{and} \quad \langle x, y \rangle = -\langle y, \Phi_\Delta x \rangle,$$

for every indecomposable non-projective A-module X, where τ_A is the Auslander-Reiten translation of A, and every pair of vectors x and y. The notation $C_\Delta = C_A$ and $\Phi_\Delta = \Phi_A$ will also be used.

4.1.3 The characteristic polynomial $\varphi_\Delta(T) = \det(T\,\mathbf{Id} - \Phi_\Delta)$ of Φ_Δ is called the **Coxeter polynomial**. The roots of $\varphi_\Delta(T)$, counting multiplicities, form the **spectrum**, or multi-set of eigenvalues $\mathrm{Spec}(\Phi_\Delta)$ of Φ_Δ.

Lemma *The Coxeter polynomial φ_Δ is self-reciprocal and does not depend on the enumeration of the vertices of the quiver Δ.*

Proof If Δ' is obtained from Δ by reordering its vertices, then there is a permutation matrix P such that $M_{\Delta'} = P^t M_\Delta P$, and therefore $\Phi_{\Delta'}$ and Φ_Δ are similar matrices. For the first claim take $\Phi_\Delta = -C^t C^{-1}$, which yields

$$\begin{aligned} T^n \varphi_\Delta(T^{-1}) &= \det(T\mathbf{Id}) \det(T^{-1} + C^t C^{-1}\mathbf{Id}) \\ &= \det(\mathbf{Id} + T C^t C^{-1}) = \det[C^t (T\mathbf{Id} + C^{-t} C) C^{-1}] \\ &= \det[(T\mathbf{Id} + C^{-t} C)^t = \varphi_\Delta(T). \end{aligned}$$

□

More generally, the Coxeter polynomial of a K-algebra A of finite global dimension does not depend on the choice of complete system of orthogonal idempotents of A.

4.1.4 Explicit formulas, special values. Here we discuss various instances where an explicit formula for the Coxeter polynomial is known. For instance, if Δ is of Dynkin type (which happens if and only if A is of finite representation type), then the Coxeter polynomial decomposes in the following way:

$$\varphi_\Delta(T) = \prod_{j=1}^{n}(T - \exp(2i\pi m_j/h))$$

where $m_1, \ldots, m_n$ are integers such that $1 = m_1 \leq m_2 \leq \ldots \leq m_{n-1} \leq m_n = h - 1$.

If Δ is of Euclidean type (which happens if and only if A is of tame type), then

$$\varphi_\Delta(T) = (T-1)^2 \prod^{n-2}(T - \exp(2i\pi m_j/h))$$

where $1 \leq m_1 \leq m_2 \leq \ldots \leq m_{n-2} < h$ are integers. Moreover, in this case, $\rho(\Delta) = 1$ is a double eigenvalue of Φ_Δ (where the **spectral radius** of Φ_Δ is denoted by $\rho(\Delta) = \rho(\Phi_\Delta)$. If Δ is neither Dynkin nor of Euclidean type, we say that Δ is a **wild quiver** (since A is of wild representation type).

4.1.5 Let A be the path algebra of a hereditary star $[p_1, \ldots, p_t]$ with respect to the standard orientation,

[2, 3, 3, 4] :

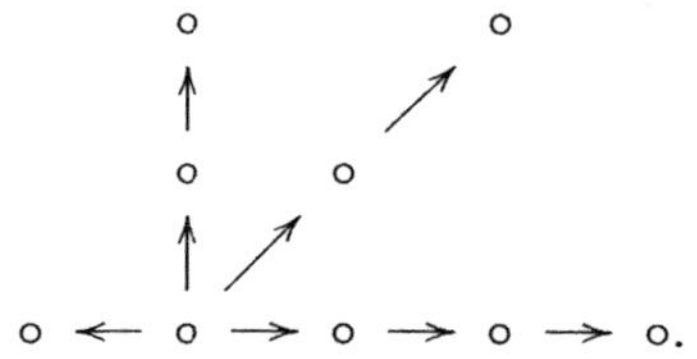

Since the Coxeter polynomial φ_A does not depend on the orientation of the arrows of Δ (see for instance [28]), we will denote it by $\varphi_{[p_1,\ldots,p_t]}$. It follows from [23, prop. 9.1] or [8] that

$$\varphi_{[p_1,\ldots,p_t]} = \prod_{i=1}^{t} v_{p_i}\left((x+1) - x\sum_{i=1}^{t}\frac{v_{p_i-1}}{v_{p_i}}\right).$$

In particular, we have an explicit formula for the sum of coefficients of $\varphi = \varphi_{[p_1,\dots,p_t]}$ as follows:

$$\varphi(1) = \prod_{i=1}^{t} p_i \left(2 - \sum_{i=1}^{t}(1 - \frac{1}{p_i})\right).$$

4.1.6 This special value of φ has a specific mathematical meaning: up to the factor $\prod_{i=1}^{t} p_i$ this is just the orbifold-Euler characteristic of a weighted projective line $K\mathbb{P}$ of weight type $(p_1, \dots, p_t)$, cf. [26]. Moreover,

1. $\varphi(1) > 0$ if and only if the star $[p_1, \dots, p_t]$ is of Dynkin type, correspondingly the algebra A is representation-finite.
2. $\varphi(1) = 0$ if and only if the star $[p_1, \dots, p_t]$ is of extended Dynkin type, correspondingly the algebra A is of tame (domestic) type.
3. $\varphi(1) < 0$ if and only if $[p_1, \dots, p_t]$ is not Dynkin or extended Dynkin, correspondingly the algebra A is of wild representation type.

The above deals with all the Dynkin types and with the extended Dynkin diagrams of type $\widetilde{\mathbb{D}}_n$, $n \geq 4$, and $\widetilde{\mathbb{E}}_n$, $n = 6, 7, 8$. To complete the picture, we also consider the extended Dynkin quivers of type $\widetilde{\mathbb{A}}_n$ $(n \geq 2)$ restricting, of course, to quivers without oriented cycles. Here, the Coxeter polynomial depends on the orientation: If p (resp. q) denotes the number of arrows in clockwise (resp. anticlockwise) orientation $(p, q \geq 1, p + q = n + 1)$, that is, the quiver has type $\mathbb{A}(p, q)$, the Coxeter polynomial φ is given by

$$\varphi_{(p,q)} = (x-1)^2\, v_p v_q.$$

Hence $\varphi(1) = 0$, fitting into the above picture.

The next table displays the v-factorization of extended Dynkin quivers

Extended Dynkin type	Star symbol	Weight symbol	Coxeter polynomial
$\widetilde{\mathbb{A}}_{p,q}$	–	(p, q)	$(x-1)^2 v_p\, v_q$
$\widetilde{\mathbb{D}}_n$, $n \geq 4$	[2,2,n-2]	$(2, 2, n-2)$	$(x-1)^2 v_2^2 v_{n-2}$
$\widetilde{\mathbb{E}}_6$	[3, 3, 3]	$(2, 3, 3)$	$(x-1)^2 v_2 v_3^2$
$\widetilde{\mathbb{E}}_7$	[2, 4, 4]	$(2, 3, 4)$	$(x-1)^2 v_2 v_3 v_4$
$\widetilde{\mathbb{E}}_8$	[2, 3, 6]	$(2, 3, 5)$	$(x-1)^2 v_2 v_3 v_5$

4.1.7 A quiver Δ is said to have a **bipartite orientation** if every vertex in Δ is either a source or a sink. In this case, there is a close relationship between the Coxeter matrix and the (symmetric) adjacency matrix of the underlying graph of Δ, described by A'Campo in [1, 1976]. We will give a detailed proof below in 4.1.8.

Theorem (A'Campo) *For a bipartite quiver Δ with n vertices and underlying graph $\overline{\Delta}$, we have*

$$\varphi_\Delta(T^2) = T^n \kappa(T + T^{-1}),$$

where κ is the characteristic polynomial of the (symmetric) adjacency matrix M of $\overline{\Delta}$. Moreover,

$$\mathrm{Spec}(\Phi_\Delta) = T^2\left(\mathrm{Spec}(M_\Delta + M^t_\Delta)\right).$$

where $T(t) = \frac{1}{2}(t + \sqrt{t^2 - 4})$.

Notice that $T^2(\mathbb{R}) = \mathbb{S} \cup \mathbb{R}^+$, where $\mathbb{S}$ denotes the unit complex sphere. In particular, the result above implies $\mathrm{Spec}(\Phi_\Delta) \subset \mathbb{S} \cup \mathbb{R}^+$.

4.1.8 Let Δ be a bipartite quiver. Consider the **adjacency matrix** A_Δ of Δ, defined as the $n \times n$-matrix $A_\Delta = (a_{ij})$ with $a_{ij} = 1$, if there is an arrow in Δ joining i and j and 0, else.

Choosing $1, \dots, m$ as sources of Δ and $m + 1, \dots, n$ as sinks, the matrix A_Δ takes the form

$$A_\Delta = \begin{bmatrix} 0 & M \\ M^t & 0 \end{bmatrix}$$

Let $\nu_1, \dots, \nu_m$ be the eigenvalues of the symmetric positive matrix $D = MM^t$. Let $x_1, \dots, x_m$ be an orthonormal basis of eigenvectors for D, with ν_i the eigenvalue of x_i. Assume that $\nu_i \neq 0, 4$, for $1 \le i \le p$; $\nu_i = 4$ for $p + 1 \le i \le q$ and $\nu_i = 0$ for $q + 1 \le i \le m$. We obtain the eigenvalues λ_{ij} and eigenvectors y_{ij} of Φ_Δ in the following way:

$$\begin{array}{l} \lambda_{i1} = \frac{1}{2}\nu_i - 1 + \frac{1}{2}\sqrt{\nu_i(\nu_i - 4)}\ ,\ y_{i1} = (x_i, \frac{1}{\lambda_{i1}+1} x_i M) \\ \lambda_{i2} = \frac{1}{2}\nu_i - 1 - \frac{1}{2}\sqrt{\nu_i(\nu_i - 4)}\ ,\ y_{i2} = (x_i, \frac{1}{\lambda_{i2}+1} x_i M) \end{array}$$

with $y_{ij}\Phi_\Delta = \lambda_{ij} y_{ij}$, for $j = 1, 2$ and $1 \le i \le p$;

$$\lambda_{i1} = 1\ ,\ y_{i1} = (x_i, \frac{1}{2} x_i M) \quad ; \quad \lambda_{i2} = 1\ ,\ y_{i2} = \frac{1}{4}(x_i, -\frac{1}{2} x_i M),$$

with $y_{i1}\Phi_\Delta = y_{i1}$ and $y_{i2}\Phi_\Delta = y_{i2} + y_{i1}$, for $p + 1 \le i \le q$;

$$\lambda_{i1} = -1\ ,\ y_{i1} = (x_i, 0) \quad \text{with} \quad x_i M = 0 \quad \text{for} \quad q + 1 \le i \le m,$$

$$\lambda_{i2} = -1\ ,\ y_{i2} = (0, x'_i) \quad \text{with} \quad 0 \neq x'_i,\ x'_i D = 0 \quad \text{for} \quad q + 1 \le i \le n - m$$

with $y_{ij}\Phi_\Delta = \lambda_{ij}y_{ij}$. We have shown:

Lemma *For* $1 \neq \lambda \in \mathrm{Spec}(\Phi_\Delta)$, *the algebraic and the geometric multiplicity of* λ *is the same. The algebraic multiplicity of* 1 *is twice its geometric multiplicity.*

4.1.9 A square matrix M is said to be **sharp** if the following conditions are satisfied:

(i) The spectral radius $\rho = \rho(M)$ is a simple root of the characteristic polynomial of M.
(ii) For any other eigenvalue λ of M we have $|\lambda| < \rho$.

Here we are interested in hereditary algebras A having sharp Coxeter matrix Φ_A. Consider a quiver Δ with a sink or source vertex i, and take a new quiver $\sigma_i(\Delta)$ obtained from Δ by reversing the orientation of all arrows incident to i. This is called a **sink-source quiver reflections**, and we say that two quivers Δ and Δ' have **equivalent orientations** if there is a sequence of sink-source quiver reflections taking Δ to Δ' (cf. [6]). In that case, the Cartan matrices C_Δ and $C_{\Delta'}$ are congruent, and therefore Δ and Δ' have the same Coxeter polynomial.

Lemma *Assume that* Δ *is a connected quiver whose underlying graph* $\overline{\Delta}$ *is not a Dynkin nor an Euclidean diagram. If* Δ *has an orientation equivalent to a bipartite orientation, then* Φ_Δ *is sharp.*

Proof Let $N = M_\Delta + M_\Delta^t$ be the adjacency matrix of $\overline{\Delta}$, which is a symmetric matrix with non-negative coefficients. Since Δ is connected, N is an irreducible matrix. By Frobenius Theorem (cf. [15, XIII §2.1]), the spectral radius r of N is a simple root of the characteristic polynomial $\kappa_N(T)$ of N. Since $\overline{\Delta}$ is neither Dynkin nor Euclidean, by 2.1.3.7 we have $r > 2$.

By the discussion above, we may assume that Δ is a bipartite quiver. Then A'Campo's Theorem 4.1.7 yields the result, for $\rho(\Phi_\Delta) = \gamma(r)$ is a simple root of $\varphi_\Delta(T)$, and since $\mathrm{Spec}(\Phi_\Delta) \subset \mathbb{S} \cup \mathbb{R}+$, then any other root λ of $\varphi_\Delta(T)$ satisfies $|\lambda| < \rho(\Phi_\Delta)$. This shows that Φ_Δ is sharp. □

Remark Ringel has shown [35] that every generalized Cartan matrix which is neither of finite nor affine type, has sharp Coxeter matrix.

4.2 Coxeter Spectrum in the Study of Indecomposable Modules

4.2.1 Take a finite dimensional hereditary algebra $A = K\Delta$. Consider the set of rays $\mathbb{R}^+(\Phi_\Delta^{-r} p_i)$ for $i = 1, \ldots, n$ and $r \geq 0$ (where p_i is the dimension of the i-th indecomposable projective A-module), and denote by $K_{\mathcal{P}}$ its convex hull (see [12]). Then $K_{\mathcal{P}}$ is a cone in $\mathbb{R}^n$ with non-empty interior, and we have

$\Phi_\Delta^{-1}(K_{\mathscr{P}}) \subset K_{\mathscr{P}}$. Applying Vandergraft's Theorem [39], there is an eigenvector $y^- \in K_{\mathscr{P}}$ of the matrix Φ_Δ with eigenvalue the spectral radius $\rho = \rho(\Phi_\Delta)$. A similar construction with the cone $K_{\mathscr{I}}$ determined by the Φ_Δ-orbits of injective vectors yields an eigenvector $y^+ \in K_{\mathscr{I}}$ of Φ_Δ with eigenvalue ρ. Since Φ_Δ is self-reciprocal (Lemma 4.1.3), notice that $\Phi_\Delta y^- = \rho^{-1} y^-$, and that $\Phi_\Delta^{-1} y^+ = \rho^{-1} y^+$.

4.2.2 Proposition *Let Δ be a wild quiver with bipartite orientation of its arrows, and with eigenvectors y^- and y^+ as defined above. Then both vectors y^- and y^+ have all positive entries.*

Proof Since all vectors in $K_{\mathscr{I}}$ have non-negative entries, so does the vector y^+. Assume that $y_s^+ = 0$, and that s is a sink in Δ (in case s is a source we consider the opposite quiver of Δ). Since Δ is connected, we may also assume that there is a vertex r connected to s such that $y_r^+ \neq 0$. Observe first that the s-th indecomposable projective $A = K\Delta$-module P_s satisfies $\dim_K P_s = C_A e_s = e_s$. By the description of C_A^{-t} given in 4.1.2, we have

$$0 = y_s^+ = (y^+)^t e_s = \frac{1}{\rho}(y^+)^t \Phi_A^t e_s = -\frac{1}{\rho}(y^+)^t C_A^{-t} C_A e_s$$

$$= -\frac{1}{\rho}(y^+)^t C_A^{-t} e_s = \frac{1}{\rho}(m_{is} y_i^+ - y_s^+) \geq \frac{1}{\rho} m_{rs} y_r^+ > 0,$$

where m_{is} is the number of arrows in Δ from vertex i to vertex s. This contradiction shows the claim. □

4.2.3 The argument for the proof of the following result is due to Ringel (cf. [5, Lemma 1.1]).

Lemma *Let Δ be a wild quiver. Then for any regular indecomposable $A = K\Delta$-module X, there is a non-zero integer m and a non-zero morphism*

$$\nu : \tau_A^{-m} X \longrightarrow X.$$

4.2.4 Theorem *Let Δ be a wild quiver with bipartite orientation, and X an indecomposable $K\Delta$-module. Then,*

(a) X is postprojective if and only if $\langle y^-, \mathbf{dim}X\rangle < 0$.
(b) X is preinjective if and only if $\langle \mathbf{dim}X, y^+\rangle < 0$.
(c) For any indecomposable $K\Delta$-module X we have

$$\langle y^-, \mathbf{dim}X\rangle \neq 0, \quad \text{and} \quad \langle \mathbf{dim}X, y^+\rangle \neq 0.$$

Proof We show (a), the proof of (b) is similar. Assume first that X is postprojective, that is, that $\tau_A^{-m} X$ is projective for some $m \geq 0$, say $\tau_A^{-m} X = P_s$ with $s \in \Delta_0$. Then $\mathbf{dim} P_s = C_\Delta \mathbf{e}_s$, and

$$\begin{aligned}\langle y^-, \mathbf{dim} X\rangle &= \langle \Phi_\Delta^{-m} y^-, \Phi_\Delta^{-m} \mathbf{dim} X\rangle = \rho^m \langle y^-, \mathbf{dim} P_s\rangle \\ &= \rho^m (y^-)^t C^{-t} C \mathbf{e}_s = -\rho^m (\Phi_\Delta y^-)^t \mathbf{e}_s = -\rho^{m-1} y_s^- < 0,\end{aligned}$$

according to the proposition above.

Assume now that $\langle y^-, \mathbf{dim} X\rangle < 0$. Since $y^- \in K_{\mathscr{P}}$, there is a sequence of vectors $y^1, y^2, y^3, \ldots$ such that $\lim_{m\to\infty} y^m = y^-$, with

$$y^m = \sum_{i=1}^{\ell_m} \mu_i^m \mathbf{dim} V_i^m, \quad \text{for some } V_i^m \in \mathscr{P} \text{ and } \mu_i^m > 0.$$

By continuity, there is some $m > 0$ such that $\langle u^m, \mathbf{dim} X\rangle < 0$, and therefore there is $1 \leq i \leq \ell_m$ with

$$\dim_K \mathrm{Hom}_A(V_i^m, X) - \dim_K \mathrm{Ext}_A^1(V_i^m, X) = \langle \mathbf{dim} V_i^m, \mathbf{dim} X\rangle < 0$$

Therefore $\mathrm{Ext}_A^1(V_i^m, X) \neq 0$, and X is postprojective.

To prove (c), assume that there is an indecomposable module X with $\langle y^-, X\rangle = 0$, and take X with minimal dimension. Then X is a regular module by (a) and (b), and by minimality, X has no proper regular submodule. Indeed, if Y is a regular submodule of X, and $0 \longrightarrow Y \longrightarrow X \longrightarrow C \longrightarrow 0$ is an exact sequence, then

$$0 \leq \langle y^-, Y\rangle = -\langle y^-, C\rangle \leq 0,$$

where the inequalities follow from (a), since C has no postprojective direct summand. Then $\langle y^-, Y\rangle = 0$, and by minimality $Y = X$. In particular, by the lemma above there is a non-zero integer m and a non-zero morphism $\nu : \tau_A^{-m} X \to X$. Since both τ_A^{-m} and X are regular modules, then $\mathrm{Im}(\nu)$ is also regular. Therefore ν is an epimorphism, for X has no regular submodules. We have an exact sequence

$$0 \longrightarrow K \longrightarrow \tau_A^{-m} X \stackrel{\nu}{\longrightarrow} X \longrightarrow 0.$$

Notice that if K_r is a regular direct summand of K, then $\tau_A^m K_r$ is a proper regular submodule of X (for τ_A preserves monomorphisms in hereditary algebras), which is impossible. Then all summands of K are postprojective, and by (a) we have

$$0 < \langle y^-, \mathbf{dim} K\rangle = \langle y^-, \mathbf{dim} \tau_A^{-m} X\rangle - \langle y^-, \mathbf{dim} X\rangle = 0,$$

a contradiction. □

4.2.5 Theorem *If the Coxeter matrix Φ_A is sharp with spectral radius ρ, then the following happens:*

(a) Let X be an indecomposable regular or preinjective A-module. Then the limit

$$\lim_{k\to\infty} \frac{\mathbf{dim}\tau_A^k X}{\rho^k} = \lambda_X^+ y^+,$$

exists, and $\lambda_X^+ > 0$.

(b) Let X be an indecomposable regular or postprojective A-module. Then the limit

$$\lim_{k\to\infty} \frac{\mathbf{dim}\tau_A^{-k} X}{\rho^k} = \lambda_X^- y^-,$$

exists, and $\lambda_X^- > 0$.

These results illustrate the relations between the Coxeter polynomial $\varphi_\Delta(T)$ and the behavior of indecomposable A-modules. Notice that Theorem 4.1.7 and 2.1.3.7 imply that if Δ is either a Dynkin or an Euclidean quiver, then Φ_Δ fails to be sharp.

Proof By Proposition 4.2.1(a), all entries of $y^+ = (y_1, \dots, y_n)$ are positive real numbers . Take $D = \mathrm{diag}(y_1, \dots, y_n)$ and define

$$P = \frac{1}{\rho} D^{-1} \Phi_\Delta D.$$

Then P is a sharp matrix with spectral radius 1. In the lemma below we will show that the limit $\lim_{k\to\infty} P^k = P^\infty$ exists. Then, for $v = \mathbf{dim}X$,

$$\lim_{k\to\infty} \frac{\Phi_\Delta^k v}{\rho^k} = \lim_{k\to\infty} DPD^{-1}v = DP^\infty D^{-1}v.$$

Also by the Lemma 4.2.6, since $\mathbb{1} = (1, \dots, 1)$ is eigenvector of P with eigenvalue 1, then $P^\infty = [\alpha_1 \mathbb{1} | \dots | \alpha_n \mathbb{1}]$, which implies that $DP^\infty D^{-1} = (\alpha_j y_i y_j^{-1})_{i,j=1}^n$. That is,

$$DP^\infty D^{-1} v = \lambda_X^+ y^+, \quad \text{where} \quad \lambda_X^+ = \sum_{\ell=1}^n v_\ell \alpha_\ell y_\ell^{-1}.$$

Clearly $\lambda_X^+ \geq 0$. That $\lambda_X^+ > 0$ follows from Theorem 4.2.1 and Proposition 4.2.1(b), since

$$\lambda_X^+ \langle y^-, y^+ \rangle = \lim_{k\to\infty} \frac{1}{\rho^k} \langle y^-, \Phi_\Delta^k y \rangle = \lim_{k\to\infty} \frac{1}{\rho^k} \langle \Phi_\Delta^{-k} y^-, v \rangle = \langle y^-, v \rangle > 0.$$

□

4.2.6 Lemma *Let P be a sharp matrix with spectral radius $\rho = 1$ and corresponding eigenvector x. Then the limit*

$$\lim_{n\to\infty} P^n = P^\infty$$

exists, and all column vectors of P^∞ are (real) scalar multiples of x.

Proof Take a linear decomposition of the minimal polynomial $\mu(T)$ of P as

$$\mu(T) = (T - \lambda_1)^{m_1} \cdots (T - \lambda_s)^{m_s},$$

with $\lambda_i \neq \lambda_j$ for $i \neq j$. Take $C(T)$ the *reduced adjoint matrix* of $(T\mathbf{Id} - P)$ (that is, the polynomial matrix satisfying $\mu(T)\mathbf{Id} = (T\mathbf{Id} - P)C(T)$, see [15, IV§6.2]). From [15, V§3.3], for an analytic function $f(T)$ we have

$$f(P) = \sum_{k=1}^{s} \frac{1}{(m_k - 1)!} \left(\frac{C(T)}{\mu(T)} (T - \lambda_k)^{m_k} f(T) \right)^{(m_k - 1)}_{T=\lambda_k},$$

where $(-)^{(m)}$ denotes the m-th derivative with respect to T. In particular, assuming that λ_1 is the simple eigenvalue 1, then

$$P^n = \frac{C(1)}{m'(1)} + \sum_{k=2}^{s} \frac{1}{(m_k - 1)!} \left(\frac{C(T)}{\mu(T)} (T - \lambda_k)^{m_k} T^n \right)^{(m_k - 1)}_{T=\lambda_k}.$$

Since $|\lambda_k| < 1$ for $k = 2, \ldots, s$, then $\lim_{n\to\infty} P^n = C(1)/m'(1)$. Now, since $PP^\infty = P^\infty$, then every non-zero column of P^∞ is an eigenvector of P with eigenvalue 1, that is, every column of P^∞ is a real multiple of x. □

4.2.7 The above criteria for indecomposable modules over hereditary algebras provide a shortcut into deep results in the representation theory, an example (cf. [4] and [19]):

Theorem *Let X, Y be indecomposable regular modules over a wild hereditary algebra A. Then there is a number N such that for every $m > N$ we have:*

(a) $\mathrm{Hom}_A(X, \tau_A^m Y) \neq 0$;
(b) $\mathrm{Hom}_A(X, \tau_A^{-m} Y) = 0$.

In particular, given two regular components C_1, C_2 we have $\mathrm{Hom}_A(C_1, C_2) \neq 0$. *Inside a regular component, most of the morphisms go in the direction opposite to the arrows.*

Proof Using Theorem 4.2.5, we have $\lim_{m\to\infty} \frac{1}{\rho(A)^m}\mathbf{dim}\tau_A^m Y = \lambda_Y^+ y^+$. Therefore $0 < \lambda_Y^+\langle \mathbf{dim}X, y^+\rangle = \lim_{m\to\infty} \frac{1}{\rho(A)^m}\langle \mathbf{dim}X, \mathbf{dim}\tau_A^m Y\rangle$. Hence the result follows by the homological formula of $\langle -, -\rangle$ of 4.1.2. This shows (a), and (b) follows similarly. □

4.3 The Canonical Representation of a Group of Symmetries

A **symmetry** g of the quiver Δ is a permutation of the set of vertices Δ_0 inducing an automorphism of Δ. By $\mathrm{Aut}(\Delta)$ we denote the group of symmetries of Δ. Each symmetry g gives rise to a matrix $g \in \mathrm{G}\ell(n)$ (identified with the group of automorphisms $\mathrm{G}\ell(K_0(A))$ of the Grothendieck group $K_0(A)$ of $A = K\Delta$), sending S_i to $S_{g(i)}$. This representation $\gamma : \mathrm{Aut}(\Delta) \to \mathrm{G}\ell(n)$ is called the **canonical representation**. Clearly, for any vertex i and any automorphism g of Δ, we have $g\mathbf{dim}P_i = \mathbf{dim}P_{g(i)}$, and similarly $g\mathbf{dim}I_i = \mathbf{dim}I_{g(i)}$ (since g takes paths from vertices i to j, to paths from vertices $g(i)$ to $g(j)$).

Lemma *The Coxeter matrix Φ_Δ is an automorphism of the canonical representation, that is, $g\Phi_\Delta = \Phi_\Delta g$.*

Proof Since $g\Phi_\Delta(\mathbf{dim}P_i) = -\mathbf{dim}I_{g(i)} = \Phi_\Delta(\mathbf{dim}P_{g(i)}) = \Phi_\Delta g(\mathbf{dim}P_i)$, we are done. □

4.3.1 Let G be any subgroup of $\mathrm{Aut}(\Delta)$. The set of G-invariant vectors $\mathrm{Inv}_G(\Delta) = \{v \in K_0(A)\otimes_{\mathbb{Z}}\mathbb{Q} : gv = v \text{ for all } g \in G\}$ is a $\mathbb{Q}$-vector space of dimension $t_0(G)$, the number of orbits of Δ under the action of G.

Proposition *Assume that Δ is a wild quiver such that the Coxeter transformation Φ_Δ is sharp. Let $m_\rho(T)$ be the minimal polynomial of the spectral radius ρ of Φ_Δ. Then*

$$\deg\ m_\rho(T) \leq t_0(\mathrm{Aut}(\Delta)).$$

In particular, $\varphi_\Delta(T)$ is not irreducible if $\mathrm{Aut}(\Delta)$ *is non-trivial.*

This result was originally proved using representation of algebras methods. Here we give a short direct proof.

Proof Since Δ has no oriented cycles we may find a source j of Δ. The simple module S_j is injective and the vector

$$v = \sum_{g\in\mathrm{Aut}(\Delta)} \mathbf{dim}S_{g(j)}\ ,$$

belongs to the space of invariants $V = \mathrm{Inv}_{\mathrm{Aut}(\Delta)}(\Delta)$. The space V is invariant under the action of Φ_Δ, hence the matrix Φ_Δ takes the following form:

$$\begin{pmatrix} \Phi_1 & * \\ 0 & \Phi_2 \end{pmatrix}$$

where Φ_1 is the restriction of Φ_Δ to V. Correspondingly, we get a factorization

$$\varphi(T) = P_{\Phi_1}(T) P_{\Phi_2}(T) ,$$

where $P_{\Phi_i}(T) \in \mathbb{Z}[T]$ is the characteristic polynomial of Φ_i, $i = 1, 2$. Since Φ_Δ is sharp, the limit

$$\lim_{n\to\infty} \frac{\Phi_\Delta^n v}{\rho^n} = \lambda y^+$$

exists and $\lambda > 0$ (Theorem 4.2.1). Therefore, the vector y^+ belongs to $V \underset{\mathbb{Q}}{\otimes} \mathbb{R}$ and ρ is a root of $P_{\Phi_1}(T)$. Hence $m_\rho(T)$ is a divisor of $P_{\Phi_1}(T)$ and degree $m_\rho(T) \leq \dim_{\mathbb{Q}} V = t_0(\mathrm{Aut}(\Delta))$ as desired. □

4.3.2 Let G be a subgroup of $\mathrm{Aut}(\Delta)$. The restriction of the canonical representation to G will also be denoted by $\gamma : G \to \mathrm{G}\ell(n)$.

Let $R_1, \ldots, R_\ell$ be a set of representatives of the **irreducible $\mathbb{Q}$-representations** of G. Let R_1 be the trivial representation. By Maschké's theorem, there exists an invertible rational matrix L such that the conjugate representation γ^L has a decomposition $\gamma^L = \bigoplus_{\alpha=1}^{\ell} R_\alpha^{r(\alpha)}$. The dimension of R_α is denoted by $\dim R_\alpha$ (that is, $R_\alpha : G \to \mathrm{GL}(\dim R_\alpha))$. Then we have $n = \sum_{\alpha=1}^{\ell} r(\alpha) \dim R_\alpha$.

Since Φ_Δ is an automorphism of γ, by Schur's lemma, the conjugate $\Phi_\Delta^L (:= L^{-1}\Phi_\Delta L)$ takes the block diagonal form, $\Phi_1 \oplus \cdots \oplus \Phi_\ell$, where $\Phi_\alpha : R_\alpha^{r(\alpha)} \to R_\alpha^{r(\alpha)}$ is an automorphism (i.e. $\Phi_\alpha \in \mathrm{G}\ell(r(\alpha)\dim R_\alpha)$). Therefore we get a factorization $\varphi_\Delta(T) = P_{\Phi_1}(T) \ldots P_{\Phi_\ell}(T)$ where $P_{\Phi_\alpha}(T) \in \mathbb{Z}[T]$ is the characteristic polynomial of Φ_α. In particular, $\varphi_\Delta(T)$ has at least $|\{\alpha : r(\alpha) > 0\}|$ factors.

4.3.3 Let $S_1, \ldots, S_m$ be a set of representatives of the irreducible $\mathbb{C}$-representations of G. Let S_1 be the trivial representation. Then m is the number of conjugacy classes of G. There is a conjugate γ^M of γ with a decomposition

$$\gamma^M = \bigoplus_{\beta=1}^{m} S_\beta^{n(\beta)} .$$

Let χ_β be the **character** corresponding to S_β (that is $\chi_\beta : G \to \mathbb{C}^*, g \mapsto \operatorname{Tr} S_\beta(g)$). The characters $1 = \chi_1, \ldots, \chi_m$ form an orthonormal basis of the class group $X(G)$, with the scalar product

$$(\chi, \chi') = \frac{1}{|G|} \sum_{g \in G} \chi(g)\overline{\chi'(g)}.$$

The following is well-known.

4.3.4 Lemma *Let G be a subgroup of* $\operatorname{Aut}(\Delta)$, *and $\gamma : G \to G\ell(n)$ the canonical representation as before.*

(a) For any $g \in G$, $\chi_\gamma(g)$ is the number of fixed points of g on Δ.
(b) ***Burnside's lemma:***

$$t_0(G) = \frac{1}{|G|} \sum_{g \in G} \chi_\gamma(g).$$

(c) $n(1) = t_0(G) = r(1)$.

Proof For (a), consider the matrix $\gamma(g) \in LG(n)$ arising from a permutation of vertices g. Then $\chi_\gamma(g) = \operatorname{Tr} \gamma(g)$ is the number of fixed vertices under g. A proof of (b) may be found in [7]. Now, since $\chi_\gamma = \sum_{\beta=1}^{m} n(\beta)\chi_\beta$, then

$$n(\beta_0) = \Big(\sum_{\beta=1}^{\ell} n(\beta)\chi_\beta, \chi_{\beta_0}\Big) = (\chi_\gamma, \chi_{\beta_0}) = \frac{1}{|G|} \sum_{g \in G} \chi_\gamma(g)\overline{\chi_{\beta_0}(g)}.$$

In particular, for $\beta_0 = 1$, we get by (b),

$$n(1) = \frac{1}{|G|} \sum_{g \in G} \chi_\gamma(g) = t_0(G)$$

Since $R_1 = S_1$, then $n(1) = r(1)$, which shows (c). □

4.3.5 Proposition *If the group* $\operatorname{Aut}(\Delta)$ *is not trivial, then $\varphi_\Delta(T)$ is not irreducible (over $\mathbb{Z}[T]$).*

Proof If $\operatorname{Aut}(\Delta) \neq (1)$, then $t_0(\operatorname{Aut}(\Delta)) < n$. Hence for the rational decomposition of the canonical representation $\gamma : \operatorname{Aut}(\Delta) \to G\ell(n)$ we get

$$n = \sum_{\alpha=1}^{\ell} r(\alpha) \dim R_\alpha = t_0(\operatorname{Aut}(\Delta)) + \sum_{\alpha=2}^{\ell} r(\alpha) \dim R_\alpha,$$

which means that there is some $\alpha \geq 2$ with $r(\alpha) > 0$. By (4.3.2), $\varphi_\Delta(T)$ is not irreducible. □

Examples

(a) The converse of the Proposition 4.3.5 is not true. Any bipartite quiver with an odd number of vertices n has a Coxeter polynomial which accepts $(T + 1)$ as a factor. Indeed, for any bipartite quiver Δ, $\varphi_\Delta(T)$ factors as

$$\varphi_\Delta(T) = (T + 1)^p (T - 1)^q Q(T),$$

with $p \equiv n \mod (2)$ and $q \equiv 0 \mod (2)$. Here is an argument: since Δ is bipartite, $\mathrm{Spec}(\Phi_\Delta) \subset \mathbb{S} \cup \mathbb{R}^+$. Moreover, the Coxeter polynomial $\varphi_\Delta(T)$ is self-reciprocal (Lemma 4.1.3), that is, if $\varphi_\Delta(\lambda) = 0$, then $\lambda \neq 0$ and λ^{-1} is a root of $\varphi_\Delta(T)$ with the same multiplicity than λ. Hence $\varphi_\Delta(T)$ has an even number of roots on $\mathbb{R}^+ \setminus \{1\}$; obviously, it has an even number of roots on $\mathbb{S} \setminus \{-1, 1\}$. Moreover, $\det(\Phi_\Delta) = (-1)^n$ and the product of the roots of $\varphi_\Delta(T)$ is

$$1 = (-1)^n \det(\Phi_\Delta) = \varphi_\Delta(0) = (1)^p (-1)^q.$$

Therefore, $q \equiv 0 \mod (2)$ and $p \equiv n \mod (2)$.

(b) The wild star quiver Δ of type (2, 3, 7),

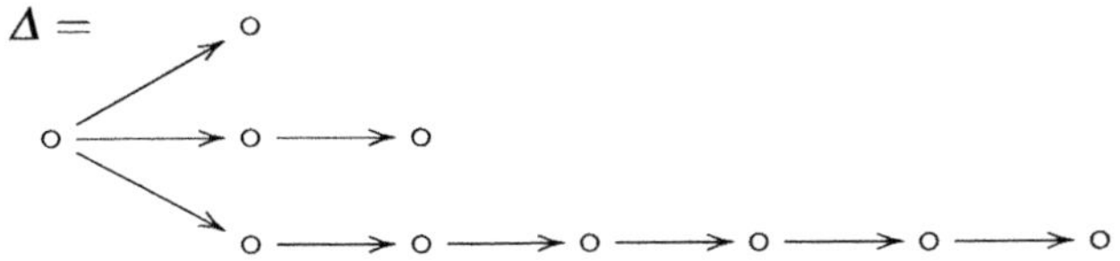

has Coxeter polynomial $\varphi_\Delta(T) = 1 + T - T^3 - T^4 - T^5 - T^6 - T^7 + T^9 + T^{10}$, which is irreducible.

4.3.6 We say that an eigenvalue λ of Φ_Δ has **algebraic multiplicity** s if λ appears exactly s times as root of $\varphi_\Delta(T)$. The number λ has **geometric multiplicity** s if there are exactly s Jordan λ-blocks in the Jordan decomposition of Φ_Δ. The **degree** $d(\lambda)$ of an eigenvalue $\lambda \in \mathrm{Spec}(\Phi_\Delta)$ is the maximal size of a Jordan λ-block of Φ_Δ. Assume that Δ is a wild quiver, then for any $\lambda \in \mathrm{Spec}(\Phi_\Delta)$ with $|\lambda| = \rho$ we have $d(\rho) \geq d(\lambda)$.

Proposition *Let G be a subgroup of* Aut(Δ) *and $\gamma : G \to G\ell(n)$ be the canonical representation. Consider the irreducible $\mathbb{C}$-decomposition*

$$\gamma^M = \bigoplus_{\beta=1}^{m} S_\beta^{n(\beta)}.$$

Let $\lambda_1 \dots, \lambda_s$ be the set of all real eigenvalues of Φ_Δ with geometric multiplicity one, and let $d(\lambda_i)$ be the degree of λ_i. Then

$$\sum_{i=1}^{s} d(\lambda_i) \le \sum_{\beta \in U_1} n(\beta),$$

where $U_1 = \{\beta \mid \dim S_\beta = 1$ and $\chi_\beta^2 = 1\}$.

Proof Let $\Phi = \Phi_\Delta$ and consider the block partition

$$\Phi^M = \begin{bmatrix} \Phi_1 & & 0 \\ & \ddots & \\ 0 & & \Phi_n \end{bmatrix}$$

as in 4.3.2. Let $V(\lambda)$ be the eigenspace corresponding to λ, that is $V(\lambda) = \{v \in \mathbb{C}^n \mid (\lambda I - \Phi)^s v = 0$ for some $s \ge 0\}$. Let $\lambda = \lambda_i$ for some $i \in \{1, \dots, s\}$. We claim that for each $g \in G$ there is some $\epsilon(g) \in \{-1, 1\}$ with $gv = \epsilon(g)v$ (here we simply write gv for $\gamma^M(g)v$) for every $v \in V(\lambda)$.

[We give the following simple argument: consider a basis $v_1, \dots, v_t$ of $V(\lambda)$ satisfying $\Phi v_i = \lambda v_i + v_{i-1}$, for $i = 1, \dots, t$ (defining $v_0 = 0$). Since gv_1 is an eigenvector of Φ with eigenvalue λ, there exists a $\mu \in \mathbb{C}$ with $gv_1 = \mu v_1$. Since v_1 is a solution of the system $(\lambda I - \Phi)x = 0$ of real equations, $v_1 \in \mathbb{R}^n$. Hence $\mu \in \mathbb{R}$. On the other hand $g^s = 1$ for some $s \in \mathbb{N}$ and therefore $\mu^s = 1$. This gives $\mu \in \{1, -1\}$. Define $\epsilon(g) = \mu$. By induction on i we show that $gv_i = \epsilon(g)v_i$.

Indeed, assume $gv_j = \epsilon(g)v_j$, for $j = 1, \dots i$. Since $\Phi g v_{i+1} = \lambda g v_{i+1} + \epsilon(g)v_i$, then $gv_{i+1} = \eta v_{i+1} + v'$, where v' belongs to the space V' generated by $v_1, \dots, v_i$.

Since $g^s = 1$, then

$$\eta^s v_{i+1} + \left(\sum_{\ell+j=s-1} \eta^\ell \epsilon(g)^j \right) v' = g^s v_{i+1} = v_{i+1},$$

and $\eta^s = 1$, $(\sum_{\ell+j=s-1} \eta^\ell \epsilon(g)^j)v' = 0$. Moreover, $\lambda\eta v_{i+1} + \eta v_i + \Phi v' = \Phi g v_{i+1} = v_{i+1}\Phi g = \lambda\eta v_{i+1} + \lambda v' + \epsilon(g)v_i$. Therefore, $\eta = \epsilon(g)$ and $v' = 0$.]

Let $0 \neq v \in V(\lambda)$ and consider the partition of the vector $v = (v_1, \ldots, v_m)$ with v_β of size $n(\beta) \dim S_\beta$. Then $\epsilon(g)v_\beta = v_\beta S_\beta^{n(\beta)}(g)$, for $g \in G$, $\alpha = 1, \ldots \ell$.

If $v_\beta = (v_\beta^{(1)}, \ldots, v_\beta^{n(\beta)})$ is a partition with each block $v_\beta^{(i)}$ of size $\dim S_\beta$, we get $\epsilon(g)v_\beta^{(i)} = v_\beta^{(i)} S_\beta(g)$. Hence $\mathbb{C}v_\beta^{(i)}$ is an invariant subspace of S_β, which is irreducible. Therefore for each $v_\beta^{(i)} \neq 0$, $\dim S_\beta = 1$ and $S_\beta : G \to \mathbb{C}^*$, $g \mapsto \epsilon(g)$. This implies that λ uniquely determines a representation $S_{\beta(\lambda)}$ such that $V(\lambda)$ is contained in the space of the representation $S_{\beta(\lambda)}^{n(\beta(\lambda))}$. Therefore,

$$\sum_{i=1}^{s} d(\lambda_i) = \sum_{i=1}^{s} \dim V(\lambda_i) \leq \sum_{i=1}^{s} n(\beta(\lambda_i)),$$

where each $\beta(\lambda_i) \in U_1$. □

4.3.7 The last proof also gives:

Proposition *With the notation of 4.3.6, let $\lambda_1, \ldots, \lambda_t$ be the set of eigenvalues of ϕ_Δ with geometric multiplicity one. Then*

$$\sum_{i=1}^{t} d(\lambda_i) \leq \sum_{\beta \in U_0} n(\alpha)\ ,$$

where $U_0 = \{\beta \mid \dim S_\beta = 1\}$.

4.3.8 If Δ is a wild quiver, there are always eigenvalues of ϕ_Δ with geometric multiplicity one. Namely the spectral radius ρ of ϕ_Δ has this property (cf. 4.1.9).

Let us consider an example. Let Δ be the quiver

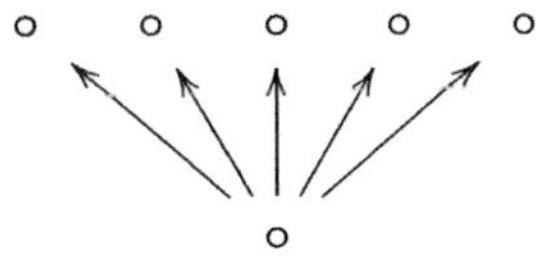

and consider the alternating group $\mathbf{A}_5$ as subgroup of $\mathrm{Aut}(\Delta)$. Let $\gamma : \mathbf{A}_5 \to G\ell(6)$ be the canonical representation with irreducible $\mathbb{C}$-decomposition $\gamma^T = \bigoplus_{\beta=1}^m S_\beta^{n(\beta)}$. It is known (see [18]) that $m = 5$ and the character table of $\mathbf{A}_5$ is

Representative of conjugacy class:	x_1	x_2	x_3	x_4	x_5
Order of the representative:	1	2	3	5	5
$1 = \chi_1$	1	1	1	1	1
χ_2	4	0	1	1	1
χ_3	5	1	1	0	0
χ_4	3	1	0	α_1	α_2
χ_5	3	1	0	α_2	α_1

with $\alpha_1 = (1 + \sqrt{5})/2$ and $\alpha_2 = (1 - \sqrt{5})/2$. We calculate

$$n(1) = (X_\gamma, X_1) = \frac{1}{60} \sum_{g \in G} \chi_\gamma(g) = 2(= t_0(\mathbf{A}_5))$$

$$n(2) = (X_\gamma, X_2) = \frac{1}{60} \sum_{g \in G} \chi_\gamma(g)\overline{\chi_2(g)} = 1, \quad \text{and} \quad \dim S_2 = 4.$$

Thus $n(3) = n(4) = n(5) = 0$ and $\gamma^T = S_1 \oplus S_1 \oplus S_2$ is also a $\mathbb{Q}$-decomposition. The Coxeter polynomial $\varphi_\Delta(T) = (1 - 3T + T^2)(1 + T)^4$ has roots $\rho = \frac{(3+\sqrt{5})}{2}$, $\rho^{-1} = \frac{(3-\sqrt{5})}{2}$ and -1 with geometric (= algebraic) multiplicity 4.

In this case the bounds given in 4.3.1, 4.3.6 and 4.3.7 are optimal.

4.4 The Automorphism Group of a Graph

4.4.1 Let Δ be a quiver as before and let n be the number of vertices of Δ.

Proposition *Let $\{\lambda_1, \ldots, \lambda_t\}$ be the different eigenvalues of Φ_Δ, where the eigenspace of λ_i has dimension m_i. Then* $\mathrm{Aut}(\Delta)$ *is a subgroup of the product*

$$U(m_1) \times \ldots \times U(m_t),$$

where $U(m_i)$ denotes the group of unitary $m_i \times m_i$-complex matrices.

Proof Let T be a matrix such that the conjugate Φ_Δ^T takes the Jordan form

$$\Phi_\Delta^T = \begin{bmatrix} J_1 & & 0 \\ & \ddots & \\ 0 & & J_t \end{bmatrix},$$

where J_i is the sum of all the Jordan λ_i-blocks.

If $\gamma : \mathrm{Aut}(\Delta) \to \mathrm{G}\ell(n)$ is the canonical representation, then for each $g \in \mathrm{Aut}(\Delta)$ the matrix $\gamma(g)^T$ has the diagonal block form

$$\gamma(g)^T = \begin{bmatrix} \gamma_1(g) & & 0 \\ & \ddots & \\ 0 & & \gamma_t(g) \end{bmatrix},$$

where each $\gamma_i(g)$ is a $m_i \times m_i$-matrix. [To see this, it is enough to show that the equation $J_i X = X J_j$, for $i \neq j$, implies that $X = 0$.] Clearly, each $\gamma_i : \mathrm{Aut}(\Delta) \to GL(m_i)$ is a representation and the image $G_i = \gamma_i(\mathrm{Aut}(\Delta))$ is a subgroup of the orthogonal matrices $O(m_i)$. Indeed, for the usual scalar product, the image under g of the canonical basis $(e_i)_i$ is an orthogonal basis:

$$(e_i g, e_j g) = (e_{g(i)}, e_{g(j)}) = \delta_{ij} = (e_i, e_j).$$

It is well known that finite subgroups of $O(m_i)$ are subgroups of $U(m_i)$, hence G is a subgroup of $U(m_1) \times \ldots \times U(m_t)$. □

One may obtain better results taking into account the number of Jordan blocks of different sizes. This does not play a very important role in practice as we will remark in the next paragraphs.

4.4.2 We get some immediate consequcnces of 4.4.1.

Proposition *Assume that all the eigenvalues* $\lambda \in \mathrm{Spec}(\Phi_\Delta)$ *have algebraic multiplicity one. Then* $\mathrm{Aut}(\Delta)$ *is abelian. Moreover for every* $g \in \mathrm{Aut}(\Delta)$, *we have* $g^2 = 1$.

Proof In this case $\mathrm{Aut}(\Delta)$ is a subgroup of $U(1) \times \ldots \times U(1)$ (n copies). □

Proposition *Let $\gamma : \mathrm{Aut}(\Delta) \to GL(n)$ be the canonical representation with irreducible $\mathbb{C}$-decomposition $\gamma^T = \bigoplus_{\beta=1}^m S_\beta^{n(\beta)}$. Then*

$$|\mathrm{Spec}(\Phi_\Delta)| \leq \sum_{\beta=1}^{m} n(\beta).$$

Proof With the notation introduced in the proof of (3.1), each representation $\gamma_i : \mathrm{Aut}(\Delta) \to G\ell(m_i)$ has an irreducible $\mathbb{C}$-decomposition,

$$\gamma_i^{T_i} = \bigoplus_{\beta=1}^{m} S_\beta^{n(i,\beta)}.$$

Hence $n(\beta) = \sum_{i=1}^{t} n(i, \beta)$ and

$$|\mathrm{Spec}(\Phi_\Delta)| \leq \sum_{i=1}^{t} (\sum_{\beta=1}^{m} n(i, \beta)) = \sum_{\beta=1}^{m} n(\beta).$$

□

In some cases this bound is optimal as shown in the example above.

4.4.3 Proposition *Let Δ be a bipartite quiver. Let $\varphi_\Delta(T) = q_1(T)^{s_1} \ldots q_\ell(T)^{s_\ell}$ be an irreducible factorization of the Coxeter polynomial $\varphi_\Delta(T)$ of Φ_Δ (we assume mcd $(q_i, q_j) = 1$ for $i \neq j$). Let $\gamma : \mathrm{Aut}(\Delta) \to GL(n)$ be the canonical representation with irreducible $\mathbb{C}$-decomposition $\gamma^T = \bigoplus_{\beta=1}^m S_\beta^{n(\beta)}$. Then*

(a) $\sum_{s_i=1} \deg q_i(T) \leq \sum_{\dim S_\beta = 1} n(\beta)$.

(b) If Δ is of wild type and $q_1(\rho) = 0$, where ρ is the spectral radius of Φ_Δ, then $\deg q_1(T) \leq n(1)$.

(c) Let $s_\ell = 2t$ be the algebraic multiplicity of 1 as root of $\varphi_\Delta(T)$. Then $\mathrm{Aut}(\Delta)$ is a subgroup of

$$\prod_{i=1}^{\ell-1} U(s_i)^{\deg q_i(T)} \times U(t).$$

Proof Part (a) is a reformulation of 4.3.6. Part (b) follows from 4.3.1 since Φ_Δ is sharp (see also 4.3.3 y 4.3.4).

Part (c) follows from 4.4.1 by observing that the group of orthogonal matrices commuting with the $\mathbb{Q}[T]$-representation

$$J_1 = \begin{bmatrix} J_1(2) & & 0 \\ & \ddots & \\ 0 & & J_1(2) \end{bmatrix}$$

with t diagonal copies for the Jordan block $J_1(2) = \begin{bmatrix} 1 & 1 \\ 0 & 1 \end{bmatrix}$, is $U(T)$. □

□

4.4.4 Let Δ be a tree. Then the above considerations may be made more precise.

First let us recall that there is a vertex $w \in \Delta_0$ which is left invariant by every $g \in \mathrm{Aut}(\Delta)$. [Since the positive vector y^+, with $y^+\Phi_\Delta = \rho y^+$, is invariant under $\mathrm{Aut}(\Delta)$ and admits a vertex w with $y^+(w) > y^+(i)$ for every $i \in \Delta_0 \setminus \{w\}$, then $g(w) = w$ for every $g \in \mathrm{Aut}(\Delta)$].

Consider the components $\Delta_1, \ldots, \Delta_s$ of $\Delta \setminus \{w\}$. The following results are left as easy exercises for the interested reader.

4.4.5 Proposition $\mathrm{Aut}(\Delta)$ *is a subgroup of* $\prod_{i=1}^{t} U(k_i) \times \mathrm{Aut}(\Delta_i)$.

4.4.6 Let Δ be a tree and $\widehat{\Delta}$ be a convex connected subtree of Δ. We say that $\widehat{\Delta}$ is a **symmetry-subtree** of Δ if there is some $g \in \mathrm{Aut}(\Delta)$ such that $\widehat{\Delta} \cap g(\widehat{\Delta}) = \emptyset$ and $\widehat{\Delta}$ is maximal with this property.

Proposition *Let* $\widehat{\Delta}$ *be a symmetry-subtree of* Δ, *then* $\Phi_{\widehat{\Delta}}(T)$ *is a factor of* $\Phi_\Delta(T)$.

4.5 Canonical Algebras

In this section we overview other types of algebras with known Coxeter polynomials.

4.5.1 Canonical Algebras A canonical algebra $\Lambda = \Lambda(\mathbf{p}, \overline{\lambda})$ is determined by a **weight sequence** $\mathbf{p} = (p_1, \ldots, p_t)$ of t integers $p_i \geq 2$ and a **parameter sequence** $\overline{\lambda} = (\lambda_3, \ldots, \lambda_t)$ consisting of $t-2$ pairwise distinct non-zero scalars from the base field K. (We may assume $\lambda = 1$ such that for $t \leq 3$ no parameters occur). Then the algebra Λ is defined by the quiver

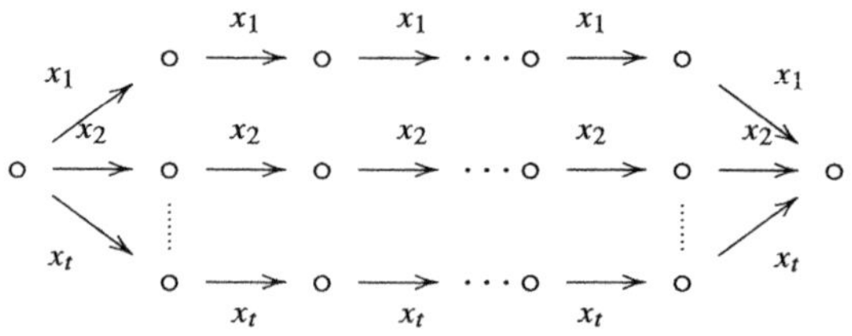

having p_i arrows with label x_i, and satisfying the $t-2$ equations:

$$x_i^{p_i} = x_1^{p_1} - \lambda_i x_2^{p_2}, \qquad i = 3, \ldots, t.$$

For more than two weights, canonical algebras are not hereditary. Instead, their representation theory is determined by a hereditary category, the category $\mathbf{coh}(K\mathbb{P})$ of coherent sheaves on a weighted projective line $K\mathbb{P} = \mathbb{X}$, naturally attached to Λ, see [16].

Proposition *Let Λ be a canonical algebra. Then Λ is the endomorphism ring of a tilting object in the category* $\mathbf{coh}(K\mathbb{P})$ *of coherent sheaves on the weighted projective line $K\mathbb{P} = \mathbb{X}$. The category* $\mathbf{coh}(K\mathbb{P})$ *is hereditary and satisfies Serre duality in the form $D\mathrm{Ext}^1(X, Y) = \mathrm{Hom}(Y, \tau_A X)$ for a self-equivalence τ_A which serves as the Auslander-Reiten translation.*

Canonical algebras were introduced by Ringel [34]. They form a key class to study important features of representation theory. In the form of tubular canonical algebras they provide the standard examples of tame algebras of linear growth. Up to tilting, canonical algebras are characterized as the connected K-algebras with a separating exact subcategory or a separating tubular one-parameter family (see [24] and [37]). That is, the module category mod-Λ accepts a separating tubular family $\mathcal{T} = (T_\lambda)_{\lambda \in \mathbb{P}K}$, where T_λ is a **homogeneous tube** for all λ with the exception of t tubes $T_{\lambda_1}, \ldots, T_{\lambda_t}$ with T_{λ_i} of rank p_i $(1 \le i \le t)$.

Canonical algebras constitute an instance, where the explicit form of the Coxeter polynomial is known, see [23] or [21].

Proposition *Let Λ be a canonical algebra with weight and parameter data* $(\mathbf{p}, \overline{\lambda})$. *Then the Coxeter polynomial of Λ is given by*

$$\varphi_\Lambda(x) = (x-1)^2 \prod_{i=1}^{t} v_{p_i}(x),$$

where $v_n(x) = x^{n-1} + x^{n-2} + \ldots + x + 1$.

The Coxeter polynomial therefore only depends on the weight sequence **p**. Conversely, the Coxeter polynomial determines the weight sequence (up to a reordering of its elements).

A finite dimensional algebra isomorphic to the endomorphism algebra of a tilting object in a (connected) hereditary abelian Hom-finite K-category $\mathbb{H}$ is called a **tilted algebra**. By a result of Happel [17] each such category is derived equivalent to the module category mod-H over a hereditary algebra H or to the category $\mathbf{coh}(K\mathbb{P})$ of coherent sheaves on a weighted projective line $K\mathbb{P}$. The Coxeter polynomials of tilted algebras are therefore the Coxeter polynomials of hereditary or canonical algebras.

4.5.2 Super Canonical Algebras Following [25], see also [38], a supercanonical algebra is defined as follows: The **double cone** $\hat{S}$ of a finite poset S is the poset obtained from S by adjoining a unique source α and a unique sink ω, like in the picture below,

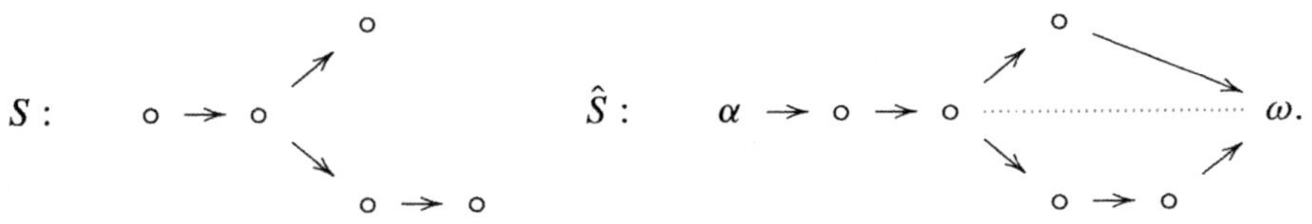

Due to commutativity there is a unique path κ_S from α to ω in $\hat{S}$. Let now $S_1, \ldots, S_t$ be finite posets, $t \geq 2$ and $\lambda_3, \ldots, \lambda_t$ pairwise different elements from $K \setminus \{0\}$. The supercanonical algebra $A = A(S_1, \ldots, S_t; \lambda_3, \ldots, \lambda_t)$ is obtained from the fully commutative quivers $\hat{S}_1, \ldots, \hat{S}_t$ by identification of the sources and sinks, respectively, and requesting additionally the $t-2$ relations $\kappa_i = \kappa_2 - \lambda_i \kappa_1$, $i = 3, \ldots, t$, where $\kappa_i = \kappa_{S_i}$. The next figure yields an example of a supercanonical algebra with three arms

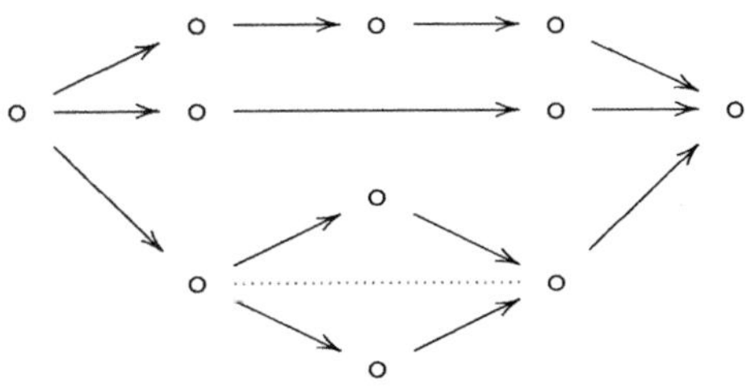

where we assume that $\kappa_3 = \kappa_2 - \kappa_1$. In case, $S_1, \ldots, S_t$ are linear quiver $S_i = [p_i]\colon 1 \to 2 \to \cdots \to p_{i-1}$, the algebra $A(S_1, \ldots, S_t; \lambda_1, \ldots, \lambda_t)$ is just the canonical algebra $\Lambda(p_1, \ldots, p_t; \lambda_3, \ldots, \lambda_t)$.

Returning to the general case, a simple calculation shows that

$$\varphi_\Lambda(x) = (x-1)^2 \prod_{i=1}^{t} \varphi_{S_i}(x),$$

where φ_{S_i} is the Coxeter polynomial of the poset algebra S_i, for $i = 1, \dots, t$.

Supercanonical algebras form a natural generalization of canonical algebras, since they arise from a canonical algebra Λ by replacing the linear arms of Λ by finite posets, that is, fully commutative quivers. There are many reasons making supercanonical algebras an interesting class (cf. [25]):

(1) The K-theory of supercanonical algebras is very similar to the K-theory of canonical algebras. There is a Riemann-Roch theorem, moreover one has an explicit formula for the Coxeter polynomial in terms of the characteristic polynomials of the posets S_i.
(2) One-point extensions of canonical algebras with exceptional modules of derived finite quasi-length are supercanonical.

4.5.3 Extended Canonical Algebras For an algebra A and a right A-module M we call $A[M] = \left(\begin{smallmatrix} A & 0 \\ M & K \end{smallmatrix}\right)$ the **one-point extension** of A by M. We call an algebra of the form $\Lambda[P]$, with P indecomposable projective, an **extended canonical algebra** of type $\langle \mathbf{p}, \overline{\lambda} \rangle$ or just of type $\langle p_1, \dots, p_t \rangle$. We use the notation $\widehat{\Lambda}\langle \mathbf{p}, \overline{\lambda} \rangle$ if P is the indecomposable projective of the sink vertex of the quiver of Λ. Note that $\widehat{\Lambda} = \widehat{\Lambda}\langle \mathbf{p}, \overline{\lambda} \rangle$ is given by the quiver

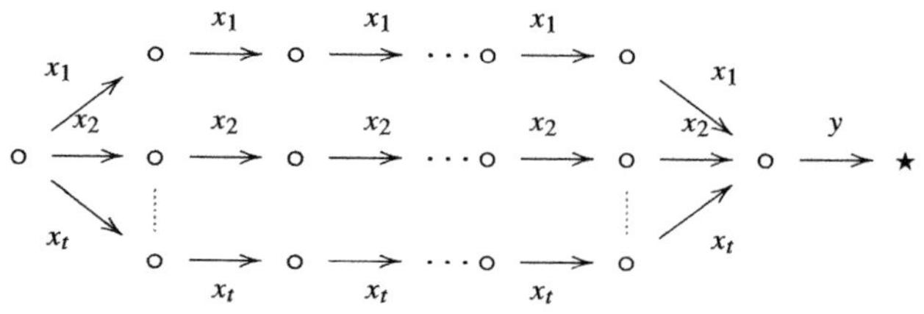

keeping all the relations for the canonical algebra Λ and not introducing new ones. In particular, there are no relations involving the new vertex $\star$. For the Coxeter polynomial $\varphi_{\widehat{\Lambda}}$, which only depends on the numbers $p_1, \dots, p_t$, we write $\widehat{\varphi}\langle p_1, \dots, p_t \rangle$ or $\varphi_{\langle p_1, \dots, p_t \rangle}$.

Remark The extended canonical algebra $\widehat{\Lambda}$ of type $\langle \mathbf{p}, \overline{\lambda} \rangle$ has Coxeter polynomial

$$\widehat{\varphi}\langle p_1, \dots, p_t \rangle = (x+1)(x-1)^2 \prod_{i=1}^{t} v_{p_i}(x) - x\varphi_{[p_1,\dots,p_t]}(x).$$

Recall that the explicit form of the Coxeter polynomial for the hereditary star $[p_1, \dots, p_t]$ is given in 4.1.5.

4.6 Self-Injective Algebras

4.6.1 Let A be a finite dimensional algebra, defined over a field k. The category of finite dimensional left A-modules and the set of isoclasses of simple A-modules will be denoted by mod-A and $\mathscr{S}(A)$, respectively. We will occasionally identify $\mathscr{S}(A)$ with a complete set of its representatives. Given a simple A-module S, we consider its projective cover $P(S)$ and its injective envelope $E(S)$. Recall that $\operatorname{top}(P(S)) \cong S$ and $\operatorname{soc}(E(S)) \cong S$. Moreover, for every projective (injective) indecomposable A-module Q there exists exactly one $S \in \mathscr{S}(A)$ with $Q \cong P(S)$ ($Q \cong E(S)$) (cf. [3, I.4,II.4]. What can be said about the structure of $\operatorname{soc}(P(S))$ (or $\operatorname{top}(E(S))$)? For instance,

(1) Let $A = k[1 \to 2]$. Setting $P_i := P(S_i)$, we obtain $\operatorname{soc}(P_1) = S_2 = \operatorname{soc}(P_2)$.
(2) If $A = k[1 \leftarrow 2 \to 3]$, then $\operatorname{soc}(P_1) = S_1$, $\operatorname{soc}(P_2) = S_1 \oplus S_3$, and $\operatorname{soc}(P_3) = P_3$.

4.6.2 In general, one can thus not hope for a fixed pattern. We are going to introduce a class of algebras where such a pattern exists and show that this class is in fact determined by the presence of a correspondence between tops and socles of the principal indecomposable modules $P(S)$. These so- called quasi-Frobenius algebras were introduced and studied by Nakayama [29, 30], whose work was inspired by results of Brauer and Nesbitt [9]. Nowadays, the following notion is commonly used.

Definition The algebra A is self-injective if the regular module ${}_AA$ is injective.

4.6.3 The principal result of this section was proved by Nakayama in the context of basic algebras.

Theorem ([14]) *The following statements are equivalent:*

(1) The algebra A is self-injective.
(2) The rule $[S] \mapsto [\operatorname{soc}(P(S))]$ defines a permutation $\nu : \mathscr{S}(A) \to \mathscr{S}(A)$.

A proof is given in 4.6.8 below. The permutation ν is referred to as the *Nakayama permutation* of the self-injective algebra A.

4.6.4 Given $M \in$ mod-A, its dual $M^* := \operatorname{Hom}_k(M, k)$ has the structure of a right A-module. Thus, $M \mapsto M^*$ is a duality between the categories mod-A and mod-A^{op}, where A^{op} denotes the opposite algebra of A. In particular, $?^*$ takes projectives to injectives and vice versa.

Lemma *Let A be self-injective.*

(1) A A-module M is projective if and only if it is injective.
(2) $\operatorname{soc}(P(S))$ is simple for every $S \in \mathscr{S}(A)$.

(3) $P(S) \cong E(\operatorname{soc}(P(S)))$ *for every* $S \in \mathscr{S}(A)$.
(4) Let S, T *be simple* A*-modules. If* $\operatorname{soc}(P(S)) \cong \operatorname{soc}(P(T))$, *then* $S \cong T$.

Proof If M is projective, then M is a direct summand of a free module A^n. Thus, M is, as a direct summand of an injective module, injective.

It follows that $\{P(S) \mid [S] \in \mathscr{S}(A)\}$ is a set of isoclasses of injective indecomposable A-modules of cardinality $|\mathscr{S}(A)|$. It thus coincides with the set of isoclasses of indecomposable injectives, and there exists a permutation $\nu : \mathscr{S}(A) \to \mathscr{S}(A)$ such that

$$P(S) \cong E(\nu(S)), \qquad \text{for all } [S] \in \mathscr{S}(A).$$

In particular, we have

- Every indecomposable injective module is projective, so that (1) follows.
- $[\operatorname{soc}(P(S))] = [\operatorname{soc}[E(\nu(S))]] = [\nu(S)]$ for all $[S] \in \mathscr{S}(A)$. Thus, we obtain (2)–(4).

□

Remark 4.6.5

(a) Owing to (1), *the class of self-injective algebras is stable under Morita equivalence.*
(b) Since the right module $A \in \text{mod-}A^{op}$ *is projective, its dual is injective, hence projective, so that* A_A *is injective. In other words, the algebra* A^{op} *is self-injective.*

By general theory, the principal indecomposable A modules are of the form $P = Ae$, where $e \in A$ is a primitive idempotent. If $M \in \text{mod-}A$, then

$$\operatorname{Hom}_A(P, M) \to eM; \quad f \mapsto f(e),$$

is an isomorphism of vector spaces, which is right A-linear in case M is a (A, A)-bimodule.

4.6.6 The **Nakayama permutation** is a combinatorial tool that does not provide any information concerning the endomorphism rings of S and $\nu(S)$ (these are actually isomorphic). For algebras over algebraically closed fields, this is of course not a problem. However, a better understanding of ν necessitates a module theoretic (functorial) description.

Definition Let A be a k-algebra. The functor

$$\mathcal{N} : \text{mod-}A \longrightarrow \text{mod-}A$$

$$M \longmapsto \text{Hom}_A(M,A)^*$$

is called the **Nakayama functor** of A.

4.6.7 The above observations yield

$$\mathcal{N}(Ae) \cong (eA)^*,$$

so that $\mathcal{N}$ sends indecomposable projectives to indecomposable injectives. (However, it does in general not send indecomposables to indecomposables).

Lemma *Let A be a k-algebra that affords a Nakayama permutation $\nu : \mathcal{S}(A) \to \mathcal{S}(A)$. Then the following statements hold:*

(1) $\mathcal{N}(P(S)) \cong E(S)$ for all $[S] \in \mathcal{S}(A)$.
(2) $\mathcal{N}(S) \cong \nu^{-1}(S)$ for all $[S] \in \mathcal{S}(A)$.

Proof

(1) Let S be a simple A-module. Pick a primitive idempotent $e_S \in A$ with $P(S) = Ae_S$. Since

$$\text{Hom}_{A^{op}}(e_S A, S^*) \cong (S^* e_S) \cong (e_S S)^* \cong \text{Hom}_A(P(S), S)^* \neq 0,$$

it follows that the principal indecomposable right A-module $e_S A$ has top S^*. As the duality ?* maps tops to socles, we obtain

$$S \cong \text{top}(e_S A) \cong \text{soc}((e_S A)^*) \cong \text{soc}(\mathcal{N}(Ae_S)) \cong \text{soc}(\mathcal{N}(P(S))),$$

so that the indecomposable injective A-module $\mathcal{N}(P(S))$ is isomorphic to $E(S)$.

(2) Let S, T be simple A-modules, e_S, e_T the corresponding primitive idempotents. Given any A-module M, we have

$$\dim_k \text{Hom}_A(P(T, M) = [M : T] \dim_k \text{End}_A(T),$$

where $[M : T]$ denotes the multiplicity of T in M. From the k-vector space isomorphisms

$$\begin{aligned}\mathrm{Hom}_A(P(T), \mathscr{N}(S)) &\cong e_T\mathscr{N}(S) \cong e_T\mathrm{Hom}_A(S, A)^* \cong (\mathrm{Hom}_A(S, A)e_T)^* \\ &\cong \mathrm{Hom}_A(S, Ae_T)^* \cong \mathrm{Hom}_A(S, \nu(T))^* \\ &\cong \mathrm{End}_A(S),\end{aligned}$$

if $T \cong \nu^{-1}(S)$, and $\mathrm{Hom}_A(P(T), \mathscr{N}(S)) = 0$ otherwise. We see that $\nu^{-1}(S)$ is the only composition factor of $\mathscr{N}(S)$. By the same token, we have

$$[\mathscr{N}(S) : \nu^{-1}(S)] \dim_k \mathrm{End}_A(\nu^{-1}(S)) = \dim_k \mathrm{Hom}_A(P(\nu^{-1}(S)), \mathscr{N}(S)) = \dim_k \mathrm{End}_A(S).$$

By applying this formula successively to the modules $\nu^{-i}(S)$, we obtain, observing $\nu^{-n}(S) \cong S$ for some $n \in \mathbb{N}$, a natural number $m \in \mathbb{N}$ such that

$$\dim_k \mathrm{End}_A(S) = m[\mathscr{N}(S) : \nu^{-1}(S)] \dim_k \mathrm{End}_A(S).$$

Thus, $[\mathscr{N}(S) : \nu^{-1}(S)] = 1$, and $\mathscr{N}(S) \cong \nu^{-1}(S)$.

□

4.6.8 Proof of Theorem 4.6.3

Proof In view of Lemma 4.6.4, it suffices to verify that (2) implies (1). Let $\ell(M)$ denote the length of the A-module M. The Nakayama functor is right exact and Lemma 4.6.7 ensures that if takes simples to simples. Induction on $\ell(M)$ then implies that

$$\ell(\mathscr{N}(S)) \leq \ell(M), \quad \text{for all } M \in \text{mod-}A.$$

Lemma 4.6.7 now yields

$$\ell(E(S)) \leq \ell(P(S)), \quad \text{for all } [S] \in \mathscr{S}(A).$$

Since $\mathrm{soc}(P(S)) = \nu(S)$, we have an embedding $i_S : P(S) \to E(\nu(S))$. Iteration gives rise to a chain

$$\ell(E(S)) \leq \ell(P(S)) \leq \ell(E(\nu(S))) \leq \ell(P(\nu(S)))\ell(E(\nu^2(S))) \leq \cdots \leq \ell(E(S)),$$

so that $\ell(P(S)) = \ell(E(\nu(S)))$. As a result, i_S is bijective, showing that $P(S)$ is injective. This implies the self-injectivity of A. □

4.6.9 The **Nakayama functor** ν : mod-$A \to$ mod-A is defined as $D\mathrm{Hom}_A(-, A)$ and is isomorphic to the functor $(-) \otimes_A D(A)$. The inverse Nakayama functor ν^{-1} : mod-$A \to$ mod-A is defined as $\mathrm{Hom}_{A^{op}}(-, A)D$ and is isomorphic to the functor

$\mathrm{Hom}_A(-, DA)$. Nakayama functors play a prominent role in the representation theory of finite dimensional algebras, since ν : proj - $\to$ inj - is an equivalence with quasi-inverse ν^{-1}. For example they appear in the definition of the Auslander-Reiten translates τ and τ^{-1}.

Proposition *Let M be an A-module with a minimal injective presentation $0 \to M \to I_0 \to I_1$. Then the following sequence is exact:*

$$0 \to \nu^{-1}(M) \to \nu^{-1}(I_0) \to \nu^{-1}(I_1) \to \tau^{-1}(M) \to 0.$$

When talking about Nakayama algebras, we assume that they are given by quivers and relations (meaning that they are basic and split algebras). This is not really a restriction since the dominant dimension is invariant under Morita equivalence and field extensions.

4.7 Further Spectral Properties

4.7.1 Isospectral Algebras Let φ_A be the Coxeter polynomial of a finite dimensional algebra A. The set of roots together with their multiplicities is denoted $\mathrm{Spec}(\varphi_A)$, or just $\mathrm{Spec}(A)$, and is called the **spectrum** of A. Two algebras are called **isospectral** (or cospectral), if they have the same spectrum, that is, the same Coxeter polynomial. In the same spirit we speak of **isospectral graphs** if their characteristic polynomials are the same.

4.7.2 Wild hereditary tree algebras which which are isospectral but not derived equivalent: Consider the tree algebras $A_1 = K\Delta_1$ and $A_2 = K\Delta_2$ given by the displayed quivers:

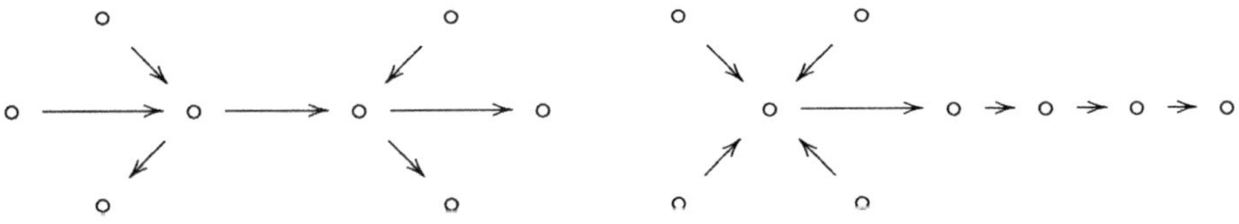

We denote the corresponding underlying graphs $\overline{\Delta}_1$ and $\overline{\Delta}_2$. In [10] the graphs $\overline{\Delta}_1$ and $\overline{\Delta}_2$ were produced as the pair of isospectral graphs with smallest number of vertices, that is, $\kappa_{\overline{\Delta}_1}(x) = \kappa_{\overline{\Delta}_2}(x)$. By A'Campo's formula (Theorem 4.1.7, see also [1]) we have $\varphi_{A_i}(x^2) = \kappa_{\overline{\Delta}_i}(x + x^{-1})$ for $i = 1, 2$, hence $\varphi_{A_1} = \varphi_{A_2}$, that is the algebras A_1 and A_2 are isospectral. Moreover, we observe that the algebras A_1 and A_2 are not derived equivalent. Indeed, the quiver Δ_i appears as a section of a transjective component of the Auslander-Reiten quiver of the derived category of mod-A_i, for $i = 1, 2$.

The tree $\overline{T} = [2, 2, 3, 5]$ and the comb $\overline{C} = [[1, 2, 2, 2, 1, 1]]$ given in the following figures,

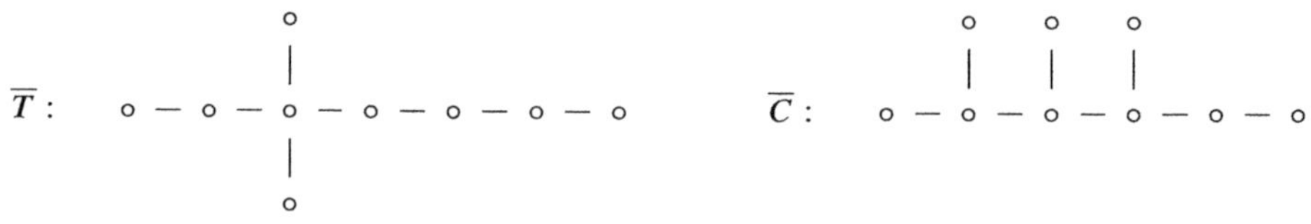

are isospectral. By the preceding argument their path algebras KT and KC will not be derived equivalent, regardless of the chosen orientation in quivers T and C.

4.7.3 Isospectral tree algebras with an arbitrary big number of vertices: Indeed, consider the algebras A and A' given by the following quivers obtained by identifying a vertex of a quiver of type $\widetilde{\mathbb{E}}_8$ with a extremal vertex of a linear quiver of type $\mathbb{A}_n$:

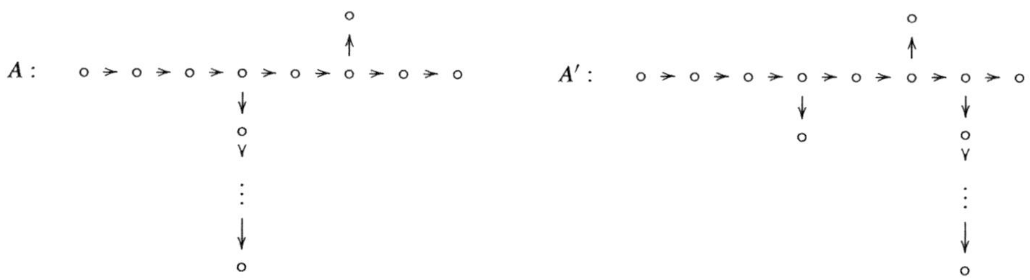

Lemma *The algebras A and A′ are isospectral.*

Proof It is enough to observe that the underlying graphs Δ and Δ' satisfy $\kappa_\Delta(x) = \kappa_{\Delta'}(x)$. This follows from the following construction at [36]:

The **coalescence** of Δ_1 and Δ_2 at vertices v_1 of Δ_1 and v_2 of Δ_2 is formed by identifying v_1 and v_2 and denoted by $\Delta_1 \bullet \Delta_2$. If Δ_2 and Δ_2' are isospectral graphs and $\Delta_2 \setminus v_2$ and $\Delta_2' \setminus v_2'$ are also isospectral, then the graphs $\Delta_1 \bullet \Delta_2$ and $\Delta_1 \bullet \Delta_2'$ are isospectral.

To show the claim, only observe that

$$\kappa_{\Delta_1 \bullet \Delta_2}(x) = \kappa_{\Delta_1}(x)\kappa_{\Delta_2 \setminus v_2}(x) + \kappa_{\Delta_1 \setminus v_1}(x)\kappa_{\Delta_2}(x) - x\kappa_{\Delta_1 \setminus v_1}(x)\kappa_{\Delta_2 \setminus v_2}.$$

In our special case $\kappa_{\Delta_2 \setminus v_2}(x) = \kappa_{\Delta_2' \setminus v_2'}(x) = x^2(x^2 - 2)(x^4 - 4x^2 + 2)$. □

Isospectral problems also illustrate the interplay between spectral graph theory and Coxeter polynomials. The following result [27], whose proof we sketch, is an example.

Proposition *Isospectral stars, with standard orientations, are isomorphic (as quivers or graphs).*

Proof Let $p = [p_1, \ldots, p_t]$ and $q = [q_1, \ldots, q_s]$ be two isospectral stars with the standard orientation. Then $n = 1 + \sum_{i=1}^{t}(p_i - 1) = 1 + \sum_{j=1}^{s}(q_j - 1)$ and we assume that $p_1 \geq p_2 \geq \ldots \geq p_t$ and $q_1 \geq q_2 \geq \ldots \geq q_s$. By (1.3), both $\kappa_p = \kappa_q$ and $\varphi_p = \varphi_q$. We shall prove that $p = q$.

First we show that $s = t$. Denote by c_i the coefficient of x^{n-i} in the polynomial κ_p. Since p a tree, the coefficient c_4 is the number of pairs of independent edges of p, see [11, Theorem 1.3]. For p, an easy computation yields

$$c_4 = \binom{n-1}{2} - (n-t-1) - \binom{t}{2} = \frac{1}{2}[(n-1)(n-4) - t(t-3)].$$

Since, by hypothesis $c_4(p) = c_4(q)$ then $t = s$.

Consider now the expression of the Coxeter polynomial of p:

$$\varphi_p = [(x+1) - x\sum_{i=1}^{t}\frac{1-x^{p_i-1}}{1-x^{p_i}}]\prod_{i=1}^{t}\frac{1-x^{p_i}}{1-x}.$$

Multiply this polynomial by $(1-x)^t$ to obtain:

$$\Gamma_p = (x+1)\prod_{i=1}^{t}(1-x^{p_i}) - x\sum_{i=1}^{t}(1-x^{p_i-1})\prod_{j\neq i}(1-x^{p_j}).$$

A simple comparison of the coefficients of Γ_p and Γ_q implies that $p = q$. □

4.7.4 Representability of Coxeter Polynomials Following [26], we say that a polynomial $p \in K[x]$ is represented by $q \in K[x]$ if $p(x^2) = q^*(x) := x^{\deg q}q(x + x^{-1})$. It follows that representable polynomials are self-reciprocal.

The concept of representability arises as a generalization of A'Campo's observation: if $A = K\Delta$ is a bipartite hereditary algebra, then φ_A is represented by κ_Δ. We shall see that there are other familiar examples of algebras with representable Coxeter polynomial and will illustrate some applications of this fact.

Example Recall that the (normalized) **Chebycheff polynomials** (of the second kind) $(u_n)_n$ may be inductively constructed by the rules:

$$u_0 = 1, \quad u_1 = x, \quad \text{and } u_{n+1} = xu_n - u_{n-1} \text{ for } n \geq 1.$$

A simple induction shows that the characteristic polynomial of the linear graph $\mathbb{A}_n = [n]$ is the polynomial u_n. Moreover, v_{n+1} is represented by u_n.

Proposition *For each $n \geq 2$, the n-th cyclotomic polynomial is representable. In fact, there is an irreducible factor f of u_{2n-1} such that $\Phi_n(x^2) = f^*(x)$.*

Proof For $n = 1$ (and $2n - 1 = 1$) we have:

$$u_1^* = x(x + x^{-1}) = x^2 + 1 = \Phi_2(x^2).$$

Assume $f_i^*(x) = \Phi_i(x^2)$ for $i < n$, then

$$\Phi_n(x^2) \prod_{\substack{1<d<n \\ d|n}} \Phi_d(x^2) = v_{n-1}(x^2) = \frac{x^{2n} - 1}{x^2 - 1} = \frac{v_{2n-1}(x)}{x + 1} = \frac{u_{2n-1}^*(x)}{x + 1}.$$

Using (1.4), consider a decomposition $u_{2n-1} = \prod_{i=1}^s g_i$ in $\mathbb{Z}[x]$ such that g_i is irreducible as a self-reciprocal polynomial. Recall that all the cyclotomic polynomials Φ_d (and hence also $\Phi_n(x^2)$) are irreducible. By induction hypothesis we may assume that for each $d|n$, we get $\Phi_d(x^2) = f_d^*(x)$ for certain $f_d = g_i$, $1 \leq i \leq s$. The result for Φ_n follows from (1.4). □

As concrete examples we calculate: $\Phi_3 = x^2 + x + 1$ with $\Phi_3(x^2) = u_2^*$; $\Phi_4 = x^2 + 1$ with $\Phi_4(x^2) = f_4^*$ for $f_4 = x^2 - 2$ which is a factor of $u_3 = x^3 - 2x$ and also of $u_7 = x^7 - 6x^5 + 10x^3 - 4x$; $\Phi_5 = x^4 + x^3 + x^2 + x + 1$ with $\Phi_5(x^2) = u_5^*(x)$.

The following remark shows the representability of the Coxeter polynomial for many algebras.

Proposition *Let $A = B[P]$ be a one-point extension of an algebra B with an indecomposable projective module P associated to a source b in B. Consider the quotient algebra $C = B/(b)$, then the following holds:*

(a) $\varphi_A(x) = (1 + x)\varphi_B(x) - x\varphi_C(x)$;
(b) Assume that φ_B and φ_C are representable, then φ_A is representable.

Proof The formula (a) is shown in [22, Prop. 18.3 and Cor. 18.2]. Therefore

$$\varphi_A(x^2) = (1+x^2)\varphi_B(x^2) - x^2\varphi_C(x^2) = x^n(x+x^{-1})q_1(x+x^{-1}) - x^n q_2(x+x^{-1}),$$

when n is the degree of φ_A. Hence $\varphi_A(x^2) = q^*(x)$ with $q(x) = xq_1(x) - q_2(x)$. □

4.7.5 Graphical Representability In general, we shall say that a polynomial $p \in \mathbb{Z}[x]$ is **graphically represented** if $p(x^2) = \kappa_\Delta(x)$ for a graph Δ. Applications of graphical representability arise as consequence of the following elementary remark. Denote by $\mathscr{R}(p)$ the set of roots of a polynomial p.

Proposition *Let p and $q \in \mathbb{Z}[x]$ be such that $p(x^2) = q^*(x)$. Then*

(a) $\mu \in \mathscr{R}(p)$, then $\mu^{-1}, \bar{\mu}, \bar{\mu}^{-1} \in \mathscr{R}(p)$.
(b) $\mathscr{R}(p) \subset \mathbb{S} \cup \mathbb{R}^+$ (where $\mathbb{S} = \{v \in \mathbb{C} \colon \|v\| = 1\}$ is the unit circle and $\mathbb{R}^+$ the positive real numbers, $0 \in \mathbb{R}^+$) if and only if $\mathscr{R}(q) \subset \mathbb{R}$.
(c) $\mathscr{R}(p) \subset \mathbb{S}$ (resp. $\mathbb{S} \setminus \{1\}$) if and only if $\mathscr{R}(q) \subset [-2, 2]$ (resp. $(-2, 2)$).

(d) If p is graphically represented, then $\mathscr{R}(q) \subset \mathbb{R}$ and $\mathscr{R}(p) \subset \mathbb{S} \cup \mathbb{R}^+$.
(e) Assume p is graphically represented by the graph Δ, that is $q = \kappa_\Delta(x)$ then

(i) Δ is a bipartite graph.
(ii) $\mathscr{R}(p) \subset \mathbb{S} \setminus \{1\}$ if and only if Δ is a union of Dynkin graphs.

Proof Observe that $0 \neq \lambda \in \mathbb{C}$, then $\mu = \lambda^2 \in \mathscr{R}(p)$ if and only if $\lambda + \lambda^{-1} \in \mathscr{R}(q)$. Assume $\lambda = r(a + ib)$ with $r \in \mathbb{R}^+$, $a^2 + b^2 = 1$, then $\lambda + \lambda^{-1} = (r + r^{-1})a + i(r - r^{-1})b$.

(a) if $\mu = \lambda^2 \in \mathscr{R}(p)$, then μ^{-1} yields $\lambda + \lambda^{-1} \in \mathscr{R}(q)$, then $\mu^{-1} \in \mathscr{R}(p)$. Since p has real coefficients, then $\bar{\mu}$ (and therefore $\bar{\mu}^{-1}$) is in $\mathscr{R}(p)$.
(b) Observe $\mu \in \mathscr{R}(p)$ lies in $\mathbb{S}$ (resp. in $\mathbb{R}^+$) iff $r = 1$ (resp. $b = 0$) iff $\lambda + \lambda^{-1} \in \mathbb{R}$.
(c) In the above case $\lambda + \lambda^{-1} \in [-2, 2]$ iff $b = 0$ iff $\mu \in \mathbb{S}$.
(d) Assume $q = \kappa_\Delta$ for Δ a graph. Then $\mathscr{R}(p) \subset \mathbb{S} \setminus \{1\}$ iff $\mathscr{R}(q) \subset (-2, 2)$, in this case Δ is a union of Dynkin graphs, see [11].

□

Example In general the Coxeter polynomial of a path algebra $A = K\Delta$ is not graphically representable. Indeed, there are wild quivers Δ where $\mathscr{R}(\varphi_A)$ is not contained in $\mathbb{R} \cup \mathbb{S}$, see [22, example 18.1]. Such a quiver must not be bipartite. Moreover, since $\mathscr{R}(\varphi_A)$ is closed under complex conjugation and taking inverses, we need at least six vertices for Δ to make the phenomenon appear. With this information at hand, many examples like

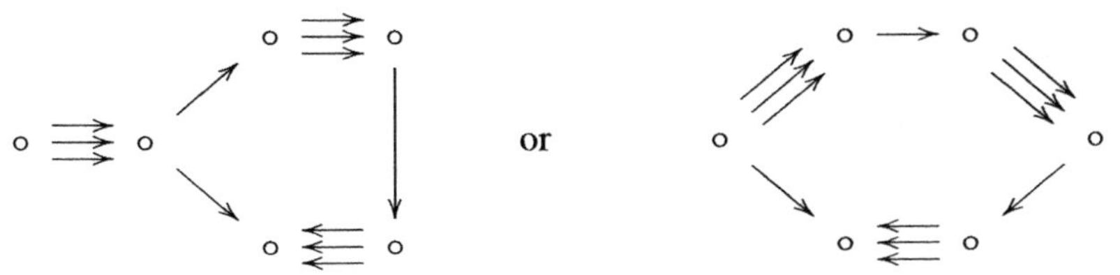

can be obtained.

The following result stresses the relationship between properties of the Coxeter transformation φ_A of an algebra A and the structure of the graph Δ in case φ_A is graphically represented by Δ.

Proposition *Let A be an algebra whose Coxeter polynomial φ_A is graphically represented by Δ, that is, $\varphi_A(x^2) = \kappa_\Delta$. Assume that the characteristic polynomial of the graph Δ' obtained by deleting a vertex a from Δ has at most $n - 1$ zeros in the interval $[2, \infty)$. Then φ_A has at most $2n$ zeros outside the complex unit circle.*

Proof The equation $\varphi_A(x^2) = \kappa_\Delta(x)$ implies that the cardinality of $\mathrm{Spec}(\varphi_A) \cap (\mathbb{R}^+ \setminus \{1\})$ is twice the cardinality of $\mathrm{Spec}(\kappa_\Delta) \cap (2, \infty)$. Hence, it is enough to prove that κ_Δ has at most n zeros in the interval $[2, \infty)$. Indeed, the adjacency matrix A

of Δ has the shape

$$A = \begin{pmatrix} A' & y^t \\ y & 0 \end{pmatrix},$$

where A' is the adjacency matrix of the graph Δ' and y is a row vector. Consider $\nu_1 \geq \nu_2 \geq \ldots \nu_{n-1}$ the roots of $\kappa_{\Delta'}$ and $\mu_1 \geq \mu_2 \geq \ldots \mu_{n-1} \geq \mu_n$ the roots of κ_Δ. The **interlacing property** of eigenvalues yields inequalities:

$$\mu_1 \geq \nu_1 \geq \mu_2 \geq \nu_2 \geq \ldots \nu_{n-1} \geq \mu_n,$$

which shows the result. □

Several results have been proved which are special cases of the above Proposition. Namely:

(a) [32, Proposition 2.6] Let $\overline{\Delta}$ be a tree graph with s ramification points and Δ be a quiver with underlying graph $\overline{\Delta}$. The Coxeter polynomial φ_A of the hereditary algebra $A = K\Delta$ has at most $2s$ roots outside the unit circle.

(b) In [20] **generalized stars** are introduced as amalgamations of linear quivers or even cycles in a selected vertex. It is shown that the Coxeter polynomial φ_S of a wild generalized star S is of the form pq where p is a product of cyclotomic polynomials and q is an irreducible polynomial with exactly one root outside the unit circle (such polynomial is called a **Salem polynomial**).

4.7.6 Fix the following notation: Let H be the star of type $[p_1, \ldots, p_t]$ with Coxeter polynomial $\varphi[p_1, \ldots, p_t]$ and let Λ be a canonical algebra of weight type $(p_1, \ldots, p_t)$ with Coxeter polynomial $\varphi(p_1, \ldots, p_t)$; let $\widehat{\Lambda}$ be the extended canonical algebra with Coxeter polynomial $\varphi\langle p_1, \ldots, p_t\rangle$.

Proposition *The above Coxeter polynomials are representable by the following procedure:*

(a) $\varphi[p_1, \ldots, p_t](x^2) = \kappa^*_{[p_1,\ldots,p_t]}$, *where* $\kappa_{[p_1,\ldots,p_t]}$ *denotes the characteristic polynomial of the adjacency matrix of star graph* $[p_1, \ldots, p_t]$*;*

(b) $\varphi(p_1, \ldots, p_t)(x^2) = \kappa_{K_2} \prod_{i=1}^t \kappa^*_{[p_i-1]}$, *where* $\kappa_{[p]}$ *denotes the characteristic polynomial of the linear path* $[p]$ *and* K_2 *is the Kronecker diagram* *;*

(c) $\Phi\langle p_1, \ldots, p_t\rangle(x^2) = \left(x\kappa_{K_2} \prod_{i=1}^t \kappa_{[p_i-1]} - \kappa_{[p_1,\ldots,p_t]}\right)^*$.

Let $q_{(p_1,\ldots,p_t)}$ be the polynomial representing $\Phi\langle p_1, \ldots, p_t\rangle$. Under the conditions shown in the above Proposition, a version of **Sturm's Theorem** [31] assures that given any interval $[\alpha, \beta] \subset \mathbb{R}$ and the roots $\lambda_1 \leq \cdots \leq \lambda_s$ of $q_{(p_1,\ldots,p_{t+1})}$ in $[\alpha, \beta]$, then $q_{(p_1,\ldots,p_t)}$ has roots $\lambda'_1 \leq \cdots \leq \lambda'_{s-1}$ in $[\alpha, \beta]$ satisfying $\lambda_1 \leq \lambda'_1 \leq \lambda_2 \leq \lambda'_2 \leq \cdots \leq \lambda_{s-2} \leq \lambda'_{s-1} \leq \lambda_s$ (interlacing property). These are the main ingredients for showing the next result [26].

Theorem *Consider the extended canonical algebra A of type* $(p_1, \ldots, p_{t-1}, p_t + 1)$ *and* A' *an extended canonical algebra of type* $(p_1, \ldots, p_t)$. *Then the following holds:*

(a) *If* $\rho(\varphi_A) = 1$, *then also* $\rho(\varphi_{A'}) = 1$.
(b) φ_A *accepts at most* 4 *eigenvalues outside* $\mathbb{S}$.

4.7.7 If the spectrum of A lies in the unit disk, then all roots of φ_A lie on the unit circle, hence A has spectral radius $\rho(A) = 1$. Clearly, for fixed degree there are only finitely many monic integral polynomials with this property. Due to the next Kronecker's theorem, these are easy to classify.

Proposition (Kronecker) *Let* f *be a monic integral polynomial whose spectrum is contained in the unit disk. Then all roots of* f *are roots of unity. Equivalently,* f *factors into cyclotomic polynomials.*

Proof Let n be the degree of f. The set of all monic polynomials with roots in the unite disc is finite. To see this, we write

$$f(z) = z^n + a_{n-1}z^{n-1} + \ldots + a_0 = \prod_{i=1}^{n}(z - z_i),$$

with $a_i \in \mathbb{Z}$ and z_j the roots in the unit circle.

Therefore, $|a_{n-1}| = |z_1 + z_2 + \ldots + z_n| \leq n$; $|a_{n-2}| \leq \sum |z_i z_j| \leq C_2^n$ (combinations of n taken 2 at a time) and $|a_{n-m}| \leq |z_1 \ldots z_n| \leq 1$.

We define $f_1(z) = f(z) = \prod_{i=1}^n (z - z_i)$ and for $k \geq 2$, $f_k(z) = \prod_{i=1}^n (z - z_i)^k$. Then:

(i) $f_k(z)$ has integral coefficients;
(ii) $f_k(z)$ has roots on the unit disc.

Therefore $f_j(z) = f_{k(j)}(z)$ for some $k(j)$.

Since the roots $\mathscr{R}(j)$ of $f_j(z)$ are z_i^j, then the sets $\mathscr{R}(j) = \mathscr{R}(k(j))$ and there is a permutation $\sigma \in S_n$ with $\sigma(\mathscr{R}(j)) = \mathscr{R}(k(j))$. Thus $z_r^j = z_{\sigma(r)}^{k(j)}$.

Consider m the order of z_r^j in $\mathscr{R}(j)$. For the root $z_1 = \lambda$ we get

$$\lambda^{j^m} = z_1^{j(j^{m-1})} = z_{\sigma(1)}^{k(j^{m-1})} = z_{\sigma^2(1)}^{k^2(j^{m-2})} = \ldots = \lambda^{k^m}.$$

Finally, $\lambda^{k^m - j^m} = 1$, that is, λ is a root of unity. □

Cyclotomic polynomials Φ_n and their products are a natural source for self-reciprocal polynomials. Clearly, $\Phi_1 = x - 1$ is not self-reciprocal, but all the remaining Φ_n (with $n \geq 2$) are. Hence, exactly the polynomials $(x-1)^{2k} \prod_{n \geq 2} \Phi_n^{e_n}$ with natural numbers k, e_n are self-reciprocal of spectral radius one.

4.7.8 Let A be a triangular finite dimensional K-algebra, that is, we assume that the quiver Δ_A of A has no oriented cycles. There is a number of measures associated to the absolute values $|\lambda|$ for λ in the spectrum $\mathrm{Spec}(\Phi_A)$ of A. Three important measures are the following: the spectral radius $\rho(A)$ of A, the **Mahler measure** of Φ_A is $M(\Phi_A) = a_n \prod_{i=1}^{n} \max\{1, |\lambda_i|\}$ and the **energy function** of Φ_A is $e(\Phi_A) = \sum_{i=1}^{n} |\lambda_i(A)|$. Due to Kronecker's theorem, we have the following result.

Theorem *Let A be a triangular algebra. The following statements are equivalent:*

1. $|\lambda_i(A)| \le 1$, *for all* $1 \le i \le n$;
2. $|\lambda_i(A)| = 1$, *for all* $1 \le i \le n$;
3. $\rho(A) = 1$;
4. $M(\Phi_A) = 1$;
5. $e(\Phi_A) = n$
6. Φ_A *factorizes as product of cyclotomic polynomials.*

An algebra A satisfying these equivalent conditions is said to be of **cyclotomic type**. The following finite dimensional algebras are known to be of cyclotomic type:

(i) Hereditary algebras of finite or tame representation type.
(ii) Tensor products of algebras of cyclotomic type.
(iii) All canonical algebras.
(iv) (Some) Extended canonical algebras.
(v) Algebras whose homological bilinear form is non-negative.

References

1. A'Campo, N. *Sur les valeurs propres de la transformation de Coxeter.* Inventiones Math. **33**, 61–67 (1976)
2. Assem, I. and Simson, D. and Skowronski, A., *Elements of the representation theory of associative algebras*, Cambridge University Press (2006), London Mathematical Society Student Texts **65**
3. Auslander, M. and Reiten, I. and Smalø, S. *Representation theory of Artin algebras*, Cambridge University Press **36** Cambridge Studies in Advanced Mathematics (1995)
4. Baer, D. *Wild hereditary Artin algebras and linear methods*, Manuscripta Mathematica Vol. **55** (1986) 69–82
5. Baer, D. *Homologlcal Properties of Wild Hereditary Artin Algebras*, Representation Theory I – Finite Dimensional Algebras. LNM**1177**, 1–12, Berlin-Heidelberg-New York (1986)
6. Bernstein, I.N. and Gelfand, I.M. and Ponomarev, V.A. *Coxeter functors and Gabriel's theorem*, Russian Math. Surveys **28** (1973) pp. 17–32
7. Biggs, N.L. and White, A.T., *Permutation groups and Combinatorial Structures*. London Math. Soc. Lecture Notes, Vol. **33**, London Math. Soc., London, 1975.
8. Boldt, A. *Coxetertransformationen*. Diplomarbeit. Paderborn (1992)
9. R. Brauer and C. Nesbitt. *On the regular representations of algebras*. Proc. Nat. Acad. Sci. U.S.A. **23** (1937), 236–240

10. Collatz, L. and Sinogowitz, U. *Spektren endlicher grafen*. Abhandlungen aus dem Mathematischen Seminar der Universität Hamburg, Vol. **21**, (1957) 63–77
11. Cvetković, D., Doob, M. and Sachs, H., *Spectra of Graphs: Theory and Application*. Academic Press, New York, (1980)
12. Dlab, V. and Ringel, C. M. *Indecomposable representations of graphs and algebras*, Mem. Amer. Math. Soc. **173**, (1976)
13. Dlab, V. and Ringel, C. M. *Eigenvalues of Coxeter transformation and the Gelfand-Kirillov dimension of the preprojective algebras*, Proceedings of the AMS, Vol.**83** (1981) 228–232.
14. Farnsteiner, R., *Self-injective algebras: the Nakayama permutation*. (2005)
15. Gantmacher, F.R., *The Theory of Matrices*, Vols. **1** and **2**, Chelsea Publishing Company, New York, N.Y. (1960)
16. Geigle, W. and Lenzing, H., *A class of weighted projective curves arising in representation theory of finite dimensional algebras*. In Singularities, representations of algebras, and vector bundles. *Springer Lect. Notes Math.* **1273** (1987), 265–297.
17. Happel, D. *Derived Equivalences and Tilting Theory*. International Symposium on Ring Theory (2001) Springer
18. Isaacs, I.M. *Character theory of finite groups*. Ac. Press (1976)
19. Kerner, K., *Representations of wild quivers*, In: CMS Conf. Proc., vol. **19**, Amer. Math. Soc., Providence, RI (1996) pp. 65–107
20. Lakatos, P., *Salem numbers, PV numbers and spectral radii of Coxeter transformations* C.R. Math. Acad. Sci. Soc. R. Can., **23** (3) (2001), pp. 71–77
21. Lenzing, H. *Wild canonical algebras and rings of automorphic forms*. In Finite Dimensional Algebras and Related Topics, V. Dlab and L.L. Scot (eds.), 191–212, Kluwer Academic Publishers, Dordrecht (1994).
22. Lenzing, H. *Coxeter transformations associated with finite-dimensional algebras*. In Computational methods for representations of groups and algebras, 287–308, Progr. Math., **173**, Birkhäuser, Basel (1999)
23. Lenzing, H. and de la Peña, J.A. *Wild canonical algebras*. Math. Z. **224** (1997) 403–425.
24. Lenzing, H. and de la Peña, J.A. *Concealed canonical algebras and separating tubular families*. Lon. Math. Soc. Vol **78**, Issue 3 (1999) 513–540
25. Lenzing, H and de la Peña, J.A. *Supercanonical algebras* J. of Algebra **282** (2004), 298–348
26. Lenzing, H. and de la Peña, J.A. *Spectral analysis of finite dimensional algebras and singularities*, inTrends in Representation Theory of Algebras and Related Topics, ed. A. Skowroński, EMS Publishing House, Zürich (2008) 541–588.
27. Lepović, M. and Gutman, I., *No starlike trees are cospectral*, Discrete Mathematics Vol. **242**, Issues 1–3 (2002) 291–295
28. Mróz, A., *Coxeter energy of graphs*. Linear Algebra and its Applications **506** (2016) 279–307
29. T. Nakayama. *On Frobeniusean algebras I*. Ann. of. Math. **40** (1939), 611–633
30. T. Nakayama. *On Frobeniusean algebras II*. Ann. of. Math. **42** (1941), 1–21
31. Obreshkoff, N., *Verteilung und Berechnung der Nullstellen reeller Polynome*. VEB Deutscher Verlag der Wissenschaften, 1963
32. de la Peña, J.A., *Coxeter transformations and the representation theory of algebras*, in: Finite Dimensional Algebras and Related Topics, in: NATO ASI Ser. C Math. Phys. Sci., vol. **424**, Kluwer Academic Publishers, Dordrecht, 1994, pp.223–253.
33. de la Peña, J.A. and Takane, M. *The spectral radius of the Galois covering of a finite graph*, Lin. Alg. Appl. **160**: 175–188 (1992).
34. Ringel, C.M. *Tame algebras and integral quadratic forms*, Springer LNM, **1099** (1984)
35. Ringel, C.M. *The spectral radius of the Coxeter transformations for a generalized Cartan matrix*, Math. Annalen **300** (1994) 331–339.
36. Schwenk, A.J., *The spectrum of a graph*, University of Michigan, 1973
37. Skowroński, A. *Simply connected algebras in Hochschild cohomologies*. In Representations of Algebras., Canad. Math. Soc. Conf. Proc. **14**, Amer. Math. Soc., Providence, RI (1993) 431–447.

38. Skowroński, A. *Generalized canonical algebras and standard stable tubes*, Colloq. Math., **90** (2001), pp. 77–93
39. Vandergraft, J.S. *Spectral properties of matrices which have invariant cones*, SIAM J. Matrix Appl. Math., **16**, 1208–1222 (1968).

Chapter 5
Group Actions on Algebras and Module Categories

It is convenient to consider $A = KQ/I$ as a K-linear category with objects Q_0 (= vertices of Q) and morphisms given by linear maps $Q(x, y) = e_y A e_x$, where e_x is the trivial path at x (for $x, y \in Q_0$). In this categorical approach we do not need to assume that Q is finite (therefore the K-algebra KQ/I may not have unity). Occasionally we write $A_0 = Q_0$ if we do not need to explicit the quiver Q.

A **symmetry** of the quiver Q is a permutation of the set of vertices Q_0 inducing an automorphism of Q. We denote by $\mathrm{Aut}(Q)$ the group of all symmetries of Q. Those symmetries $g \in \mathrm{Aut}(Q)$ inducing a morphism $g: KQ \to KQ$ such that $g(I) \subset I$ form the group $\mathrm{Aut}(Q, I)$. In natural way, any $g \in \mathrm{Aut}(Q, I)$ induces an automorphism of the module category mod-A and on the Auslander-Reiten quiver Γ_A of A (since the action commutes with the Auslander-Reiten translation τ_A of Γ_A).

In Sect. 5.1, we recall some basic facts about the actions of groups $G \subset \mathrm{Aut}(Q, I)$ on $A = KQ/I$ (orbits, stabilizers, Burnside's lemma) and show that for a representation-finite standard algebra A, we have $\mathrm{Aut}(Q, I) \cong \mathrm{Aut}(\Gamma_A)$, where $\mathrm{Aut}(\Gamma_A)$ is formed by the symmetries of Γ_A commuting with the translation τ_A. We recall that A is **standard** if for a choice of representatives of the iso-classes of indecomposables, the induced full subcategory of mod-A (denoted by $\mathrm{Ind}_A/\cong$) is equivalent to $K(\Gamma_A)$ which is the quotient of the path algebra $K\Gamma_A$ by the ideal generated by the **meshes** $\sum_{i=1}^{s} (\sigma\alpha_i) \cdot \alpha_i$ of the almost split sequences

$$0 \longrightarrow \tau_A X \xrightarrow{(\sigma\alpha_i)} \bigoplus_{i=1}^{s} Y_i \xrightarrow{(\alpha_i)} X \longrightarrow 0\ ,$$

Recall that for $\mathrm{chr}\, K \neq 2$, any representation-finite algebra $A = KQ/I$ is standard. We also consider relations between the structure of $\mathrm{Aut}(Q, I)$ and the **Coxeter polynomial** of A, that is, the characteristic polynomial of the Coxeter transformation of A.

J.-A. de la Peña, *Representations of Algebras*, Algebra and Applications 30,
https://doi.org/10.1007/978-3-031-12288-0_5

In Sect. 5.2 we present the main constructions associated to groups acting on algebras:

- If G acts freely on A (that is, $G \subset \mathrm{Aut}(Q, I)$ and $g(x) = x$ for a vertex x, implies $g = 1$), then the **Galois covering** $F: A \to A/G$ is a G-invariant functor of K-categories.
- If B is a G-graded K-category, the **smash product** $B \# G$ is a K-category accepting the free action of G. Moreover, $(B \# G)/G \cong B$.

Galois coverings were introduced by Bongartz and Gabriel [9] for the study of representation type of algebras. Smash product is a well-known construction in ring theory (see [6]), but only recently was observed that it yields the inverse operation to Galois coverings [17]. Indeed, if G acts freely on A, then A/G is a G-graded category such that $(A/G) \# G \cong A$. Following [3], we introduce functors relating the module categories of A and A/G when G acts freely on A.

In Sect. 5.3 we give examples of results [30, 38] showing that for a Galois covering $F: A \to A/G$, the category A is tame if and only if A/G is tame, provided certain restrictions on the group G or on the category A/G are satisfied.

The main results of the chapter relate the representation theory of A and A/G. Indeed, given a Galois covering $F: A \to A/G = B$ with B a finite dimensional K-algebra, then B is representation-finite if and only if A is **locally representation-finite** (that is, for each $i \in A_0$, there are only finitely many indecomposable A-modules X, up to isomorphism, with $X(i) \neq 0$). The proof of this result was partially given in [26] and completed in [36], and provides an efficient tool to deal with representation-finite algebras.

The representation-infinite situation is more involved. We recall that A is said to be **tame** if for every $d \in \mathbb{N}$ there are finitely many $A - K[t]$-bimodules $M_1, \ldots, M_{s(d)}$ which are free finitely generated as right $K[t]$-modules and such that any indecomposable A-module X with dimension d is of the form $M_i \otimes_{K[t]} (K[t]/(t - \lambda))$ for some $1 \leq i \leq s(d)$ and some $\lambda \in K$. We say that a tame algebra A is **domestic** (resp. of **polynomial growth**) if $s(d)$ can be chosen $\leq c$ (resp. $\leq d^m$ for some m) for all d. It is not hard to show that A is tame if A/G is tame for a group acting freely on A. The converse is false, as shown in [39]; nevertheless there are many interesting, general situations where it holds true.

We denote by $K_0(A)$ the **Grothendieck group** of A, freely generated by representatives $S_1, \ldots, S_n$ of the simple A-modules, where $n = n(Q)$ is the number of vertices of Q. We denote P_i (resp. I_i) the projective cover (resp. injective envelope) of S_i. With the categorical approach an A-module is a functor $X: A \to$ Mod-K, and a morphism $f: X \to Y$ is a natural transformation. In case A has finite global dimension, gl.dim$A < \infty$, the Coxeter matrix $\varphi_A: K_0(A) \to K_0(A)$ is defined by $\varphi_A([P_i]) = -[I_i]$.

5.1 The Group of Automorphisms of an Algebra

5.1.1 Let $A = KQ/I$ be a finite dimensional K-algebra: $\mathrm{Aut}(A)$ denotes the group of automorphisms of A. By $\mathrm{Aut}(Q)$ we denote the group of symmetries of Q and by $\mathrm{Aut}(Q, I)$ the group of symmetries of Q fixing I (that is, $g \in \mathrm{Aut}(Q)$ induces $g\colon KQ \to KQ$ such that $g(I) = I$). Clearly, $\mathrm{Aut}(Q, I)$ is a subgroup of $\mathrm{Aut}(A)$.

5.1.2 Each symmetry $g \in \mathrm{Aut}\,(Q)$ gives rise to a matrix $g \in G\ell_{\mathbb{Z}}(n(Q))$ sending S_i to $S_{g(i)}$. This representation $\gamma\colon \mathrm{Aut}(Q) \to G\ell_{\mathbb{Z}}(n(Q))$ is called the **canonical representation**. Let $G \subset \mathrm{Aut}(Q, I)$ be a subgroup. For a given $i \in Q_0$, Gi denotes the **orbit** of i and subgroup of G, $G_i = \{g \in G\colon gi = i\}$ denotes the **stabilizer** of i. Clearly G_i is a subgroup of G.

Lemma *The mapping $Gi \to G/G_i$, $gi \mapsto gG_i$ is a bijection from the orbit to the set of left cosets G/G_i.*

$$n(Q) = |G| \sum \frac{1}{|G_i|}$$

where the sum runs over representatives of the orbits.

5.1.3 Let $K_0(A) \cong \mathbb{Z}^{n(Q)}$ be the Grothendieck group of A. Consider

$$\mathrm{Inv}_G(A) = \{v \in K_0(A) \otimes_{\mathbb{Z}} \mathbb{Q}\colon v^g = v \text{ for } g \in G\},$$

the $\mathbb{Q}$-space of G-invariant vectors. Then $t_0(G)$, the number of orbits of G in Q_0, equals $\dim_{\mathbb{Q}} \mathrm{Inv}_G(Q)$. Let $S_1, \dots, S_m$ be a set of representatives of the **irreducible $\mathbb{C}$-representations** of G (here m is the number of conjugacy classes of G). Let S_1 be the trivial representation. Consider χ_β the character corresponding to S_β (that is, $\chi_\beta\colon G \to \mathbb{C}^*, g \mapsto \mathrm{Tr}\, S_\beta(g)$). The characters $1 = \chi_1, \dots, \chi_m$ form an orthonormal basis of the class group $X(G)$, with the scalar product $(\chi, \chi') = \frac{1}{|G|} \sum\limits_{g \in G} \chi(g)\overline{\chi'(g)}$.

Lemma (Burnside's Lemma) *We have $t_0(G) = \frac{1}{|G|} \sum\limits_{g \in G} \chi_\gamma(g)$ where $\chi_\gamma\colon G \to \mathbb{Q}^*$, is the character of the canonical representation γ.*

Proof Observe that $\chi_\gamma(g)$ is the number of fixed points of g.

$$\sum_{g \in G} \chi_\gamma(g) = \sum_{g \in G} \sum_{i \in Q_g} 1 = \sum_{i} \sum_{g \in G_i} 1 = \sum_{i \in Q_0} |G_i| = |G| \sum_{i \in Q_0} |Gi|^{-1} = |G| t_0(G),$$

where $Q_g = \{j \in Q_0\colon gj = j\}$. □

5.1.4 If G is a subgroup of $\mathrm{Aut}(Q, I)$, then G acts on Mod-A as follows:

$$X \in \text{Mod-}A \text{ and } g \in G, \text{ then } X^g \in \text{Mod-}A,$$

such that for $i \xrightarrow{\alpha} j$, we have $X^g(i) = X(gi) \xrightarrow{X(g\alpha)} X(gj)$. Similarly, for $f \in \mathrm{Hom}_A(X, Y)$, we define $f^g \in \mathrm{Hom}_A(X^g, Y^g)$.

Clearly, this action preserves indecomposable modules and induces an action of G of the Auslander-Reiten quiver Γ_A, satisfying:

(a) The action preserves projective, injective and simple modules.
(b) The action preserves Auslander-Reiten sequences (in particular, $(\tau_A X)^g = \tau_A X^g$).
(c) G is a subgroup of $\mathrm{Aut}(\Gamma_A)$, the group of automorphisms of the quiver Γ_A (commuting with the τ_A-structure).

For (c), let $g \in G$ be an element inducing a trivial action on $\mathrm{Aut}(\Gamma_A)$; we shall prove $g = 1$. Indeed, $P_i = P_i^g = P_{gi}$ implies $gi = i$ for every vertex $i \in Q_0$.

Let $i \xrightarrow{\alpha_s} j$, $s = 1, \ldots, m$ be all arrows between i and j, then g establishes a permutation of the α_s. Let X_s be the 2 dimensional A-module with $X_s(\alpha_i) = \delta_{is} \colon K \to K$. Since X_s is indecomposable, $X_s^g = X_s$ or $g\alpha_s = \alpha_s$. Therefore $g = 1$.

5.1.5 Proposition *Let* $A = KQ/I$ *be an algebra satisfying:*

(a) A *is representation finite,*
(b) A *is standard,*

then $\mathrm{Aut}(Q, I) = \mathrm{Aut}(\Gamma_A)$.

Proof We already know that $\mathrm{Aut}(Q, I)$ is a subgroup of $\mathrm{Aut}(\Gamma_A)$. Let $g \in \mathrm{Aut}(\Gamma_A)$ and consider the induced automorphism $\bar{g}$ of the mesh category $K(\Gamma_A)$. Clearly, $\bar{g}$ restricts to an automorphism of A (considering the full embedding $i \mapsto P_i$, $i \in Q_0$). By definition there is an automorphism $h \in \mathrm{Aut}(Q, I)$ inducing $\bar{g}$, and therefore inducing g. □

5.1.6 Let $A = KQ/I$ and assume that $\mathrm{gl.dim}(A) < \infty$. For example, this happens if A is triangular, that is, Q has no oriented cycles.

Recall that $\varphi_A : K_0(A) \xrightarrow{\sim} K_0(A)$, given by $\mathbf{dim}(P_i) \mapsto -\mathbf{dim}(I_i)$, is the **Coxeter matrix** of A, which is $\mathbb{Z}$-invertible. In case $A = KQ$ (i.e. $I = 0$), then $(\mathbf{dim}(X))\varphi_A = \mathbf{dim}(\tau_A X)$ for any non-projective indecomposable X.

The characteristic polynomial $p_A(t) = \det(t\mathbf{Id} - \varphi_A)$ is called the **Coxeter polynomial** of A.

5.1.7 Proposition

(a) *φ_A is an automorphism of the representation γ.*
(b) *If* Aut(Q, I) *is not trivial, then $p_A(t)$ is not irreducible.*

Proof

(a) It suffices to observe that $g\varphi_A = \varphi_A g$ for any $g \in G$.
(b) Let $R_1, \ldots, R_m$ be a set of representatives of the irreducible $\mathbb{Q}$-representations of $G = \mathrm{Aut}(Q, I)$. Let R_1 be the trivial representation. Up to conjugation (with L)

$$\gamma^L = \bigoplus_{\alpha=1}^{m} R_\alpha^{r(\alpha)}$$

$R_\alpha \colon G \to GL_{\mathbb{Q}}(\dim R_\alpha)$, then $n = \sum_{\alpha=1}^{m} r(\alpha) \dim R_\alpha$.

By Schur's Lemma, $\varphi_A^L = \begin{bmatrix} \varphi_1 & & 0 \\ & \ddots & \\ 0 & & \varphi_m \end{bmatrix}$, where $\varphi_\alpha \colon R_\alpha^{r(\alpha)} \to R_\alpha^{r(\alpha)}$ is an automorphism. Hence $p_A(t) = p_{\varphi_1}(t) \ldots p_{\varphi_m}(t)$, and if $p_A(t)$ is irreducible then $r(\alpha) = 0$ for $\alpha \geq 2$. If $G \neq (1)$, then $t_0(G) < n$. Moreover, the characters

$$\chi_\alpha \colon G \to \mathbb{Q}^*, \quad g \mapsto \mathrm{Tr}\ R_\alpha(g), \quad \alpha = 1, \ldots, m.$$

form an orthonormal basis of $X(G)$ with scalar product

$$(\chi, \chi') = \frac{1}{|G|} \sum_{g \in G} \chi(g)\overline{\chi'(g)}.$$

Then

$$r(\alpha) = (\chi_\gamma, \chi_\alpha) = \frac{1}{|G|} \sum_{g \in G} \chi_\gamma(g)\overline{\chi_\alpha(g)},$$

for $\alpha = 1$, $r(1) = \frac{1}{|G|} \sum_{g \in G} \chi_\gamma(g) = t_0(G)$. Finally, $n = \sum_{\alpha=1}^{m} r(\alpha) \dim R_\alpha = t_0(G) + \sum_{\alpha=2}^{m} r(\alpha) \dim R_\alpha$, which implies the existence of $\alpha \geq 2$ with $r(\alpha) > 0$. Therefore $p_A(t)$ is not irreducible. □

5.1.8 Consider $G = A_5 \subset \text{Aut}(Q)$ and $\gamma\colon A_5 \to G\ell(6)$ the canonical representation. It is not hard to calculate the character table of A_5:

conjugacy classes	$\{1\}$	(123)	$(12)(34)$	(12345)	(13524)
$\lvert Gx_i\rvert$	1	20	15	12	12
$1 = \chi_1$	1	1	1	1	1
χ_2	4	1	0	-1	-1
χ_3	5	-1	1	0	0
χ_4	3	0	-1	α_1	α_2
χ_5	3	0	-1	α_2	α_1

$\alpha_1 = (1+\sqrt{5})/2$

$\alpha_2 = (1-\sqrt{5})/2$

where χ_i corresponds to S_i irreducible $\mathbb{R}$-representation of G (with S_1 the trivial representation).

Then $\gamma^T = \bigoplus_{\beta=1}^{s} S_\beta^{n(\beta)}$, $n(1) = t_0(G) = 2$

$$n(2) = (\chi_\gamma, \chi_2) = \frac{1}{60}\sum_{g\in G} \chi_\gamma(g)\overline{\chi_2(g)} = \frac{1}{60}[24+60-24] = 1$$

Since $\dim\gamma = 6$, $\gamma^T = S_1 \oplus S_1 \oplus S_2$ which is also a $\mathbb{Q}$-decomposition. Moreover, $p_A(t) = (1-3t+t^2)(1+t)^4$ is an irreducible factorization.

5.2 Constructions of Algebras Associated to Groups of Automorphisms (Coverings and Smash Products)

5.2.1 Let A be a K-category given as $A = KQ/I$. Let G be a subgroup of $\text{Aut}(Q, I) \subset \text{Aut}(A)$. We say that G **acts freely** on A if $gi = i$ for some $i \in Q_0$ implies $g = 1$.

Lemma *Let G be a group acting freely on $A = KQ/I$. Then there exists a K-category $B = K\bar{Q}/\bar{I}$ and a functor $F\colon A \to B$ satisfying:*

(a) (G-invariant): $Fg = F$ for every $g \in G$;
(b) (universal G-invariant): for any functor $F'\colon A \to B'$ which is G-invariant, there exists a unique functor $\bar{F}\colon B \to B'$ such that $F' = \bar{F}F$;
(c) $\bar{Q}_0$ is formed by the G-orbits of vertices in Q_0 and for any $a = Gi$ and $b = Gj$

$$B(a, b) = \bigoplus_{g\in G} A(i, gj).$$

Proof Let B be defined as in (c) with composition maps

$$B(a,b)\otimes B(b,c) = \Big(\bigoplus_{g\in G} A(i,gj)\Big) \otimes \Big(\bigoplus_{h\in G} A(j,h\ell)\Big) \longrightarrow \bigoplus_{g'} A(i,g'j) = B(a,c)$$

$$(f_g)\otimes(f'_h) \longmapsto \Big(\sum_{gh=g'} (f'_h)^g f_g\Big)$$

Define $F\colon A \to B$, $f \in A(i,j) \subset \bigoplus_g A(i,gj) = B(a,b)$ in the unique possible way (since $gj = j$ implies $g = 1$). □

5.2.2 If G acts freely on $A = KQ/I$, the functor $F\colon A \to B$ as in 5.2.1 is called a **Galois covering** defined by G, and $B = A/G$ is a **Galois quotient** of A.

We say that a K-category B is G-**graded** if for each pair of objects a, b there is a vector space decomposition $B(a,b) = \bigoplus_{g\in G} B^g(a,b)$ such that the composition induces linear maps

$$B^g(a,b)\otimes B^h(b,c) \to B^{gh}(a,c).$$

Lemma *$B = A/G$ is a G-graded K-category.*

5.2.3 For a G-graded K-category B we define the **smash product** as the K-category $B\#G$ with objects $B_0 \times G$ and for any pair $(a,g), (b,h) \in B_0 \times G$, the morphisms are

$$(B\#G)((a,g),(b,h)) = B^{g^{-1}h}(a,b),$$

with composition given by:

$$\begin{array}{ccc}(B\#G)((a,g),(b,h))\otimes(B\#G)((b,h),(c,t)) & \longrightarrow & (B\#G)((a,g),(c,t)) \\ \| & & \| \\ B^{g^{-1}h}(a,b)\otimes B^{h^{-1}t}(b,c) & \longmapsto & B^{g^{-1}t}(a,c)\end{array}$$

Theorem ([17]) *The category $B\#G$ accepts a free action of G such that $(B\#G)/G \cong B$. Moreover, if G acts freely on A, then $(A/G)\#G \cong A$.*

Proof Clearly there is a G-invariant functor $F\colon B\#G \to B$ inducing a functor $\bar{F}\colon (B\#G)/G \to B$ which is the identity on objects. Check that $\bar{F}$ is an isomorphism.

Assume G acts freely on A and let $F\colon A \to A/G$ be the induced Galois covering. Since G acts freely, there is a bijection $f : (A/G)_0 \times G \to A_0$

which commutes with the G-action. Moreover, if $a, b \in (A/G)_0$, $f(a, 1) = i$, $f(b, 1) = j \in A_0$ then

$$
\begin{aligned}
(A/G\#G)((a, g), (b, h)) = A/G^{g^{-1}h}(a, b) = A(i, g^{-1}hj) = A(gi, hj) \\
= A(f(a, g), f(b, h))
\end{aligned}
$$

is compatible with the composition. □

In other words, smash products and Galois quotients are inverse operations in the class of K-categories.

5.2.4 Let B be a K-category. There is a $1 - 1$ correspondence ($A \xrightarrow{F} B$ Galois covering defined by G) $\mapsto$ (G group grading B), such that for any two Galois coverings $A \xrightarrow{F} B$ defined by G and $A' \xrightarrow{F'} B$ defined by H, there exists a commutative diagram

$$
\begin{array}{ccc}
A & \xrightarrow{F} & B \\
\uparrow{\scriptstyle \bar{F}} & \nearrow{\scriptstyle F'} & \\
A' & &
\end{array}
$$

if and only if there exists a normal subgroup $H \trianglelefteq G'$, such that $G'/H = G$.

5.2.5 A **Galois triple** (A, F, G) of B is a Galois covering $F: A \to B$ defined by the (free) action of a group G. There is a universal object in this set of Galois triples. Indeed, let $B = KQ/I$ and consider W the set of all walks in Q starting and ending at a fixed vertex b. Let $\sim$ be the equivalence relation induced by the following elementary relations:

(i) $\alpha\alpha^{-1} \sim e_y$ and $\alpha^{-1}\alpha \sim e_x$ for any arrow $x \xrightarrow{\alpha} y$;

(ii) if $\sum_{i=1}^{s} \lambda_i w_i \in I(x, y)$ with $\lambda_i \in K^*$, such that for any $L \subsetneq \{1, \ldots, s\}$ we have $\sum_{i \in L} \lambda_i \notin I(x, y)$, then $w_i \sim w_j$ for i, j;

(iii) if $w \sim w'$, then $ww'' \sim w'w''$, whenever the products are defined.

Then $\tilde{G} = W/\sim$ has a group structure such that B is $\tilde{G}$-graded. The group $\tilde{G}$ is called the **fundamental group** of B.

Proposition *Let $\tilde{B} = B \# \tilde{G}$ and $\tilde{F}: \tilde{B} \to B$ be defined by $\tilde{G}$. The triple $(\tilde{B}, \tilde{F}, \tilde{G})$ is a **universal Galois covering**, that is, for any Galois covering $F: A \to B$ defined by the action of a group G, there exists a covering $\bar{F}: \tilde{B} \to A$ defined by $H \lhd \tilde{G}$ such that $\tilde{G}/H = G$.*

5.2.6 Let $F: A \to B$ be a Galois covering of K-categories defined by the action of G. We shall denote by

(a) MOD-A the category of left A-modules;
(b) Mod-A those $X \in$ MOD-A with $\dim_K X(i) < \infty$ for every $i \in A_0$;
(c) mod-A those $X \in$ MOD-A with $\sum_{i \in A_0} \dim_K X(i) < \infty$.

There are naturally defined functors:

$$f_\bullet : \text{MOD-}B \to \text{MOD-}A, \quad (Y : B \to \text{Mod-}K) \mapsto (Y \circ F : A \to \text{Mod-}K),$$

called the **pull-up functor**;

$$F_\lambda : \text{MOD-}A \to \text{MOD-}B, (X : A \to \text{Mod-}K) \mapsto F_\lambda X(a) = \bigoplus_{g \in G} X(gi),$$

and such that for $f = (f_g) \in B(a, b) = \bigoplus_{g \in G} A(i, gj)$, then $F_\lambda X(f) : F_\lambda X(a) \to F_\lambda X(b)$, sends (a_g) to $\left(\sum_h X(f^h_{h_g^{-1}})(a_h)\right)_g$, called the **push-down functor**. We observe that F_λ is a left adjoint to $F_\bullet$. Similarly, there is a right adjoint $F_\rho : \text{MOD-}A \to \text{MOD-}B$ to $F_\bullet$. For modules $X \in$ mod-A, the modules $F_\lambda X$ and $F_\rho X$ coincide.

5.2.7 Recall that G acts on MOD-A and $X \in$ MOD-A is G-**stable** if $X^g = X$ for every $g \in G$. The category of G-stable A-modules is denoted by MOD-A^G.

Lemma *Let $F: A \to B$ be a Galois covering defined by the action of G. Then the following happens:*

(a) The categories MOD-A^G *and* MOD-B *are equivalent.*
(b) For any $X \in$ MOD-A and $g \in G$, we have $F_\lambda X^g \cong F_\lambda X$. Moreover,
$F_\bullet F_\lambda X \xrightarrow{\sim} \bigoplus_{g \in G} X^g$ *as A-modules.*
(c) Let H be a subgroup of G and $X \in$ MOD-A^H. Then $F_\lambda X$ has a natural structure as KH-module. If H is a finite group and chr $K \not| \, |H|$, *then $F_\lambda X$ decomposes as a direct sum of at least $|H|$ factors.*

Proof

(a) Is clear.
(b) Observe that $F_\lambda X(a) = \bigoplus_{h \in G} X(hi) \xrightarrow{\sim} F_\lambda X^g(a) = \bigoplus_{h \in G} X(hgi)$ canonically.
Hence $F_\bullet F_\lambda X(i) = F_\lambda X(Fi) = \bigoplus_{h \in G} X(hi) = \bigoplus_{h \in G} X^h(i)$ and correspondingly in morphisms.

(c) Choose a set W of representatives in G of the right cosets G/H. Then $F_\lambda X(a) = \bigoplus_{g\in G} X(i) = KH \otimes_K \left(\bigoplus_{w\in W} X(wi)\right)$ and correspondingly in morphisms.

Hence we get $\varphi\colon KH \to \mathrm{End}_B(F_\lambda X)$ a group homomorphism such that for each idempotent e of KH, $\varphi(e)$ is idempotent and there is a factorization of $F_\lambda X$.

In case H is a finite group with chr $K \not| \, |H|$, by Mashké theorem, the group algebra KH is semisimple (with $|H|$ idempotents). The result follows.

□

5.2.8 Assume G acts freely on A and $F\colon A \to B = A/G$ is the corresponding Galois covering.

Let $X \in$ MOD-A, the stabilizer G_X is the subgroup of G formed by those $g \in G$ such that $X^g \cong X$. That is, $X \in$ MOD-A^H if $H \subset G_X$.

Proposition

(a) If $X \in \mathrm{Ind}_A$ and G is torsion free, then $G_X = (1)$.
(b) If $X \in \mathrm{Ind}_A$ and $G_X = (1)$, then $F_\lambda X$ is indecomposable and for any module $Y \in$ mod-A with $F_\lambda X \simeq F_\lambda Y$, then $Y \simeq X^g$ for some $g \in G$.

Proof

(a) Let $g \in G_X$ for some $X \in \mathrm{Ind}_A$, then g establishes a permutation of supp X (a finite set). Then for some $s \in \mathbb{N}$, $1 = g^s$ on supp X. Since G acts freely on A, then $g^s = 1$. Since G is torsion free, then $g = 1$ and $G_X = (1)$.
(b) Assume $F_\lambda X \simeq Z \oplus Z'$, then $\bigoplus_{g\in G} X^g = F_\bullet F_\lambda X \simeq F_\bullet Z \oplus F_\bullet Z'$. Assume X is a direct summand of $F_\bullet Z \in$ MOD-A^G, then $\bigoplus_{g\in G} X^g \subset F_\bullet Z$ and $F_\bullet Z' = 0$. Therefore $F_\lambda X$ is indecomposable.
If $F_\lambda X \simeq F_\lambda Y$, then Y is indecomposable and $Y \simeq X^g$ for some $g \in G$.

□

5.3 Coverings and the Representation Type of an Algebra

5.3.1 A full subcategory B' of $B = KQ/I$ is **convex** in B if $B' = KQ'/I'$ for a quiver Q' which is path closed in Q (i.e. $i_0 \to i_1 \cdots \to i_s \to i_{s+1}$ with $i_0, i_{s+1} \in Q'$ implies $i_1, \ldots, i_s \in Q'$).

Theorem *Let B be a representation-finite K-algebra. Then*

(*a*) [29] *There exists a universal Galois covering $\tilde{\Gamma}_B \to \Gamma_B$ defined by the action of a free group G. Moreover $K(\tilde{\Gamma}_B) = \mathrm{Ind}_{\tilde{B}}$ for a K-category $\tilde{B}$. If B is standard, then $\tilde{B} \to B$ is a universal Galois covering.*
(*b*) [13] *For every finite convex subcategory C of $\tilde{B}$, we have $\tilde{\Gamma}_C = \Gamma_C$ and C is representation-directed (i.e. C is representation-finite and Γ_C is a preprojective component).*

5.3.2 We say that the group G **acts freely on indecomposable classes** of A-modules if $X^g \simeq X$ for $X \in \mathrm{Ind}_A$ implies $g = 1$. Observe that for the algebra B with quiver

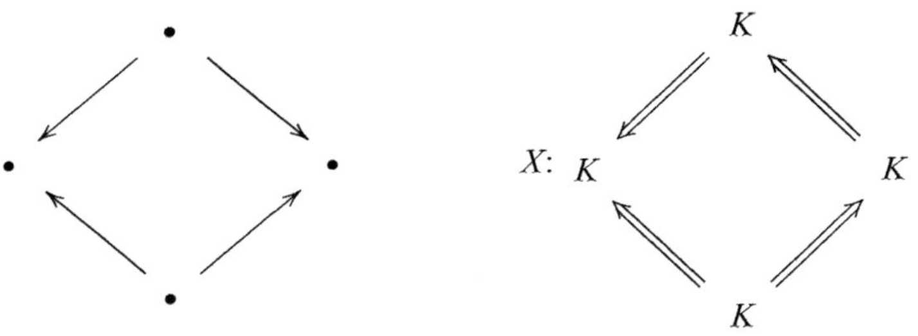

the indecomposable representation X has non-trivial stabilizer.

Theorem *Let $F : A \to A/G = B$ be a Galois covering defined by a group G,*

(*a*) [3] *If G acts freely on indecomposable classes of A-modules, then F_λ induces an injection $(\mathrm{Ind}_A/\cong)/G \hookrightarrow \mathrm{Ind}_B/\cong$ and preserves Auslander-Reiten sequences.*
(*b*) [3, 36] *If B is finite (hence a K-algebra), then B is representation-finite if and only if* (i) *G acts freely on $\mathrm{Ind}_A/\cong$ and* (ii) *A is locally representation-finite (i.e. for each $i \in A_0$, $\{X \in \mathrm{Ind}_A : X(i) \neq 0\}$ is finite). In that case, $\Gamma_A/G \simeq \Gamma_B$.*

Proof

(*a*) Let $X \in \mathrm{Ind}_A$, then we know $G_X = (1)$ and $F_\lambda X \in \mathrm{Ind}_B$. Moreover $F_\lambda X \simeq F_\lambda Y$ implies $Y \simeq X^g$ for some $g \in G$. For an almost split sequence

$$\eta : 0 \longrightarrow \tau_A X \xrightarrow{\alpha} E \xrightarrow{\beta} X \longrightarrow 0,$$

with $X \in \mathrm{Ind}_A$ we get an exact sequence

$$F_\lambda \eta : 0 \longrightarrow F_\lambda \tau_A X \longrightarrow F_\lambda E \longrightarrow F_\lambda X \longrightarrow 0,$$

with $F_\lambda X$, $F_\lambda \tau_A X$ indecomposable. For $f\colon Y \to F_\lambda X$ a non-invertible epimorphism in mod-B, we get a map

$$\begin{array}{ccc}
 & & F_\bullet Y \\
 & & \downarrow{\scriptstyle F_\bullet f} \\
\bigoplus_{g\in G} E^g = F_\bullet F_\lambda E & \xrightarrow{F_\bullet F_\lambda \beta} & F_\bullet F_\lambda X = \bigoplus_{g\in G} X^g \\
{\scriptstyle \pi_1}\downarrow & & \downarrow{\scriptstyle \pi_1} \\
E & \xrightarrow{\quad\beta\quad} & X
\end{array}$$

There is some $g \in \operatorname{Hom}_A(F_\bullet Y, E)$ such that $\beta g = \pi_1 F_\bullet (f)$. From the adjunction $(F_\bullet, F_\rho)$ we get

$$\operatorname{Hom}_A(F_\bullet Y, E) \xrightarrow{\sim} \operatorname{Hom}_B(Y, F_\rho E) = \operatorname{Hom}_B(Y, F_\lambda E),\ g \mapsto g'.$$

Then $\beta g' = f$ and $F_\lambda \eta$ is almost split sequence in mod-B.

(b) Assume B is representation-finite and $Y_1, \dots, Y_s$ are representatives of $\operatorname{Ind}_B/\cong$. Let $X \in \operatorname{Ind}_B$ and suppose $g \in G$ is such that $X^g \simeq X$. We shall prove that $g = 1$. Let $\tilde{B} \xrightarrow{F} B$ be a covering defined by a free group H as in (5.1), that is, we get commutative diagrams:

$$\begin{array}{ccc}
\tilde{B} & \hookrightarrow & \operatorname{Ind}_{\tilde{B}} = K(\tilde{\Gamma}_B) \\
{\scriptstyle F}\downarrow & & \downarrow{\scriptstyle K(\pi)} \\
B & \hookrightarrow & \operatorname{Ind}_B = k(\Gamma_B)
\end{array}
\qquad
\begin{array}{ccc}
\tilde{\Gamma}_B & \overset{g}{- - -} & \tilde{\Gamma}_B \\
{\scriptstyle \pi}\downarrow & & \downarrow{\scriptstyle \pi} \\
\Gamma_B & \overset{g}{- - -} & \Gamma_B
\end{array}$$

Then g induces an automorphism $\tilde{g}$ on $\tilde{\Gamma}_B$ such that $g\pi = \pi\tilde{g}$ and $\tilde{g}X' = X'$ for some $X' \in \tilde{\Gamma}_B$ with $\pi X' = X$. We get an automorphism g' of $\tilde{B}$ acting freely (since $g = Fg'$ acts freely on B).

Let C be the full subcategory of $\tilde{B}$ formed by supp X. It is easy to see that C is convex in $\tilde{B}$ and therefore Γ_C is a preprojective component. Since $g'X' = X'$, then we get $h \in \operatorname{Aut}(C)$ with $(X')^h \simeq X'$. Since h commutes with τ_C, there is a projective C-module P_x with $P_x^h = P_x$, which is a contradiction unless $h = 1$.

We check that A is locally representation-finite: let $i \in A_0$ and let $(X_\alpha)_{\alpha\in\Delta}$ be all indecomposable A-modules which are direct summands of $F_\bullet Y_j$ for some $1 \le j \le s$ and $X_\alpha(i) \neq 0$. This set is finite [indeed, $\bigoplus_{\Delta'} X_\alpha = F_\bullet Y_j$, then $\bigoplus F_\lambda X_\alpha = F_\lambda F_\bullet Y_j = \bigoplus Y_\ell^{(I_\ell)}$ and $X_\alpha \in \operatorname{Ind}_A$, for all α. Since $\dim_k F_\bullet Y_j(i) = \dim_k Y_j(Fi) < \infty$, the set is finite].

Finally, let $Y \in \mathrm{Ind}_B$. Then $F_\lambda F_\bullet Y = \oplus F_\lambda X_\alpha$ and

$$\mathrm{Hom}_A(F_\bullet Y, F_\bullet Y) \xrightarrow{\cong} \mathrm{Hom}_B(Y, F_\rho F_\bullet Y) = \mathrm{Hom}_B(Y, F_\lambda F_\bullet Y)$$

$$1_{F_\bullet Y} \longmapsto e\colon Y \mapsto F_\lambda F_\bullet Y$$

hence $Y \simeq F_\lambda X_\alpha$ for some α. Therefore $\Gamma_B = \Gamma_A/G$. □

Let $F\colon A \to A/G$ be a Galois covering with group G. In case A/G is a representation-finite K-algebra, we have seen that F_λ covers all indecomposable modules. The situation is different for representation-infinite algebras. We shall briefly discuss the new occurring phenomena. By $\mathrm{Ind}_1 A/G$ we denote the full subcategory of $\mathrm{Ind}_{A/G}$ formed by all objects isomorphic to $F_\lambda M$ for some $M \in \mathrm{Ind}_A$ (these modules are called A/G-modules of the **first kind**). The remaining indecomposable modules (called of the **second kind**) form the category $\mathrm{Ind}_2 A/G$.

If $M \in \text{mod-}A^H$ is such that $\mathrm{supp}\, M$ is not finite but $\mathrm{supp}\, M/H$ is finite, then M is called a **weakly G-periodic** module.

Lemma ([30]) *Let $F\colon A \to A/G$ be a Galois covering such that G acts freely on $\mathrm{Ind}_A/\cong$. For $X \in \mathrm{Ind}_{A/G}$ the following conditions hold:*

(1) $X \in \mathrm{Ind}_1 A/G$ if and only if $F_\bullet X \cong \bigoplus_{i\in I} Z_i$, where all $Z_i \in \text{mod-}A$.

(2) $X \in \mathrm{Ind}_2 A/G$ if and only if $F_\bullet X \cong \bigoplus_{i\in I} Y_i$, where each Y_i is weakly G-periodic.

Proof

(1) If $X \cong F_\lambda M$ for some $M \in \text{mod-}A$, then $F_\bullet X \cong \bigoplus_{g\in G} M^g$. Assume now that $Z \in \mathrm{Ind}_A$ is a direct summand of $F_\bullet X$. Let $j \in \mathrm{Hom}_A(Z, F_\bullet X)$ and $p \in \mathrm{Hom}_A(F_\bullet X, Z)$ be such that $pj = 1_Z$. We get morphisms $j^g\colon Z^g \to (F_\bullet X)^g \cong F_\bullet X$ and $p^g\colon (G_\bullet X)^g \to Z^g$ which yield a pair of maps $j'\colon \bigoplus_{g\in G} Z^g \to F_\bullet X$ and $p'\colon F_\bullet X \to \bigoplus_{g\in G} Z^g$. Since $Z^g \ncong Z$ for $g \neq h$, then the endomorphism $p' \circ j'$ of $\bigoplus_{g\in G} Z^g$ is invertible.

Since $F_\bullet$ induces an equivalence of categories $\text{MOD-}A/G \cong \text{MOD-}A^G$, then $F_\lambda Z$ is a direct summand of X. Thus $X \cong F_\lambda Z$.

(2) Let $X \in \text{mod}_2\text{-}A/G$. By the proof of (1), $F_\bullet X = \bigoplus_{i\in I} Y_i$ where $Y_i \in \mathrm{IND}_A \setminus \mathrm{Ind}_A$. Then we have to show that an indecomposable direct summand Y of $F_\bullet X$ is weakly G-periodic. Indeed, let $\mathscr{U}$ be a set of representatives of the cosets of G with respect to G_Y. Then $\bigoplus_{g\in\mathscr{U}} Y^g$ is a direct summand of

$F_\bullet X$. Since $\operatorname{supp} Y \subset \operatorname{supp} X$ is contained in a finite number of G-orbits, then $(\operatorname{supp} Y)/G_Y$ is finite. The converse follows from (1).

□

5.3.3 A category $A = KQ/I$ is called **locally support finite** if for each $x \in Q_0$, the full subcategory A_x of A, determined by the set of vertices of all $\operatorname{supp} M$ with $M \in \operatorname{Ind}_A$ and $M(x) \neq 0$, is finite.

Proposition *Let $F: A \to A/G$ be a Galois covering. Then:*

(a) If G acts freely on $\operatorname{Ind}_A/\cong$, then F_λ: mod-$A \to$ mod-A/G induces a bijection $F_\lambda: (\operatorname{Ind}_A/\cong)/G \to (\operatorname{Ind}_{A/G}/)\cong$.

(b) If A is locally support finite, then $\operatorname{Ind}_A = \operatorname{ind}_A$. In particular if G acts freely on $\operatorname{Ind}_A/\cong$, then F_λ induces a bijection between $(\operatorname{Ind}_A/\cong)/G$ and $(\operatorname{Ind}_A/G)/\cong$. Moreover, in this case A/G is tame (resp. domestic, polynomial growth) if and only if so is A. For the Auslander-Reiten quivers we have $\Gamma_{A/G} \cong \Gamma_A/G$.

Proof

(*a*) Follows from (4.2) since there are no weakly G-periodic indecomposable A-modules.

(*b*) Let $Y \in \operatorname{Ind}_A$ and consider $x \in Q_0$ with $Y(x) \neq 0$. Let $\hat{A}_x$ be the (finite) full subcategory of A consisting in the objects $y \in Q_0$ such that $A(y, z) \neq 0$ or $A(z, y) \neq 0$ for some $z \in A_x$. Let Z be an indecomposable direct summand of the restriction $Y|\hat{A}_x$ such that $Z(x) \neq 0$. Hence $\operatorname{supp} Z \subset A_x$ and it is easy to check that Z is a direct summand of Y in mod-A.

Thus $Y = Z \in \operatorname{Ind}_A$. The last claim is easy to prove.

□

5.3.4 The category $A_{\alpha,\beta}$ given by the quiver with relations

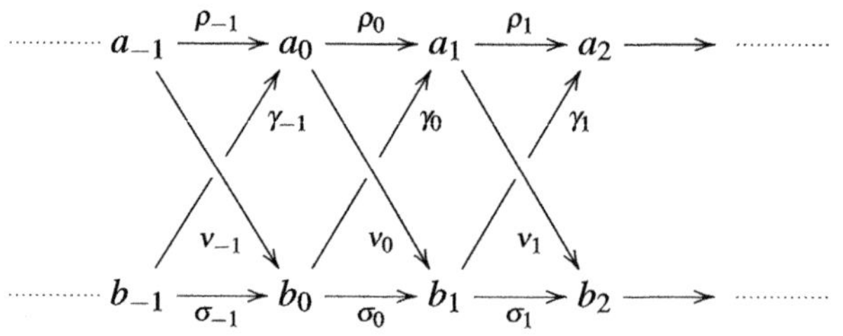

$$\left.\begin{aligned} \sigma_{i+1}\sigma_i &= \alpha\nu_{i+1}\gamma_i \\ \nu_{i+1}\rho_i &= \beta\sigma_{i+1}\nu_i \\ \gamma_{i+1}\nu_i &= \rho_{i+1}\rho_i \\ \rho_{i+1}\gamma_i &= \gamma_{i+1}\sigma_i \end{aligned}\right\}$$

with $(\alpha, \beta) \neq (1, 1)$, is locally support finite (why?). Moreover the group $\mathbb{Z}$ generated by the action $(a_i \mapsto a_{i+1}, b_i \mapsto b_{i+1})$ acts freely on $A_{\alpha,\beta}$ and on $\operatorname{Ind}_{A_{\alpha,\beta}}/\cong$. Hence the Galois covering $F: A_{\alpha,\beta} \to \bar{A}_{\alpha,\beta}/\mathbb{Z}$ yields a bijection

$$F_\lambda: (\operatorname{Ind}_{A_{\alpha,\beta}}/\cong)/\mathbb{Z} \to (\operatorname{Ind}_{\bar{A}_{\alpha,\beta}})/\cong.$$

The algebra $\bar{A}_{\alpha,\beta}$ is given by the quiver with relations

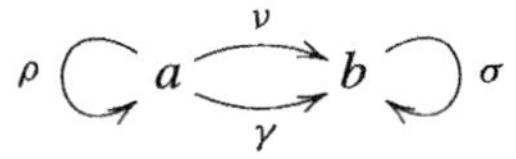

$$\left.\begin{aligned} \sigma^2 &= \alpha\gamma\nu \\ \rho\nu &= \beta\nu\sigma \\ \nu\gamma &= \rho^2 \\ \gamma\rho &= \sigma\gamma \end{aligned}\right\}$$

Since $A_{\alpha,\beta}$ is tame (resp. polynomial growth for $\alpha\beta \neq 1$), so is $\bar{A}_{\alpha,\beta}$.

5.3.5 Given a natural number $p \geq 0$, a group G is said to be p-**residually finite** if for each finite subset $S \subset G \setminus \{1\}$ there is a normal subgroup of finite index $H \triangleleft G$ such that $H \cap S = \emptyset$ and $p \not| \ (G/H)$. For example, free groups are p-residually finite.

Theorem ([38]) *Let $F\colon (\tilde{Q}, \tilde{I}) \to (Q, I)$ be a Galois covering given by the action of a p-residually finite group G, where $p =$ chr K. Assume that $\tilde{A} = K\tilde{Q}/\tilde{I}$ is locally support finite. Then $\tilde{A}$ is tame if and only if $A = KQ/I$ is tame.*

Proof Without loss of generality, we assume that Q is finite. We denote also by $F\colon \tilde{A} \to A$ the induced covering functor. We divide the proof in several steps.
Step 1. Let $F_\lambda\colon$ mod-$\tilde{A} \to$ mod-A be the push-down functor. Consider a sequence $(\Gamma_n)_{n\in\mathbb{N}}$ of finite full subcategories of $\tilde{A}$ such that $\bigcup\limits_n \Gamma_n = \tilde{A}$ and if $\tilde{A}(x, \Gamma_n) \neq 0$ or $\tilde{A}(\Gamma_n, x) \neq 0$, then $x \in \Gamma_{n+1}$. The restriction functor $\varepsilon^n_\bullet$: mod-$\tilde{A} \to$ mod-Γ_n has a left adjoint $\varepsilon^n_\lambda\colon$ mod-$\Gamma_n \to$ mod-$\tilde{A}$ such that $\varepsilon^n_\bullet\, \varepsilon^n_\lambda = \mathbf{Id}_{\text{mod-}\Gamma_n}$. Therefore, the functor

$$F_n = F_\lambda \varepsilon^n_\lambda\colon \text{mod-}\Gamma_n \to \text{mod-}A,$$

is right exact, $n \in \mathbb{N}$.
Step 2. Let $Y \in \text{Ind}_A$, we shall prove that Y is a direct summand of $F_\lambda F_\bullet\, Y$. Indeed, since $\tilde{A}$ is locally support-finite, by (5.3), the pull-up $F_\bullet\, Y$ decomposes as $F_\bullet\, Y \simeq \bigoplus\limits_{i\in L} X_i$, where $X_i \in \text{Ind}_{\tilde{A}}$. Thus $F_\lambda F_\bullet\, Y \simeq \bigoplus\limits_{i\in L} F_\lambda X_i$. We proceed in two steps.

Step 2.1 There exists a normal subgroup H of G with finite index not divisible by p, such that the Galois covering $\bar{F}\colon \tilde{A} \to \tilde{A}/H$ induces a bijection

$$\bar{F}_\lambda\colon (\text{Ind}_{\tilde{A}}/\simeq)/H \to (\text{Ind}_{\tilde{A}}/H)/\simeq .$$

Since A is finite, the set of vertices of $\tilde{Q}$ is a disjoint union of a finite number of G-orbits Gx_i, $i = 1, \ldots, s$. With the notation introduced above, we consider the full subcategory $R_i = \tilde{A}_{x_i}$ of $\tilde{A}$. Let S be the set of all elements $g \in G/\{1\}$ such that $g(\hat{\hat{R}}_i) \cap \hat{\hat{R}}_i \neq \emptyset$, for some $i \in \{1, \ldots, s\}$. Since G acts freely on $\tilde{Q}$, then S is finite and there is a normal subgroup H of G such that G/H is finite of order not divisible by p and $H \cap S = \emptyset$. Hence $h(\hat{\hat{R}}_i) \cap \hat{\hat{R}}_i = \emptyset$, $i = 1, \dots, s$. By (5.1), the induced Galois covering $\bar{F}\colon \tilde{A} \to \tilde{A}/H$ yields an injection $\bar{F}_\lambda(\mathrm{Ind}_{\tilde{A}}/\simeq)/H \to (\mathrm{Ind}_{\tilde{A}}/H)/\simeq$. We show that this map is also surjective. Let $M \in \mathrm{Ind}_{\tilde{A}}/H$ and $b = \bar{F}(x_j)$ a vertex of $\tilde{Q}/H$ such that $M(b) \neq 0$. The restriction $F' = \bar{F}| : \hat{R}_j \to F(\hat{R}_j)$ is a bijection and there is an indecomposable $\hat{R}_j$-module $Y = F'N$ with $Y(x_j) \neq 0$ and such that N is an indecomposable direct summand of the restriction M' of M to $F(\hat{R}_j)$. We conclude that $Y \in \mathrm{Ind}_{\tilde{A}}$ and $N \in \mathrm{Ind}_{\tilde{A}}/H$, showing that $M = N = F_\lambda Y$.

Step 2.2 For the proof of (2), consider a normal subgroup H of G as in (2.1) and a factorization of Galois coverings

$$\begin{array}{ccc} \tilde{A} & \xrightarrow{\;F\;} & A = \tilde{A}/G \\ {\scriptstyle F}\big\downarrow & \nearrow_{F'} & \\ \bar{A} = \tilde{A}/H & & \end{array}$$

Then $F_\lambda = F'_\lambda \bar{F}_\lambda$ and $F_\bullet = \bar{F}_\bullet F'_\bullet$. For an indecomposable $Y \in \mathrm{Ind}_A$, we get

$$F_\lambda F_\bullet Y = F'_\lambda(\bar{F}_\lambda \bar{F}_\bullet)F'_\bullet Y,$$

Consider the indecomposable decomposition $F'_\bullet Y = \bigoplus_{i=1}^{t} Y_i$ with $\bar{F}_\lambda \bar{F}_\bullet Y_i \simeq Y_i$ for $i = 1, \dots, t$. Then $F_\lambda F_\bullet Y \simeq F'_\lambda F'_\bullet Y$ and the result follows from (4.2). In particular, Y is a direct summand of $F_\lambda X_j$ for some $j \in L$.

Step 3. There is some $n \in \mathbb{N}$ and a function $f\colon \mathbb{N} \to \mathbb{N}$ such that

$$F_n\colon \text{mod-}\Gamma_n \to \text{mod-}A,$$

has the property that for every $Y \in \mathrm{Ind}_A$ with $\dim_K Y = d$, there exists some $X \in \mathrm{Ind}_{\Gamma_n}$ with $\dim_K X \leq f(d)$ and such that Y is a direct summand of $F_\bullet X$.

Indeed, the set of vertices $\tilde{Q}_0$ is the disjoint union $\bigcup_{i=1}^{s} Gx_i$, then there is some $n \in \mathbb{N}$ such that Γ_n contains all the categories $\hat{R}_i$ for $R_i = \tilde{A}_{x_i}$, $i = 1, \dots, s$. By (2), Y is a direct summand of $F_\lambda Z$ for some $Z \in \mathrm{Ind}_{\tilde{A}}$. For some $g \in G$, $\mathrm{supp}\, Z^g \subset \Gamma_n$ and $X = \varepsilon^n_\bullet Z^g \in \mathrm{Ind}_{\Gamma_n}$ satisfies that Y is a direct summand of $F_n X$. Clearly, $f(d) = |\Gamma_n| d$ (where $|\Gamma_n|$ denotes the number of objects of Γ_n) satisfies the desired property.

Step 4. Assume that $\tilde{A}$ is tame, we show that A is tame. Indeed, let $n \in \mathbb{N}$ and $f\colon \mathbb{N} \to \mathbb{N}$ be as in (3). Since F_n is right exact (1), there is a $A - F_n$-bimodule N such that $F_n = N \otimes_{\Gamma_n} (-)$. Let $z \in \mathbb{N}^{Q_0}$.

Since Γ_n is tame, there is a family $M_1, \ldots, M_t$ of $\Gamma_n - K[T]$-bimodules which are finitely generated free as $K[T]$-right modules and such that any indecomposable Γ_n-module X with $\dim_K X \leq f(d)$ is isomorphic to $M_i \otimes_{K[T]} S$ for some $1 \leq i \leq t$ and S a simple $K[T]$-module. By (3), for every $Y \in \mathrm{Ind}_A$ with $\dim_K Y = d$, there exists some $1 \leq i \leq t$ and S a simple $K[T]$-module such that Y is a direct summand of $(N \otimes_{\Gamma_n} M_i) \otimes_{K[T]} S$. It is easy to see that this (apparently) weaker property is equivalent to the tameness of A.
Step 5. Assume now that A is tame. It is enough to show that each Γ_n is tame. Consider the right exact functors

$$H_n = \varepsilon^n_{\bullet}\ F_{\bullet}\ : \text{mod-}A \to \text{mod-}\Gamma_n.$$

Let $Y \in \mathrm{Ind}_{\Gamma_n}$. Thus $X = F_\lambda \varepsilon^n_\lambda Y \in \text{mod-}A$ and

$$H_n X = \varepsilon^n \left(\bigoplus_{g \in G} (\varepsilon^n_\lambda Y)^g \right) \cong \bigoplus_{g \in S} \varepsilon^n_{\bullet}\ (\varepsilon^n_\lambda Y)^g = Y \oplus \left(\bigoplus_{g \in S \setminus \{1\}} \varepsilon^n_{\bullet}\ (\varepsilon^n_\lambda Y)^g \right),$$

where S is the finite set of $g \in G$ such that $\mathrm{supp}\,(\varepsilon^n_\lambda Y)^g \cap \Gamma_n \neq \emptyset$. As in (3), (4) we get that Γ_n is tame. □

5.3.6 We consider several examples:

(*a*) Let $F: \tilde{A} \to A$ be a Galois covering induced from a Galois covering of quivers with relations and defined by the action of a group G. Assume that A is locally support-finite.

- If G is a finite group such that $p = \mathrm{chr}\, K$ does not divide the order of G, then (5.5) applies and $\tilde{A}$ is tame if and only if so is A.
- If G is a free group, then both (5.2) and (5.7) apply.

(*b*) Consider example (5.4). For $(\alpha, \beta) \neq (1, 1)$, the category $A_{\alpha,\beta}$ is tame and locally support finite. Hence the algebra $\bar{A}_{\alpha,\beta}$ is tame.
(*c*) [39] Consider the Galois covering

$$F: A = A^{(2)}_{1,1} \to \bar{A} - A^{(2)}_{1,1}/\mathbb{Z}_2$$

given in (2.2.c) and assume that $\mathrm{chr}\, K = 2$. We Know that A is tame. We show that $\bar{A}$ is a wild algebra.

Set $x_0 = \alpha_0 + \beta_0$, $y_0 = \beta_0$, $x_1 = \alpha_1 + \beta_1$, $y_1 = \beta_1$. Then $\bar{A}$ is isomorphic to the algebra A' given by the quiver with relations.

$$A' : \bullet \underset{y_0}{\overset{x_0}{\rightrightarrows}} \bullet \underset{y_1}{\overset{x_1}{\rightrightarrows}} \qquad \left.\begin{array}{l} x_1 x_0 = 0 \\ y_1 x_0 = x_1 y_0 \end{array}\right\}$$

The universal covering $\tilde{A}'$ constructed as in (2.3) admits a full convex subcategory Δ as follows

Since Δ is wild (observe that the Tits form q_Δ takes value -1 in the indicated vector), then $\tilde{A}'$ is wild. Then A is also wild.

(d) The algebra $\bar{A}$ of example (c) provides another example of a wild algebra whose Tits form $q_{\bar{A}}$ is weakly non-negative. Indeed,

$$q_{\bar{A}}(a, b, c) = (a - b + c)^2.$$

5.3.7 We consider briefly the situation of coverings $F\colon A \to A/G$ where A is not necessarily locally support-finite.

Let $F\colon A \to A/G$ be a Galois covering. A *line* in A is a full convex subcategory isomorphic to the path category of a linear quiver (of type $\mathbb{A}_n$, $\mathbb{A}_\infty$ or $\mathbb{A}_\infty^\infty$). A **line** L is G-**periodic** if its stabilizer $G_L = \{g \in G\colon gL = L\}$ is non-trivial (then L is of type $\mathbb{A}_\infty^\infty$).

Let L be a G-periodic line in A, we construct an indecomposable weakly G-periodic A-module B_L by setting $B_L(x) = K$ for $x \in L$ and $B_L(x) = 0$ for $x \notin L$ and $B_L(\alpha) = \mathbf{Id}$ for any arrow α in L. Then $G_{B_L} \simeq G_L$ is isomorphic to $\mathbb{Z}$. Let W_x be a set of representatives of the G_{B_L}-orbits in Gx, for any object x in A. Then $F_\lambda B_L$ is a $A/G - K[T, T^{-1}]$-bimodule such that for each $x \in A$, $F_\lambda B_L(x)$ is a free $K[T, T^{-1}]$-module of rank $\sum_{y \in W_x} \dim_k B_L(y)$. We consider the functor

$$\phi^L = F_\lambda B_L \otimes_{K[T,T^{-1}]} (-)\colon \text{mod-}K[T, T^{-1}] \to \text{mod-}A/G.$$

Let $\mathscr{L}$ be the set of all lines in A and $\mathscr{L}_0$ be the set of representatives of the G-orbits in $\mathscr{L}$.

Theorem ([30]) *Let $F\colon A \to A/G$ be a Galois covering such that G acts freely on* $(\text{Ind}_A)/\simeq$. *Assume that for any weakly G-periodic A-module X,* supp X *is a line. Then*

(a) Every module in $\text{Ind}_2 A/G$ *is of the form $\phi^L(V)$ for some $L \in \mathscr{L}_0$ and some indecomposable $K[T, T^{-1}]$-module V.*

(b) $\Gamma_{A/G} = \Gamma_A/G \vee \bigvee_{L \in \mathscr{L}_0} \Gamma_{K[T,T^{-1}]}$, *where $\Gamma_{K[T,T^{-1}]}$ is the translation quiver of finite dimensional indecomposable $K[T, T^{-1}]$-modules, consisting of a K^*-family of stable tubes of rank one.*

(c) A is tame if and only if so is A/G.

5.4 Balanced Functors

5.4.1 Let A be a locally bounded K-category, that is, A has a (possibly infinite) set of objects A_0 such that

(a) $A(a, b)$ is a K-vector space and the composition corresponds to linear maps $A(a, b) \otimes_K A(b, c) \to A(a, c)$ for every a, b, c objects in A_0;
(b) $A(a, a)$ is a local ring for every a in A_0;
(c) $\sum_b A(a, b)$ and $\sum_b A(b, a)$ are finite dimensional for every a in A_0.

Let $F: A \to B$ be a K-linear functor between two locally bounded K-categories. The **pull-up** functor $F_\bullet$: B-Mod $\to$ A-Mod, $M \mapsto MF$ admits a left adjoint F_λ: A-Mod $\to$ B-Mod, called the **push-down** functor, which is uniquely defined (up to isomorphism) by the following requirements:

(i) $F_\lambda A(a, -) = B(Fa, -)$;
(ii) F_λ commutes with direct limits.

In particular, F_λ preserves projective modules. Denote by F_ρ: A-Mod $\to$ B-Mod the right adjoint to $F_\bullet$.

5.4.2 In the next paragraphs we shall develop covering theory for balanced functors. Since it is close to the theory developed for Galois coverings, we will only comment the results and leave the proofs as exercises.

Lemma *Let $F: A \to B$ be a covering functor. Assume that B is connected and a fiber $F^{-1}(i)$ is finite, for some $i \in B_0$. Then the fibers have constant cardinality.*

In case $F: A \to B$ is a covering functor with B connected and A_0 is finite, we define the *order* of F $\operatorname{ord}(F) = |F^{-1}(i)|$ for any $i \in B_0$. Thus $\operatorname{ord}(F)|B_0| = |A_0|$.

5.4.3 A covering functor $F: A \to B$ is **balanced** if ${}_b f_a^\bullet = {}_b^\bullet f_a$ for every couple of objects a, b in A.

Lemma *Let $F: A \to B$ be a balanced covering functor, then $F_\lambda = F_\rho$ as functors* A-mod $\to$ B-mod.

(*a*) Let A be a locally bounded K-category and let G be a group of K-linear automorphisms acting freely on A (that is, for $a \in A_0$ and $g \in G$ if $ga = a$, then $g = 1$). The quotient category A/G has as objects the G-orbits in the objects of A; a morphism $f: i \to j$ in A/G is a family $f: ({}_b f_a) \in \prod_{a,b} A(a, b)$, where a (resp. b) ranges in i (resp. j) and $g \cdot {}_b f_a = {}_{gb} f_{ga}$ for all $g \in G$. The canonical projection $F: A \to A/G$ is called a **Galois covering** defined by the action of G.

5.4.4 A particular situation is illustrated by the following algebras (given as quivers with relations):

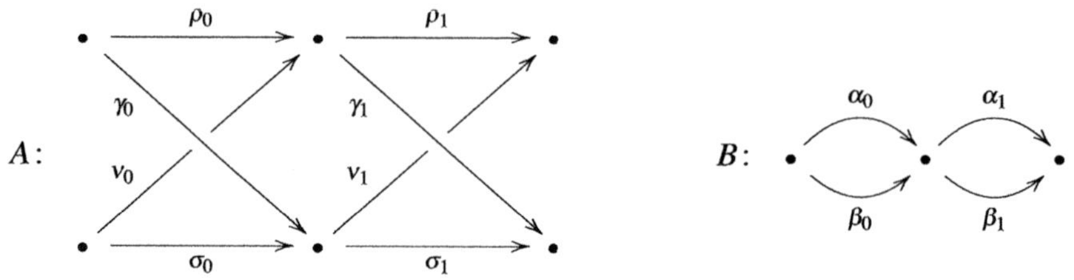

$$\begin{cases} \rho_1\rho_0 = \nu_1\gamma_0 \\ \sigma_1\sigma_0 = \gamma_1\nu_0 \\ \rho_1\nu_0 = \nu_1\sigma_0 \\ \sigma_1\gamma_0 = \gamma_1\rho_0 \end{cases} \qquad \begin{cases} \alpha_1\alpha_0 = \beta_1\beta_0 \\ \beta_1\alpha_0 = \alpha_1\beta_0 \end{cases}$$

The algebra A is tame, but B is wild when char $K = 2$ [17]. The cyclic group C_2 acts freely on A and A/C_2 is isomorphic to B.
(b) Consider the algebras given by quivers with relations and the functor F as follows:

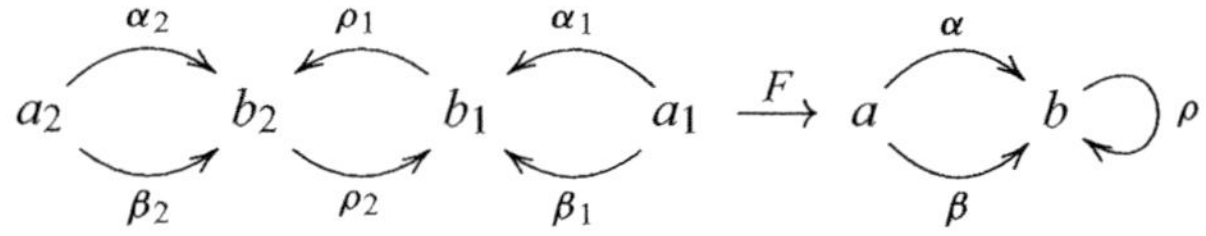

both algebras with $\text{rad}^2 = 0$ and $F\alpha_i = \alpha$, $F\beta_i = \beta$, $F\rho_i = \rho$. It is a simple exercise to check that F is a balanced covering, but obviously it is not of Galois type.
(c) Consider the functor

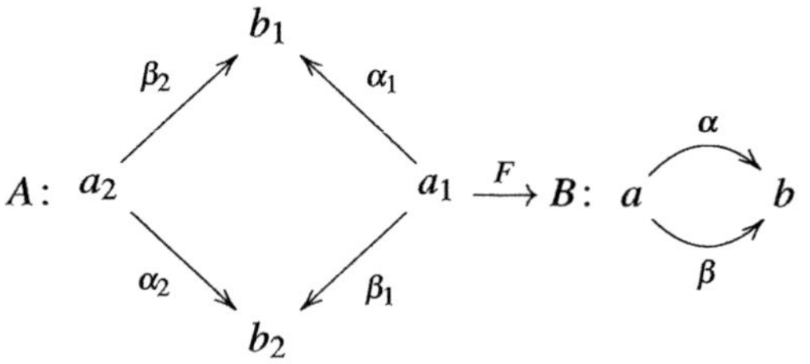

where $F\alpha_i = \alpha$, $i = 1, 2$, $F\beta_1 = \beta$, $F\beta_2 = \alpha + \beta$. Since $F(\beta_2 - \alpha_2) = \beta$ and $F(\beta_1) = \beta$, then ${}_{b_2}\beta^{\bullet}_{a_2} = -\alpha_2$ and ${}^{\bullet}_{b_2}\beta_{a_2} = 0$. Hence F is a non-balanced covering functor.

For the two dimensional indecomposable A-module $X\colon Ka_2 \to Kb_2$ it follows that $F_\bullet F_\lambda X$ is indecomposable and hence X is not a direct summand of $F_\bullet F_\lambda X$.

(*d*) As a further example, consider the infinite category A and the balanced covering functor defined in the obvious way:

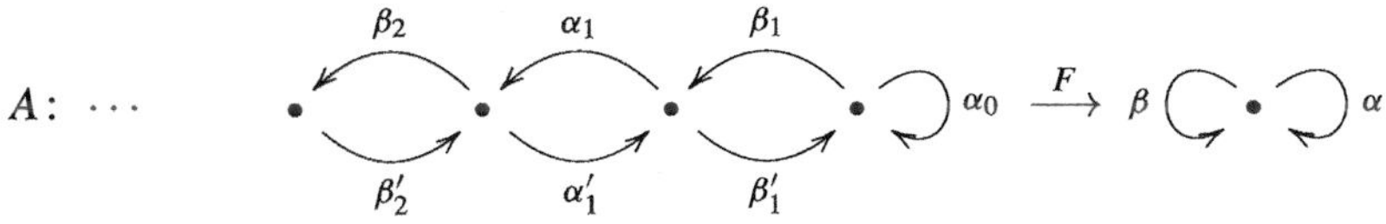

where both categories A and B have $\mathrm{rad}^2 = 0$.

5.4.5 We say that a locally bounded K-category B is **schurian** if for every $i, j \in B_0$, $\dim_K B(i, j) \le 1$.

Lemma *Let $F\colon A \to B$ be a covering functor and assume that B is schurian, then F is balanced.*

Proposition *Let $F\colon A \to B$ be a Galois covering, then F is balanced.*

5.4.6 We say that a K-category B is G-**graded** with respect to the group G if for each pair of objects i, j there is a vector space decomposition $B(i, j) = \bigoplus_{g \in G} B^g(i, j)$ such that the composition induces linear maps

$$B^g(i, j) \otimes B^h(j, m) \to B^{gh}(i, m).$$

Then the **smash product** $B \# G$ is the K-category with objects $B_0 \times G$, and for pairs $(i, g), (j, h) \in B_0 \times G$, the set of morphisms is

$$(B \# G)((i, g), (j, h)) = B^{g^{-1}h}(i, j),$$

with compositions induced in natural way.

It is clear that $B \# G$ accepts a free action of G such that

$$(B \# G)/G \xrightarrow{\sim} B.$$

Moreover, if $B = A/G$ is a quotient, then B is a G-graded K-category and

$$(A/G) \# G \xrightarrow{\sim} A.$$

Proposition *Let $F\colon A \to B$ be a covering functor and assume that B is a G-graded K-category. Then*

(i) Assume A accepts a G-grading compatible with F, that is, $F(A^g(a, b)) \subseteq B^g(Fa, Fb)$, for every pair of vertices a, b and $g \in G$. Then there is a covering

functor $F \# G \colon A \# G \to B \# G$ *completing a commutative square*

$$\begin{array}{ccc} A \# G & \xrightarrow{F\#G} & B \# G \\ \downarrow & & \downarrow \\ A & \xrightarrow{F} & B \end{array}$$

where the vertical functors are the natural quotients. Moreover F *is balanced if and only if* $F \# G$ *is balanced.*

(*ii*) *In case* B *is a schurian algebra, then* A *accepts a* G*-grading compatible with* F.

5.4.7 Theorem *Assume that* K *is algebraically closed and let* $F \colon A \to B$ *be a covering induced from a map of quivers with relations. Then* B *is representation-finite if and only if* A *is locally representation-finite. Moreover in this case the functor* $F_\lambda \colon A\text{-mod} \to B\text{-mod}$ *preserves indecomposable modules and Auslander-Reiten sequences.*

Proof Let $F \colon A \to B$ be induced from $q \colon (Q', I') \to (Q, I)$ where $A = KQ'/I'$ and $B = KQ/I$. Let $\tilde{B} = K\tilde{Q}/\tilde{I}$ be the universal cover of B and $\tilde{F} \colon \tilde{B} \to B$ the universal covering functor. By 5.2.3 there is a covering functor $F' \colon \tilde{B} \to A$ such that $\tilde{F} = FF'$.

(1) Assume that B is a connected representation-finite algebra. If B is standard, by 5.4.1, A is locally representation-finite. We may assume that B is not standard. By [40, Section 9], B has a full subcategory $C = KQ_C/I_C$ given by the **Riedtmann quiver**

Q_C:

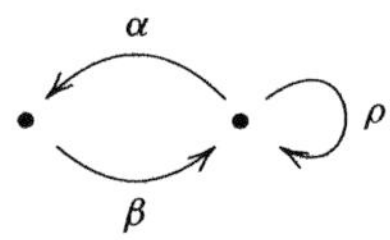

with I_C generated by $\beta\alpha - \beta\rho\alpha$, $\alpha\beta - \rho^2$ and ρ^4.

It is well-known (cf. [35]) that (Q_C, I_C) does not accept a proper covering $(Q'_C, I'_C) \to (Q_C, I_C)$. Therefore $F^{-1}(a)$ has a single point and by (5.1.3), F is a trivial covering map. In this case $A = B$ is representation-finite.

(2) Assume that A is locally representation-finite. Let X be an indecomposable A-module. We shall prove first that X is isomorphic to $F'_\lambda N$ for some indecomposable $\tilde{B}$-module N. Since indecomposable projective A-modules are of the form $A(a, -) = F'_\lambda \tilde{B}(x, -)$ for some x in $\tilde{B}$, using the connectedness of Γ_A, we may assume that there is an irreducible morphism $f : Y \to X$ such that $Y = F'_\lambda N$ for some indecomposable $\tilde{B}$-module N. Clearly, without loss of generality, we may assume that N is not injective and consider the Auslander-Reiten sequence $\xi : 0 \longrightarrow N \xrightarrow{g} N' \xrightarrow{g'} N'' \longrightarrow 0$ in $\tilde{B}$-mod. We shall prove

that the push-down

$$F'_\lambda\xi : 0 \longrightarrow F'_\lambda N \xrightarrow{F'_\lambda g} F'_\lambda N' \xrightarrow{F'_\lambda g'} F'_\lambda N'' \longrightarrow 0$$

is an Auslander-Reiten sequence in A-mod. This implies that there exists a direct summand $\bar{N}$ of N' such that $X \xrightarrow{\sim} F'_\lambda \bar{N}$ which completes the proof of the first claim.

To verify that $F'_\lambda\xi$ is an Auslander-Reiten sequence, let $h\colon F'_\lambda N \to Z$ be non-split mono in ${}_A A$-mod. Consider $\mathrm{Hom}_A(F'_\lambda N, Z) \xrightarrow{\sim} \mathrm{Hom}_{\tilde{B}}(N, F'_\bullet Z)$, $h \mapsto h'$ which is not a split mono (otherwise, then $\mathrm{Hom}_{\tilde{B}}(F'_\bullet Z, N) \xrightarrow{\sim} \mathrm{Hom}_A(Z, F'_\rho N)$, $v \mapsto v'$ with $vh' = 1_{F'_\bullet Z}$. By 5.1.4, $F'_\lambda = F'_\rho$ and $v'h = 1_Z$). Then there is a lifting $\bar{h}\colon N' \to F'_\bullet Z$ with $\bar{h}g = h'$. Hence $\mathrm{Hom}_{\tilde{B}}(N', F'_\bullet Z) \xrightarrow{\sim} \mathrm{Hom}_A(F'_\lambda N', Z)$, $\bar{h} \mapsto \bar{h}'$ with $\bar{h}' F'_\lambda g = h$.

(*a*) We show that B is representation-finite. Indeed, by case (1), since $F'\colon \tilde{B} \to A$ is a covering induced by a map of a quiver with relations, then $\tilde{B}$ is locally representation-finite. By [29], B is representation-finite.

(*b*) We show that F_λ preserves Auslander-Reiten sequences. Let X be an indecomposable A-module of the form $X = F'_\lambda N$ for an indecomposable $\tilde{B}$-module N. Then $F_\lambda X = F_\lambda F'_\lambda N = \tilde{F}_\lambda N$. Since by [29], $\tilde{F}_\lambda$ preserves indecomposable modules, then $F_\lambda X$ is indecomposable. Finally, as above, we conclude that F_λ preserves Auslander-Reiten sequences.

□

5.4.8 Let K be an algebraically closed field. We recall that A is said to be of **tame representation type** if for each dimension $d \in \mathbb{N}$ and each object $a \in A_0$, there are finitely many $A - K[t]$-bimodules $M_1, \ldots, M_t$ which satisfy:

(a) M_i is finitely generated free as right $K[t]$-module $i = 1, \ldots, t$;
(b) each indecomposable $X \in A$-mod with $X(a) \neq 0$ and $\dim_K X = d$ is isomorphic to some module of the form $M_i \otimes_{K[t]} (K[t]/(t - \lambda))$ for some $i \in \{1, \ldots, t\}$ and $\lambda \in K$.

In fact, it is shown in [36] that A is tame if (b) is substituted by the weaker condition

(b') each indecomposable $X \in A$-mod with $X(a) \neq 0$ and $\dim_K X = d$ is a direct summand of a module of the form $M_i \otimes_{K[t]} (K[t]/(t - \lambda))$ for some $i \in \{1, \ldots, t\}$ and $\lambda \in K$.

Theorem *Let $F\colon A \to B$ be a balanced covering functor. If B is tame, then A is tame.*

Proof Let $a \in A_0$ and $d \in \mathbb{N}$. Let $M_1, \ldots, M_t$ be the $B - K[t]$-bimodules satisfying (a) and (b): each indecomposable $M \in B$-mod with $M(Fa) \neq 0$ and $\dim_K M \leq d$ is isomorphic to some $M_i \otimes_{K[t]} (K[t]/(t - \lambda))$ for some $i \in \{1, \ldots, t\}$ and $\lambda \in K$.

By 5.3.2, each indecomposable $X \in A$-mod with $X(a) \neq 0$ and $\dim_K X = d$ is a direct summand of some $F_\bullet M_i \otimes_{K[t]} (K[t]/(t-\lambda))$ for some $i \in \{1, \ldots, t\}$ and $\lambda \in K$. Hence A satisfies condition (b'). □

5.5 Galois Coverings of Algebras

5.5.1 Following [9], by a **locally bounded category** we mean a K-category R which is isomorphic to a factor category KQ/I, where Q is a locally finite quiver and I is an admissible ideal of the path category KQ of Q.

An algebra A will be considered as a **finite category**, that is, a locally bounded category given by a finite quiver. A locally bounded category R is said to be **interval-finite** if, for any two vertices x and y of Q_R, the number of pairwise different vertices of Q_R lying on paths from x to y in Q_R is finite. We note that this is a much weaker property than the interval-finiteness defined in [10, (2.6)], which does not force R to be triangular. A full subcategory C of a locally bounded category R is said to be **convex** if any path in Q_R with source and target in Q_C lies entirely in Q_C. Observe that the convex hull of a finite subcategory of an interval-finite category R is a finite convex subcategory of R.

An R-module M is called **finite dimensional** (resp. **locally finite dimensional**) if $\dim M = \sum_{x \in R} \dim_K M(x) < \infty$ (respectively, $\dim_K M(x) < \infty$ for any object x of R). Recall from 5.2.6 that we denote by MOD-R the category of all R-modules, by Mod-R (respectively, mod-R) the category of all locally finite dimensional (respectively, all finite dimensional) R-modules, and by Ind$R$, (respectively, Ind$R$) the full subcategory of Mod-R (respectively, mod-R) formed by all indecomposable modules.

5.5.2 Let G be a group of K-linear automorphisms of R acting freely on the objects of R. Then following [26] we may consider the orbit category R/G defined as follows. The objects of R/G are the G-orbits of the objects of R, and, for any two objects a and b of R/G, the morphism K-space $(R/G)(a, b)$ is defined as

$$R/G(a, b) = \left\{ (f_{y,x}) \in \prod_{(x,y) \in a \times b} R(x, y) \;\middle|\; g f_{y,x} = f_{gy,gx} \text{ for all } g \in G, x \in a, y \in b \right\}$$

and the composition ef of morphisms $f : a \to b$ and $e : b \to c$ in R/G is defined by $(ef)_{zx} = \sum_{y \in b} e_{zy} f_{yx}$. Then we have a canonical **Galois covering functor**

$$F : R \longrightarrow R/G$$

which assigns to any object x of R its G-orbit Gx and maps a morphism $f \in R(x, y)$ onto the family $F(f) \in (R/G)(Gx, Gy)$ such that $F(f)_{hy,gx} = gf$ or 0 according as $h = g$ or $h \neq g$. Moreover, F induces the K-linear isomorphisms

$$\bigoplus_{F(y)=a} R(x, y) \xrightarrow{\sim} (R/G)(F(x), a) , \quad \bigoplus_{F(y)=a} R(y, x) \xrightarrow{\sim} (R/G)(a, F(x)) ,$$

for all objects x of R and a of R/G. For a full subcategory D of R we denote by gD the full subcategory of R formed by the objects gx, $x \in D$, and its stabilizer $G_D = \{g \in G \mid gD = D\}$. Then we may consider the locally bounded category D/G_D. The group G acts on MOD-R by the translations ${}^g(-)$ which assign to each R-module M the R-module ${}^gM = M \circ g^{-1}$. For each R-module M, we denote by G_M the stabilizer $\{g \in G \mid {}^gM \cong M\}$ of M. Following [24], a module Y in Ind$R$ is said to be *weakly G-periodic* if supp$Y$ is infinite and (supp$Y$)$/G_Y$ is a finite category. Observe that in such a case, G_Y is infinite. We mention that in Dowbor's papers [18–21] the weakly G-periodic modules are called infinite G-atoms.

5.5.3 Assume now that G is a group of K-linear automorphisms of R acting freely on the isoclasses of modules in Ind$R$. We point out that G acts freely on the objects of R, since G acts freely on the isoclasses of indecomposable projective R-modules $R(-, x), x \in R$. Consider the associated Galois covering functor $F : R \to R/G$. As in 5.2.6, we denote by $F_\bullet$: MOD-$R/G \to$ MOD-R the **pull-up functor**, which assigns to an R/G-module M the R-module $M \circ F$, and by F_λ : MOD-$R \to$ MOD-R/G the **push-down functor**, left adjoint to $F_\bullet$ (see [9, (3.2)]). Since G acts freely on the isoclasses in Ind$R$, F_λ induces an injection from the set (Ind$R/\cong)/G$ of G-orbits of isoclasses in Ind$R$ into the set (Ind$R/G)/\cong$ of isoclasses in IndR/G [26, (3.5)]. We denote by mod$_1$-R/G the full subcategory of mod-R/G consisting of all modules isomorphic to $F_\lambda(M)$ for some module M in mod-R, and by mod$_2$-R/G the full subcategory of mod-R/G formed by all modules without nonzero direct summands from mod$_1$-R/G. It was shown in [24, (2.2) and (2.3)] that a module X from mod-R/G belongs to mod$_1$-R/G (respectively, mod$_2$-R/G) if and only if $F_\bullet(X)$ is a direct sum of finite dimensional R-modules (respectively, weakly G-periodic R-modules). We denote by Ind$_1 R/G$ (respectively, Ind$_2 R/G$) the full subcategory of mod$_1$-R/G (respectively, mod$_2$-R/G) formed by the indecomposable modules. Following [24] the modules from Ind$_1 R/G$ (respectively, Ind$_2 R/G$) are called **indecomposable modules of the first kind** (respectively, **indecomposable modules of the second kind**). The category R is said to be G-**exhaustive** if mod-R/G = mod$_1$-R/G [24]. From [24, (2.5)] we know that R is G-exhaustive provided R is **locally support-finite**, that is, for each object x of R, the full subcategory of R consisting of the objects of all supports supp$M$, where M is a module in Ind$R$ with $M(x) \neq 0$, is a finite category. An important class of locally-support finite categories is formed by the locally representation-finite locally bounded categories playing a fundamental role in the representation theory of representation-finite algebras (see [5, 7, 9, 26]). We also mention that if R is G-exhaustive then the Auslander-Reiten quiver $\Gamma_{R/G}$ of

R/G is the orbit translation quiver Γ_R/G of the Auslander-Reiten quiver Γ_R under the induced action of G on Γ_R, because the push-down functor F_λ : mod-R $\to$ mod-R/G preserves the almost split (Auslander-Reiten) sequences [26, (3.6)]. The problem whether the G-exhaustivity of R forces R to be a locally support-finite category remains still open, although it was confirmed in some important cases (see [2, 48]). We refer also to [18–21, 24] for some results on indecomposable R/G-modules of the second kind.

5.5.4 A crucial role in the proof of our main result will be played by a reduction theorem from [24], allowing to describe the indecomposable modules of the second kind for a class of Galois coverings.

Assume that R is not G-exhaustive. Following [24, (3.1)], a family S' of full subcategories of R is called *separating* (with respect to G) if S' satisfies the following conditions:

(i) for each $L \in S'$ and $g \in G$, $gL \in S'$;
(ii) for each $L \in S'$ and each G-orbit $\mathcal{O}$ of R, $\mathcal{O} \cap L$ is contained in finitely many G_L-orbits;
(iii) for any two different $L, L' \in S'$, $L \cap L'$ is locally support-finite;
(iv) for each weakly G-periodic R-module Y, there exists an $L \in S'$ such that $\operatorname{supp} Y \subseteq L$.

For and additive category C and a full subcategory C_0 of C we denote by $C/[C_0]$ the factor category of C by the ideal $[C_0]$ of all morphisms in C factorized through a direct sum of some objects of C_0.

For a convex subcategory L of a locally bounded category R we have a canonical embedding E^L : MOD-L $\to$ MOD-R such that, for a module M in MOD-L, $E^L(N)$ is an R-module such that $E^L(N)(x) = N(x)$ for any object x of L, $E^L(N)(f) = N(f)$ for any morphism f in L, and $E^L(N)(y) = 0$ for any object y of R which is not in L. Moreover, for any group G of K-linear automorphisms of R acting freely on the objects of R, we have a commutative diagram of functors

$$\begin{array}{ccc} \text{MOD-}L & \xrightarrow{E^L} & \text{MOD-}R \\ \downarrow{\scriptstyle F^L_\lambda} & & \downarrow{\scriptstyle F_\lambda} \\ \text{MOD-}L/G_L & \xrightarrow{E^L_\lambda} & \text{MOD-}R/G \end{array}$$

where F^L_λ is the push-down functor associated to the Galois covering $F^L : L \to L/G_L$, F_λ is the push-down functor associated to the Galois covering $F : R \to R/G$, and E^L_λ assigns to a module X in MOD-L/G_L the module $E^L_\lambda(X)$ in MOD-R/G such that $F_\bullet E^L_\lambda(X) = \bigoplus_{g \in U_L} {}^g F^L_\bullet(X)$, where $F_\bullet$: MOD-R/G $\to$ MOD-R and $F^L_\bullet$: MOD-L/G_L $\to$ MOD-L are the pull-up functors associated to F and F^L, and U_L is a fixed set of representatives of the cosets of G modulo G_L (see [24, (2.4), (3.2)]).

5.5.5 The following theorem is a special (convex) version of [24, Theorem 3.1].

Theorem *Let R be a locally bounded K-category and G a group of K-linear automorphisms of R acting freely on the isoclasses in* Ind$R$*. Let S' be a separating family of convex subcategories of R with respect to G and S'_0 a fixed set of representatives of G-orbits in S'. Then the functors E^L_λ : mod-$L/G_L \to$ mod-R/G, $L \in S'_0$, induce a natural K-linear equivalence of categories*

$$E : \bigoplus_{L\in S'_0} (\text{mod-}L/G_L)/[\text{mod}_1\text{-}L/G_L] \longrightarrow (\text{mod-}R/G)/[\text{mod}_1\text{-}R/G].$$

In particular, the Auslander-Reiten quiver $\Gamma_{R/G}$ of R/G is the disjoint union of the translation quivers

$$\Gamma_{R/G} = (\Gamma_R/G) \sqcup \left(\bigoplus_{L\in S'_0} \left(\Gamma_{L/G_L}\right)_2 \right),$$

where $\left(\Gamma_{L/G_L}\right)_2$ is the union of all connected components of Γ_{L/G_L} formed by the indecomposable L/G_L-modules of the second kind.

5.5.6 In this book we are interested in coverings of algebras by locally bounded categories of tame representation type. Recall that, following Drozd [25], a finite category R is called **tame** if, for any dimension d, there exists a finite number of $K[T]$-R-bimodules M_i, $1 \le i \le n_d$, which are finitely generated and free as left $K[T]$-modules, and all but a finite number of isoclasses of indecomposable R-modules of dimension d are of the form $K[T]/(T-\lambda) \otimes_{K[T]} M_i$ for some $\lambda \in K$ and some $i \in \{1, \dots, n_d\}$. Let $\mu_R(d)$ be the least number of $K[T]$-R-bimodules satisfying the above condition for d. Then R is said to be of **polynomial growth** (respectively, **domestic**) if there is a positive integer m such that $\mu_R(d) \le d^m$ (respectively, $\mu_R(d) \le m$) for any $d \ge 1$ (see [15, 45]). An arbitrary locally bounded category R is said to be **tame** (respectively, **of polynomial growth**, **domestic**) if so is every finite full subcategory of R. We refer to [23] for results characterizing tame locally bounded categories. It has been proved in [22] that if R is a locally bounded category, G a group of automorphisms of R acting freely on the objects of R and R/G is tame, then R is tame. The inverse implication is still an open problem.

5.5.7 The following solution of the stabilizer conjecture by Dowbor in [21] is important for applications of covering techniques to tame algebras.

Theorem *Let R be a tame locally bounded K-category and G a torsion-free automorphism group of R acting freely on the objects of R. Then the stabilizer G_Y of any weakly G-periodic R-module Y is an infinite cyclic group.*

The following consequence of [24, (2.4), (3.6)] and [26, pp. 94-95] shows the importance of the above theorem for Galois coverings of algebras by tame locally bounded categories. We note that the restriction imposed on the group to be torsion-free is natural because of results of [26, 28].

Proposition *Let R be a locally bounded K-category, G a group of K-linear automorphisms of R acting freely on the objects of R, and Y a weakly G-periodic R-module whose stabilizer G_Y is an infinite cyclic group. Then $F_\lambda(Y)$ carries a canonical structure of a KG_Y-R/G-bimodule which is a free module of finite rank as left module over the group algebra KG_Y of G_Y. In particular, we have a canonical functor*

$$\Phi^Y = - \otimes_{KG_Y} F_\lambda(Y) : \text{mod}_K\text{-}G_Y \longrightarrow \text{mod-}R/G,$$

whose image is contained in $\text{mod}_2\text{-}R/G$.

5.5.8 We also note that, if G_L is infinite cyclic, then KG_L is isomorphic to the algebra $K[T, T^{-1}]$ of Laurent polynomials.

Let R be a locally bounded K-category and G a group of K-linear automorphisms of R acting freely on the objects of R. A **line** in R is a convex subcategory L of R which is isomorphic to the path category KQ of a linear quiver Q of type $\mathbb{A}_n$, $\mathbb{A}_\infty$ or ${}_\infty A_\infty$. A line L in R is said to be G-**periodic** if its stabilizer G_L is nontrivial. Clearly, in this case, the quiver Q_L of L is of type

$${}_\infty\mathbb{A}_\infty : \quad \cdots \;\text{———}\; \bullet \;\text{———}\; \bullet \;\text{———}\; \bullet \;\text{———}\; \bullet \;\text{———}\; \bullet \;\text{———}\; \cdots$$

and has a G_L-periodic orientation. With each G-periodic line L of R we may associate a canonical weakly G-periodic R-module M_L by setting $M_L(x) = K$ for any vertex x of Q_L, $M_L(y) = 0$ for all vertices y of $Q_R \setminus Q_L$, and $M_L(\gamma) = \mathbf{Id}_K$ for each arrow γ of Q_L. Since $G_{M_L} = G_L = \mathbb{Z}$, we obtain then a canonical functor

$$\Phi^L = - \otimes_{K[T,T^{-1}]} F_\lambda(M_L) : \text{mod}_K\text{-}[T, T^{-1}] \longrightarrow \text{mod-}R/G,$$

where $\text{mod}_K\text{-}[T, T^{-1}]$ denotes the category of finite dimensional modules over $K[T, T^{-1}]$.

5.6 Cycle-Finite Algebras

5.6.1 In this section we describe properties of cycle-finite algebras applied in the proof of our main theorem and its consequences.

By a **tame concealed algebra** we mean a tilted algebra $C = \text{End}_H(T)$, where H is the path algebra KQ of a quiver Q of Euclidean type $\widetilde{\mathbb{A}}_m (m \geq 1)$, $\widetilde{\mathbb{D}}_n (n \geq 4)$, or $\widetilde{\mathbb{E}}_p (6 \leq p \leq 8)$, and T is a (multiplicity-free) preprojective tilting H-module. The tame concealed algebras have been completely classified by quivers and relations in [8] and [31]. Recall also that the Auslander-Reiten quiver Γ_C of a tame concealed algebra C is of the form

$$\Gamma_C = \mathscr{P}^C \vee \mathscr{T}^C \vee \mathscr{Q}^C,$$

where $\mathscr{P}^C$ is a preprojective component containing all indecomposable projective C-modules, $\mathscr{Q}^C$ is a preinjective component containing all indecomposable injective C-modules, and $\mathscr{T}^C$ is a $\mathbb{P}_1(K)$-family $\mathscr{T}^C_\lambda$, $\lambda \in \mathbb{P}_1(K)$, of pairwise orthogonal standard stable tubes, all but finite number of them of rank one (see [43, Chapter 4]) and [44] for further properties of the module category of a tame concealed algebra).

5.6.2 By a **tubular algebra** we mean a tubular extension (equivalently, tubular coextension) of a tame concealed algebra of tubular type $(2, 2, 2, 2)$, $(3, 3, 3)$, $(2, 4, 4)$, or $(2, 3, 6)$, as defined in [43]. Recall that a tubular algebra B admits two different tame concealed convex subcategories C_0 and C_∞ such that the Auslander-Reiten quiver Γ_B of B is of the form

$$\Gamma_B = \mathscr{P}^B_0 \vee \mathscr{T}^B_0 \vee \left(\bigvee_{q \in \mathbb{Q}^+} \mathscr{T}^B_q \right) \vee \mathscr{T}^B_\infty \vee \mathscr{Q}^B_\infty,$$

where $\mathscr{P}^B_0$ is the preprojective component $\mathscr{P}^{C_0}$ of Γ_{C_0}, $\mathscr{T}^B_0$ is a $\mathbb{P}_1(K)$-family of pairwise orthogonal standard ray tubes, obtained from the stable tubes of $\mathscr{T}^{C_0}$ by ray insertions, $\mathscr{Q}^B_\infty$ is the preinjective component $\mathscr{Q}^{C_\infty}$ of Γ_{C_∞}, $\mathscr{T}^B_\infty$ is a $\mathbb{P}_1(K)$-family of pairwise orthogonal standard coray tubes, obtained from the stable tubes of $\mathscr{T}^{C_\infty}$ by coray insertions, and, for each $q \in \mathbb{Q}^+$ (the set of positive rational numbers), $\mathscr{T}^B_q$ is a $\mathbb{P}_1(K)$-family of pairwise orthogonal standard stable tubes (we refer to [43, Chapter 5] for more properties of module categories of tubular algebras).

5.6.3 Recall also that a module M over an algebra A is called **sincere** if $\operatorname{supp} M = A$. The following characterization of tame concealed and tubular algebras was established in [46, Theorem 4.1].

Theorem *Let A be an algebra. The following conditions are equivalent:*

(i) A is cycle-finite and admits a sincere indecomposable module lying in a stable tube of Γ_A.
(ii) A is either tame concealed or tubular.

5.6.4 Moreover, we have the following consequence of the above theorem and a result of Crawley-Boevey [14, Corollary E], proved in [46, Theorem 4.3].

Theorem *Let A be a cycle-finite algebra. Then A is of polynomial growth.*

5.6.5 An algebra A is said to be **minimal representation-infinite** if A is representation-infinite but every proper convex subcategory of A is representation-finite. Then we have the following characterization of minimal representation-infinite cycle-finite algebras established in [46, Corollary 4.4].

Theorem *For an algebra A the following conditions are equivalent:*

(i) A is minimal representation-infinite and cycle-finite.
(ii) A is tame concealed.

5.6.6 We also mention the following characterization of domestic cycle-finite algebras, proved in [46, Theorem 5.1].

Theorem *Let A be a cycle-finite algebra. The following conditions are equivalent:*

(i) A is domestic.
(ii) A does not contain a tubular algebra as a convex subcategory.
(iii) All but finitely many components of Γ_A are stable tubes of rank one.

5.6.7 Let A be a cycle-finite algebra. The support supp$M$ of a module M in Ind$A$ is not necessarily a convex subcategory of A. We denote by $c(M)$ the number of pairwise different tame concealed convex subcategories of the convex hull of supp$M$ in A.

The following fact was established in our joint work with Malicki [33, Theorem B].

Theorem *Let A be a cycle-finite algebra and M be an indecomposable finite dimensional A-module. Then $c(M) \leq 3$.*

We note that there are many cycle-finite algebras A having indecomposable finite dimensional modules M with $c(M) = 3$. For example, the iterated tubular algebras studied in [42] have this property. We also mention that there are (discrete) indecomposable finite dimensional modules over cycle-finite algebras with a very complicated structure, for example the modules lying in generalized multicoils of tame generalized multicoil algebras (see [34]).

5.6.8 We end this section with a result on the number of terms in the middle of almost split sequences over cycle-finite algebras. For an algebra A and a non-projective indecomposable module X in mod-A, consider an almost split sequence

$$0 \longrightarrow \tau_A X \longrightarrow \bigoplus_{i=1}^{s(X)} Y_i \longrightarrow X \longrightarrow 0,$$

where all Y_i are indecomposable A-modules. It has been proved by Bautista and Brenner [4] (see also [32]) that if A is representation-finite then $s(X) \leq 4$, and if $s(X) = 4$ then one of the modules Y_i is projective-injective. It has been conjectured by Brenner that, for a tame algebra A, $s(X) \leq 5$ for every nonprojective indecomposable finite dimensional A-module X. This has been confirmed for some classes of tame algebras but for arbitrary tame algebras it is still an open and interesting problem.

5.6.9 For cycle-finite algebras this was shown in [41, Theorem 3] as follows.

Theorem *Let A be a cycle-finite algebra and*

$$0 \longrightarrow \tau_A X \longrightarrow \bigoplus_{i=1}^{s(X)} Y_i \longrightarrow X \longrightarrow 0$$

be an almost split sequence in mod-A, *where the modules Y_i are indecomposable. Then $s(X) \leq 5$. Moreover, if $s(X) = 5$, then one of the modules Y_i is projective-injective.*

We recall that an almost split sequence in mod-A whose middle term contains an indecomposable projective-injective module P is isomorphic to the almost split sequence of the form

$$0 \longrightarrow \text{rad}P \longrightarrow (\text{rad}P/\text{soc}P) \oplus P \longrightarrow P/\text{soc}P \longrightarrow 0$$

(see [1, Proposition 3.11]).

5.6.10 A locally bounded K-category R is said to be **cycle-finite** if R is interval-finite and every finite convex subcategory of R is cycle-finite. A group G of K-linear automorphisms of a locally bounded category R is said to be **admissible** if its action on the objects of R is free and has finitely many orbits. In such a case, R/G is a finite category (algebra). For a locally bounded category R and an admissible group G of automorphisms of R, we denote by $\mathscr{A}(R, G)$ a fixed set of representatives of isoclasses of weakly G-periodic R-modules closed under the induced action of G, and by $\mathscr{A}_0(R, G)$ a fixed set of representatives of G-orbits in $\mathscr{A}(R, G)$. Moreover, we identify the group algebra $K[\mathscr{Z}]$ of the infinite cyclic group $\mathscr{Z}$ with the algebra $K[T, T^{-1}]$ of Laurent polynomials.

5.6.11 The following theorem is the main result of the section.

Theorem *Let R be a connected cycle-finite locally bounded K-category over an algebraically closed field K, G a torsion-free admissible group of K-linear automorphisms of R, $A = R/G$, $\mathscr{A} = \mathscr{A}(R, G)$, and $\mathscr{A}_0 = \mathscr{A}_0(R, G)$. Then the functors $\Phi^Y = - \otimes_{K[T,T^{-1}]} F_\lambda(Y) : \mathrm{mod}_K\text{-}[T, T^{-1}] \to \mathrm{mod}\text{-}A$, $Y \in \mathscr{A}_0$, induce a K-linear equivalence of categories*

$$\Phi : \bigoplus_{\mathscr{A}_0} \mathrm{mod}_K\text{-}[T, T^{-1}] \xrightarrow{\sim} \mathrm{mod}\text{-}A/[\mathrm{mod}_1\text{-}A] \ .$$

Moreover, the following statements hold:

(i) A is tame.
(ii) Every indecomposable finite dimensional A-module X is isomorphic either to $F_\lambda(M)$ for some indecomposable finite dimensional R-module M or to $\Phi^Y(V)$ for some $Y \in \mathscr{A}_0$ and some indecomposable finite dimensional $K[T, T^{-1}]$-module V.
(iii) The Auslander-Reiten quiver Γ_A of A has the disjoint union decomposition

$$\Gamma_A = (\Gamma_R/G) \sqcup \left(\bigoplus_{\mathscr{A}_0} \Gamma_{K[T,T^{-1}]} \right)$$

where $\Gamma_{K[T,T^{-1}]}$ is the Auslander-Reiten quiver of the category of finite dimensional $K[T, T^{-1}]$-modules.

5.6.12 We present now an example of a tame algebra A which admits a Galois covering $R \to R/G = A$ with infinite cyclic group G and R an interval-finite tame locally bounded K-category having infinitely many pairwise nonisomorphic weakly G-periodic modules with the support equal to R.

Let A be the bound quiver algebra KQ/I given by the quiver Q of the form

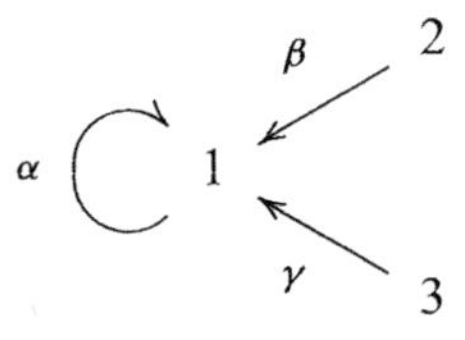

and the ideal I of the path algebra KQ generated by α^2. The algebra A admits a Galois covering $R \to R/G = A$, where R is the bound quiver category $K\Delta/J$ with Δ the quiver of the form

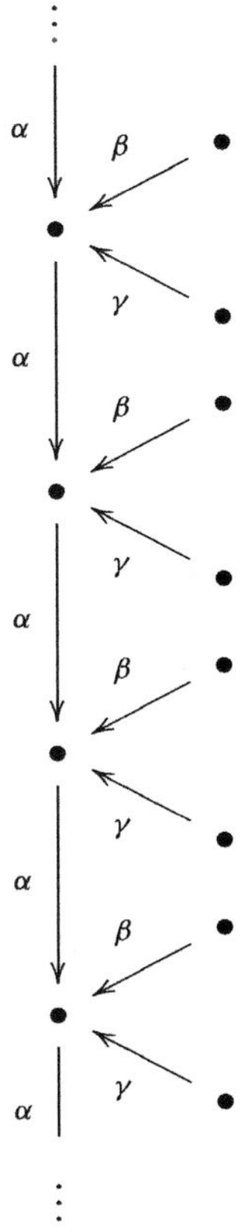

J the ideal of the path category $K\Delta$ of Δ generated by all elements of the form α^2, and G is the infinite cyclic group generated by the canonical vertical shift of $R = K\Delta/J$.

5.6.13 Then R is an interval-finite strongly simply connected locally bounded K-category which does not contain a hypercritical finite convex subcategory (see [49] for the list of hypercritical algebras), and consequently R is tame, by [12,

Corollary 1]. On the other hand, R contains the finite convex subcategory $R' = K\Delta'/J'$ given by a subquiver Δ' of Δ of the form

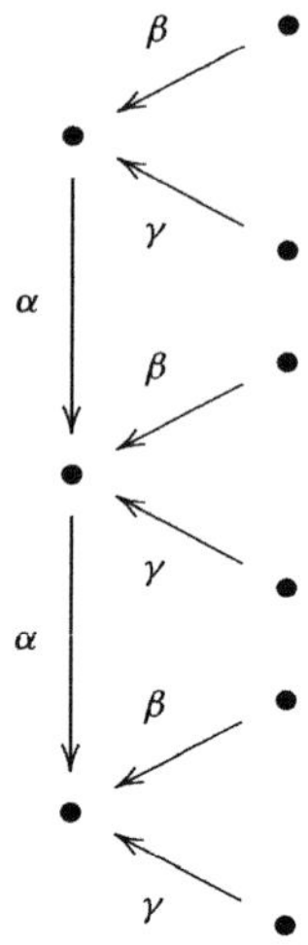

and the ideal J' of $K\Delta'$ generated by α^2, which is a pg-critical algebra in the sense of [37]. In particular, by [47, Theorem 4.1], R' is not of polynomial growth, and hence is not cycle-finite. Therefore, R is a tame strongly simply connected locally bounded K-category which is not cycle-finite. Moreover, it is not hard to show that there are infinitely many pairwise nonisomorphic weakly G-periodic R-modules with the support equal to R. We also mention that A is a tame algebra. In fact, by [11, Lemma 5.3], A degenerates (in the variety of K-algebras of dimension 8) to the special biserial bound quiver algebra $\bar{A} = K\bar{Q}/\bar{I}$ given by the quiver $\bar{Q}$ of the form

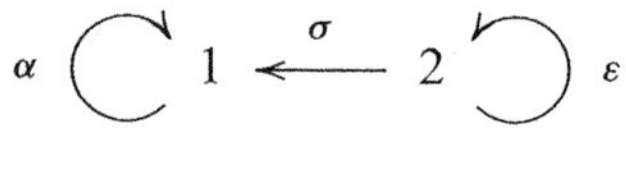

and the ideal $\bar{I}$ of the path algebra $K\bar{Q}$ generated by α^2 and ε^2. Since $\bar{A}$ is tame (see [24, (5.2)]), A is also tame, by [27] or [16].

References

1. Assem, I. and Simson, D. and Skowronski, A., *Elements of the representation theory of associative algebras*, Cambridge University Press (2006), London Mathematical Society Student Texts **65**
2. I. Assem and A. Skowroński, *On tame repetitive algebras*, Fund. Math. **142** (1993), 59–84.
3. Assem, I., Skowroński, A. and Tomé, B.: *Coil enlargements of algebras*. Tsukuba Journal of Mathematics Vol. **19**, No. 2 (December 1995), pp. 453–479

4. R. Bautista and S. Brenner, *On the number of terms in the middle of an almost split sequence*, In: Representations of Algebras, Lecture Notes in Math., Vol. **903**, Springer-Verlag, Berlin-Heidelberg, 1981, pp. 1–8.
5. R. Bautista, P. Gabriel, A. V. Roiter and L. Salmerón, *Representation-finite algebras and multiplicative bases*, Invent. Math. **81** (1985), 217–285.
6. M. Beattie. *A generalization of the smash product of a graded ring.* J. Pure Appl. Algebra 52 (1988) 219–229.
7. K. Bongartz, *A criterion for finite representation type*, Math. Ann. **269** (1984), 1–12.
8. K. Bongartz, *Critical simply connected algebras*, Manuscripta Math. **46** (1984), 117–136.
9. K. Bongartz and P. Gabriel, *Covering spaces in representation theory*, Invent. Math. **65** (1982), 337–378.
10. O. Bretscher and P. Gabriel, *The standard form of a representation-finite algebra*, Bull. Soc. Math. France **111** (1983), 21–40.
11. T. Brüstle, *Tame tree algebras*, J. Reine Angew. Math. **567** (2004), 51–98.
12. T. Brüstle, J. A. de la Peña and A. Skowroński, *Tame algebras and Tits quadratic forms*, Preprint, 2008.
13. O. Butscher and P. Gabriel. *The standard form of a representation-finite algebra.* Bull. Soc. Math. France 111 (1983) 21–40.
14. W. Crawley-Boevey, *On tame algebras and bocses*, Proc. London Math. Soc. **56** (1988), 451–483.
15. W. Crawley-Boevey, *Tame algebras and generic modules*, Proc. London Math. Soc. **63** (1991), 241–265.
16. W. Crawley-Boevey, *Tameness of biserial algebras*, Arch. Math. (Basel) **65** (1995), 399–407.
17. Dean, A. and de la Peña, J. A.: *Algorithms for weakly non-negative quadratic forms.* Linear Algebra and its Applications Volume **235**, 1 March 1996, Pages 35–46
18. P. Dowbor, *On the category of modules of second kind for Galois coverings*, Fund. Math. **149** (1996), 31–54.
19. P. Dowbor, *Galois covering reduction to stabilizers*, Bull. Polish Acad. Sci. Sér. Sci. Math. **44** (1996), 341–352.
20. P. Dowbor, *Properties of G-atoms and full Galois covering reduction to stabilizers*, Colloq. Math. **83** (2000), 231–265.
21. P. Dowbor, *Stabilizer conjecture for representation-tame Galois coverings of algebras*, J. Algebra **239** (2001), 119–149.
22. P. Dowbor and A. Skowroński, *On Galois coverings of tame algebras*, Arch. Math. (Basel) **44** (1985), 522–529.
23. P. Dowbor and A. Skowroński, *On the representation type of locally bounded categories*, Tsukuba J. Math. **10** (1986), 63–77.
24. P. Dowbor and A. Skowroński, *Galois coverings of representation-infinite algebras*, Comment. Math. Helv. **62** (1987), 311–337.
25. Y. A. Drozd, *Tame and wild matrix problems*, In: Representation Theory II, Lecture Notes in Mathematics, Vol. **832**, Springer-Verlag, Berlin-Heidelberg, 1980, pp. 242–258.
26. P. Gabriel. *The universal cover of a representation-finite algebra.* In 'Representation of Algebras'. Springer LNM 903 (1981) 68–105.
27. C. Geiss, *On degenerations of tame and wild algebras*, Arch. Math. (Basel) **64** (1995), 11–16.
28. C. Geiss and J. A. de la Peña, *An interesting family of algebras*, Arch. Math. (Basel) **60** (1993), 25–35.
29. Happel, D., Reiten, I. and Smalø, S.: *Tilting in abelian categories and quasitilted algebras.* Memoirs of the American Mathematical Society (1996) 88.
30. Happel, D., Reiten, I. and Smalø, S.: *Quasitilted algebras.* NATO Advanced Science Institute Series/Klüwer Ac. Pub.
31. D. Happel and D. Vossieck, *Minimal algebras of infinite representation type with preprojective component*, Manuscripta Math. **42** (1983), 221–243.
32. S. Liu, *Almost split sequences for nonregular modules*, Fund. Math. **143** (1993), 183–190.
33. P. Malicki, J. A. de la Peña and A. Skowroński, *Cycle-finite module categories*, Preprint 2009.

34. P. Malicki and A. Skowroński, *Algebras with separating almost cyclic coherent Auslander-Reiten components*, J. Algebra **291** (2005), 208–237.
35. R. Martínez-Villa and J. A. de la Peña. *The universal cover of a quiver with relations*. J. Pure and Appl. Algebra 30 (3) (1983) 277–292.
36. R. Martínez-Villa and J. A. de la Peña. *Automorphisms of representation-finite algebras*. Invent. Math. 72 (1983) 359–362.
37. R. Nörenberg and A. Skowroński, *Tame minimal non-polynomial growth simply connected algebras*, Colloq. Math. **73** (1997), 301–330.
38. de la Peña, J. A.: *Quadratic forms and the representation type of an algebra*. Engänzungs-reihe 90–103. Bielefeld (1990).
39. de la Peña, J. A.: *On the corank of the Tits form of an algebra*. Journal of Pure and Applied Algebra Volume **107**, Issue 1, 26 February 1996, Pages 89–105
40. de la Peña, J.A. *Tame algebras: some fundamental notions*. Ergänzungsreihe 95–010 (1995) Bielefeld
41. J. A. de la Peña and M. Takane, *On the number of terms in the middle of almost split sequences over tame algebras*, Trans. Amer. Math. Soc. **351** (1999), 3857–3868.
42. J. A. de la Peña and B. Tomé, *Iterated tubular algebras*, J. Pure Appl. Algebra **64** (1990), 303–314.
43. Ringel, C.M. *Tame algebras and integral quadratic forms*, Springer LNM, **1099** (1984)
44. D. Simson and A. Skowroński, *Elements of the Representation Theory of Associative Algebras 2: Tubes and Concealed Algebras of Euclidean Type*. London Mathematical Society Student Texts, Vol. **71**, Cambridge University Press, Cambridge, 2007.
45. A. Skowroński, *Algebras of polynomial growth*, In: Topics in Algebra, Banach Center Publ., Vol. **26**, Part 1, PWN Warsaw, 1990, pp. 535–568.
46. A. Skowroński, *Cycle-finite algebras*, J. Pure Appl. Algebra **103** (1995), 105–116.
47. A. Skowroński, *Simply connected algebras of polynomial growth*, Compositio Math. **109** (1997), 99–133.
48. A. Skowroński, *Tame algebras with strongly simply connected Galois coverings*, Colloq. Math. **72** (1997), 335–351.
49. L. Unger, *The concealed algebras of the minimal wild, hereditary algebras*, Bayreuter Math. Schriffen **31** (1990), 145–154.

Chapter 6
Reflections and Weyl Groups

6.1 Vinberg's Characterization of Dynkin Diagrams

6.1.1 Let A be an Artin algebra. Given two indecomposable modules M, N, let $\mathrm{Irr}(M, N) = \mathrm{rad}(M, N)/\mathrm{rad}^2(M, N)$ be the bimodule of **irreducible maps** and denote by a_{MN} the length of $\mathrm{Irr}(M, N)$ as an $\mathrm{End}_A(N)$-module and by a'_{MN} its length as an $\mathrm{End}_A(M)$-module. Note that in case M is not injective, then a_{MN} is equal to the multiplicity of N occurring in the middle term of the Auslander-Reiten sequence starting with M, whereas if N is not projective, then a'_{MN} is equal to the multiplicity of M occurring in the middle term of the Auslander-Reiten sequence ending with N. The Auslander-Reiten quiver Γ_A has as vertices the isomorphism classes of the indecomposable A-modules, and there is an arrow $[M] \to [N]$ provided $\mathrm{Irr}(M, N) > 0$. We endow this arrow with the valuation (a_{MN}, a'_{MN}), and in this way we consider the valuated quiver Γ_A. We denote the Auslander-Reiten translations by $\tau_A = DTr$ and $\tau_A^- = TrD$. An indecomposable module is called stable provided $\tau^n M \neq 0$ and $\tau^{-n} M \neq 0$ for all $n \geq 1$. The full subquiver $\Gamma_A^{(s)}$ of Γ_A consisting of the isomorphism classes of stable modules is called the **stable Auslander-Reiten-quiver**. Any component of the stable Auslander-Reiten quiver determines (uniquely) a Cartan matrix, and we call it its **Cartan class**. Also, a module M is called **periodic** provided $\tau^p(M) \cong M$ for some $p > 0$.

Theorem *The Cartan class of a component of the stable Auslander-Reiten quiver of an Artin algebra containing periodic modules is either a Dynkin diagram or* A_∞.

6.1.2 Let I be an index set. A **Cartan matrix** C on I is a function $C : I \times I \to \mathbb{Z}$ satisfying the following properties:

$(C1)$ $C_{ii} = 2$, for all $i \in I$.
$(C2)$ $C_{ij} \leq 0$ for all $i \neq j$ in I.
$(C3)$ $C_{ij} = 0$ if and only if $C_{ji} = 0$.

J.-A. de la Peña, *Representations of Algebras*, Algebra and Applications 30,
https://doi.org/10.1007/978-3-031-12288-0_6

Fig. 6.1 Dynkin diagrams

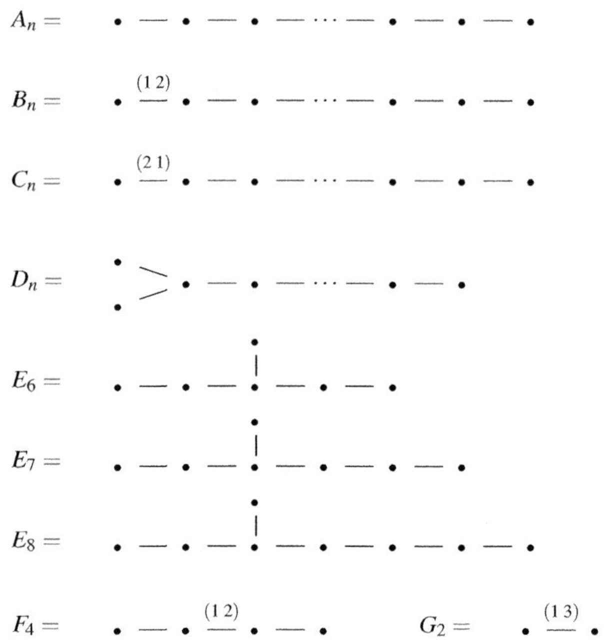

The underlying graph of C has as vertices the elements of I, and edges $\{i, j\}$ for all pairs $i \neq j$ with $C_{ij} \neq 0$. Starting with the underlying graph, we add to the edges $i — j$, a valuation i.e. a pair of numbers

$$(|C_{ij}|, |C_{ji}|),$$

in case both numbers are non-zero. The Cartan matrix C will be called **connected** in case its underlying graph is connected. In particular, we are interested in the diagrams of Figs. 6.1, 6.2 and 6.3:

6.1.3 Let C be a Cartan matrix on I. A **subadditive function** for C is a function $d : I \to \mathbb{N}$ satisfying

$$\sum_{i \in I} d_i C_{ij} \geq 0, \qquad \text{for all } j \in I.$$

Moreover, if the inequalities are always zero, then d is called an **additive function**.

Lemma *Let C be a Euclidean diagram. Then any subadditive function for C is additive.*

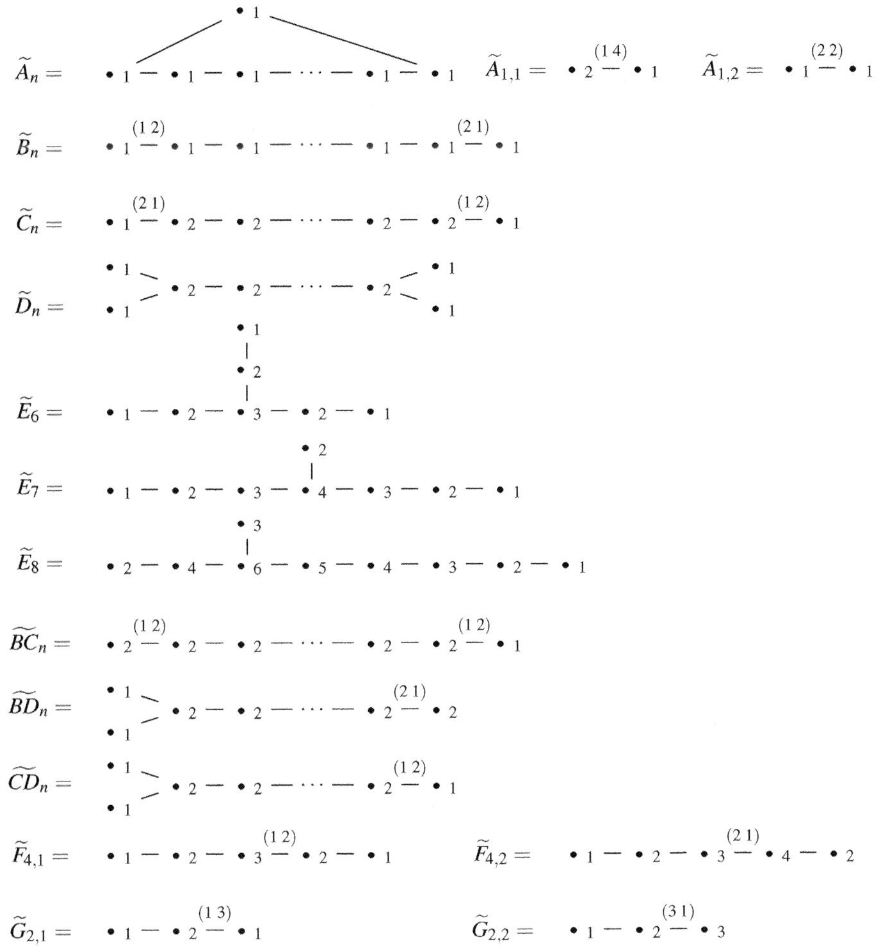

Fig. 6.2 Euclidean diagrams. The integers on the vertices determine an additive function h

Proof Let C^t be the transpose of C, that is $C^t_{ij} = C_{ji}$, for all $i, j \in I$. Clearly, C^t is an Euclidean diagram of the same type as C. For every Euclidean diagram, there is an additive function h (see Fig. 6.2).

Given the Euclidean diagram C let δ be a fixed additive function for C^t, thus $\delta C^t = 0$. Let d be a subadditive function for C. Then $(dC)\delta^t = d(C\delta^t) = 0$. By assumption, the components of dC are non-negative, while those of δ are positive. Therefore, the equality all components of dC are zero, which means that d is additive. □

6.1.4 Lemma *Every subadditive function for any one of the infinite diagrams A^{∞}_{∞}, B_{∞}, C_{∞} or D_{∞} is additive and bounded.*

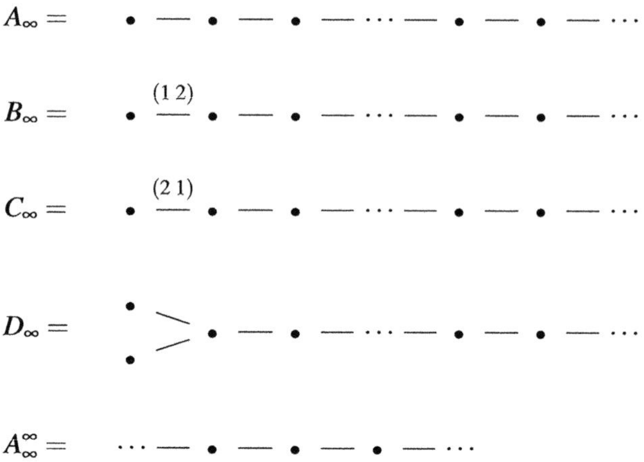

Fig. 6.3 A family of infinite diagrams

Proof Consider first the case A_∞^∞. We may assume $I = \mathbb{Z}$, with edges $\{i, i+l\}$. Given a subadditive function d on I, chose i_0 where d reaches its minimal value. Then

$$2d_i \geq d_{i-1} + d_{i+1} \quad \text{implies} \quad d_{i-1} = d_i = d_{i+1}.$$

We get that d is constant.

In the remaining cases, we get a subadditive function on A_∞^∞. Indeed, in case d is defined on B_∞, we consider

$$\cdots \quad d_2 \text{ —— } d_1 \text{ —— } d_0 \text{ —— } d_1 \text{ —— } d_2 \quad \cdots$$

In case d is defined on C_∞, we consider

$$\cdots \quad d_2 \text{ —— } d_1 \text{ —— } 2d_0 \text{ —— } d_1 \text{ —— } d_2 \quad \cdots$$

If d is defined on D_∞, we get

$$\begin{array}{lll} d_0 \text{ —— } & d_1 \text{ —— } & d_2 \cdots \\ & / & \\ d_0' & & \end{array}$$

and construct the following function on A_∞^∞:

$$\cdots \quad d_2 \;\text{——}\; d_1 \;\text{——}\; d_0 + d_0' \;\text{——}\; d_1 \;\text{——}\; d_2 \quad \cdots$$

Since the subadditive functions on A_∞^∞ are constant, then d is additive and bounded. □

6.1.5 Theorem *Let C be a connected Cartan matrix and d a subadditive function for C. Then the following holds:*

(a) C is either a Dynkin diagram, an Euclidean diagram or one of the infinite diagrams classified in Lemma 6.1.4.
(b) If d is not additive, then C is a Dynkin diagram or A_∞.
(c) If d is unbounded, then C is of type A_∞.

Proof If C is neither a Dynkin diagram nor one of the infinite cases of the above Lemma, then there exists a Euclidean diagram C' which is smaller than C (an easy verification).

Now if $C' \neq C$, then the restriction $d|C'$ of a subadditive function d' cannot be additive. This is a contradiction, since $d|C'$ must be additive, according to Lemma 6.1.3. This proves (a).

If d is not additive, then Euclidean diagrams or their infinite counterparts cannot occur according to the lemmas, this proves (b). If d is unbounded, then I has to be infinite, and only A_∞ remains according to Lemma 6.1.4. □

6.1.6 An application to AR quivers. We consider a quiver $F = (F_0, F_1)$ with F_0 the set of vertices and F_1 the set of arrow. We assume that F does not have loops or double arrows. If x is a vertex, we denote by x^+ (resp. x^-) the set of endpoints of arrows with starting (resp. ending) point x. In case the sets x^+ and x^- are finite for all x, we will call the quiver **locally finite**. A **translation quiver** $T = (T_0, T_1, \tau)$ is given by a quiver $T = (T_0, T_1)$, together with an injective function $\tau : G_0 \to T_0$ defined on a subset G_0 of T_0 satisfying $(\tau x)^+ = x^-$. Given an arrow $\alpha : y \to x$, there is a unique arrow $\sigma(\alpha) : \tau x \to y$.

A translation quiver is called **stable** provided τ is defined on all of $T_0 = G_0$ and is also surjective. Of course, any translation quiver has a unique maximal stable translation subquiver. A vertex x of a translation quiver T will be called periodic of period $p \geq 0$ provided $\tau^p(x) = x$. We will be interested in stable translation quivers containing periodic elements.

An important example is the following: let F be an oriented tree, and define $\mathbb{Z}F$ as follows: its vertices are the elements of $\mathbb{Z} \times F_0$, and given an arrow $\alpha : x \to y$, there are arrows $(n, \alpha) : (n, x) \to (n, y)$ and $\sigma(n, \alpha) : (n + l, y) \to (n, x)$ for all $n \in \mathbb{Z}$. In this way, we obtain a $\mathbb{Z}F$ stable translation quiver.

Given a quiver (F_0, F_1), a function $a : F_1 \to \mathbb{N} \times \mathbb{N}$ will be called a **valuation**, and $F = (F_0, F_1, a)$ a **valued quiver**. The image of $\alpha : x \to y$ will be denoted by (a_α, a'_α) with $a = a_{xy}$ and $a' = a_{yx}$. If F is a valued quiver, we can associate with it a **Cartan matrix** $C = C(F)$ on the index set F_0 as follows: for $x \in F_0$, let $C_{xx} = 2$ and for $x \to y$ in F_1, let $C_{xy} = -a_{xy} = a'_{yx}$, where $a_{xy} = 0 = a'_{xy}$ in case there is no arrow with starting point x and endpoint y.

A valued translation quiver $F = (F_0, F_l, \tau, a)$ is given by a translation quiver (F_0, F_l, τ) and a valuation a for (F_0, F_1) such that $a_{\sigma(\alpha)} = a'_\alpha\ a'_{\sigma(\alpha)} = a_\alpha$ for all arrows $\alpha : x \to y$. A typical example is again the following: let (F_0, F_l, a) be a valued oriented tree, and define on (F_0, F_1) a valuation by $a_{(n,\alpha)} = a_\alpha = a'_{\sigma(n,\alpha)}$ and $a'_{(n,\alpha)} = a'_\alpha = a_{\sigma(n,\alpha)}$. This valued translation quiver is denoted by $\mathbb{Z}(F_0, F_1, a)$.

6.1.7 Proposition *Let F, F' be valued oriented trees. Then $\mathbb{Z}F$ and $\mathbb{Z}F'$ are isomorphic if and only if the Cartan matrices $C(F)$ and $C(F')$ are equivalent. Given any stable valued translation quiver Γ, there is a valued oriented tree F and a group G of automorphisms of $\mathbb{Z}F$ such that Γ is isomorphic to $\mathbb{Z}F/G$.*

In case Γ is isomorphic to $\mathbb{Z}F/G$, for some valued oriented tree F, we call $C(F)$ the **Cartan class** of Γ; it is uniquely determined by Γ.

Proof Indeed, if $\Gamma = (\Gamma_0, \Gamma_1, \tau, a)$ is a valued translation quiver, and $(\Gamma_0, \Gamma_1, \tau) = \mathbb{Z}(F_0, F_1)/G$ for some oriented tree (F_0, F_1), then using the projection from $\mathbb{Z}(F_0, F_1)$ onto $(\Gamma_0, \Gamma_1, \tau)$, the valuation a of Γ gives rise to a valuation on $\mathbb{Z}(F_0, F_1)$, also denoted by a, in such a way that $\mathbb{Z}(F_0, F_1)$ becomes a valued translation quiver. The canonical embedding of (F_0, F_1) into $\mathbb{Z}(F_0, F_1)$ endows (F_0, F_1) with a valuation, again denoted by a, such that $(\Gamma_0, \Gamma_1, \tau, a) = \mathbb{Z}(F_0, F_1, a)/G$. Since (F_0, F_1) is a tree, we can choose a vertex x which is a sink. Denote by $\sigma_x(F_0, F_l, a)$ the full valued subquiver of $\mathbb{Z}(F_0, F_1, a)$ with vertices $(0, y)$ for $y \neq x$, and (l, x). It is obvious that the Cartan matrices $C(F_0, F_1, a)$ and $C(\sigma_x(F_0, F_l, a))$ are equivalent. This shows the unicity of the Cartan matrix associated to $\mathbb{Z}(F_0, F_1, a)$. □

6.1.8 If $\Gamma = (\Gamma_0, \Gamma_1, \tau, a)$ is a valued translation quiver, a **subadditive function** f for Γ satisfies, by definition,

$$f(x) + f(\tau x) \geq \sum_{y \in x^-} f(y) a'_{yx},$$

for all $x \in \Gamma_0$. Such a function is called **additive**, provided we always have equalities.

Theorem *Let $\Gamma = (\Gamma_0, \Gamma_1, \tau, a)$ be a stable valued translation quiver which is connected, and contains a periodic vertex. Assume there is a subadditive function f for Γ, then*

(a) The Cartan class of Γ is either a Dynkin diagram, an Euclidean diagram, or one of the infinite diagrams A_∞, A_∞^∞, B_∞, C_∞ or D_∞.
(b) If f is not additive, then the Cartan class of Γ is a Dynkin diagram, or A_∞.
(c) If f is unbounded, then the Cartan class of Γ is A_∞.

Proof First note that the existence of a subadditive function implies that Γ is locally finite. Let y be a periodic vertex, say $\tau^p(y) = y$. Now, since y^+ is finite, then τ^p induces a permutation on y^+, and therefore τ^{pm} is the identity on y^+, for some $m \geq 0$. Thus any $z \in y^+$ is also periodic. Similarly, vertices in y^- are periodic. But in this way, we can reach any other vertex of Γ, since it is a connected quiver.

Let Γ be a quotient of $\mathbb{Z}F$, with F a valuated oriented tree with Cartan matrix C. We can assume that $F = \{0\} \times F$ is embedded into $\mathbb{Z}F$ and denote the corresponding map $F \to \mathbb{Z}F \to \Gamma$ just by $u \mapsto \bar{u}$. By definition of C, we have

$$C_{uv} = \begin{cases} 2, & \text{if } u = v, \\ a_{\bar{u}\bar{v}}, & \text{if } u \to v, \\ a'_{\bar{v}\bar{u}}, & \text{if } v \to u, \\ 0, & \text{else.} \end{cases}$$

Assume now there is given a subadditive function f for Γ. We consider first the case where there exists a fixed number p with $\tau^p(x) = x$ for all vertices x of Γ (for instance, in case F is finite). From f we obviously obtain a τ-invariant subadditive function d for Γ, by

$$d(x) = \sum_{i=0}^{p-1} f(\tau^i(x)),$$

and d is additive if and only if f is. □

6.2 *M*-Matrices and Positivity

6.2.1 Lemma *If M is a real $n \times n$-matrix and $M + M^t$ is positive definite, then M is non-singular and $-M^{-1}M^t$ is diagonalizable over $\mathbb{C}$, all its eigenvalues having modulus one.*

Proof For $v \in \mathbb{C}^n$ define $a(v) = \overline{v}^t M v$, where the bar denotes complex conjugation. If $a(v) = 0$, then

$$0 = a(v) + \overline{a(v)} = \overline{v}^t M v + \overline{v}^t M^t v,$$

and hence $v = 0$, since $M + M^t$ is positive definite. In particular, this shows that $Mv = 0$ implies $v = 0$, and M is non-singular.

Suppose that $-M^{-1}M^t v = \lambda v$, where $v \neq 0$. Then $-\overline{v}^t M^t v = \lambda \overline{v}^t M v$, and hence $\lambda = -\overline{a(v)} a(v)^{-1}$. Thus $|\lambda| = 1$.

It remains to show that $-M^{-1}M^t$ is diagonalizable, and for this it suffices to show that if λ is a repeated eigenvalue of M and $u, v \in \mathbb{C}^n$ satisfy $(-M^{-1}M^t -$

$\lambda\mathbf{Id}_n)u = v$ and $(-M^{-1}M^t - \lambda\mathbf{Id}_n)v = 0$, then $v = 0$. Indeed, we have

$$-M^t u = \lambda M u + M v, \quad \text{and} \quad -M^t v = \lambda M v.$$

Then, $-\overline{u}^t M^t v = \lambda\overline{u}^t M v$ and, taking transpose conjugate, $-\overline{v}^t M u = \overline{\lambda}\overline{v}^t M^t u$. Since $|\lambda| = 1$, this gives $-\overline{v}^t M^t u = \lambda\overline{v}^t M u$. Therefore $\overline{v}^t M v = 0$, and thus $v = 0$.

□

6.2.2 The class of Z-matrices are those matrices whose off-diagonal entries are less than or equal to zero. Let A be a $n \times n$ real Z-matrix. That is, $A = (a_{ij})$ where $a_{ij} \leq 0$ for all $i \neq j$. Then A is an M-matrix if it can be expressed in the form $A = sI - B$, where $B = (b_{ij})$ with $b_{ij} \geq 0$, for all $1 \leq i, j \leq n$, where s is at least as large as the maximum of the moduli of the eigenvalues of B, and I is the identity matrix.

Below, $\leq$ denotes the coordinate-wise order. That is, for any real matrices A, B of size $n \times n$, we write $A \leq B$ if $a_{ij} \leq b_{ij}$, for all i, j.

Proposition *Let A be a $n \times n$ real Z-matrix, then the following statements are equivalent to A being a non-singular M-matrix:*

- $(M1)$ *All the principal minors of A are positive. That is, the determinant of each submatrix of A obtained by deleting a set, possibly empty, of corresponding rows and columns of A is positive.*
- $(M2)$ *$A + D$ is non-singular for each non-negative diagonal matrix D.*
- $(M3)$ *Every real eigenvalue of A is positive.*
- $(M4)$ *All the leading principal minors of A are positive.*
- $(M5)$ *There exist lower and upper triangular matrices L and U respectively, with positive diagonals, such that $A = LU$.*
- $(P1)$ *A is **inverse-positive**. That is, A^{-1} exists and $A^{-1} \geq 0$.*
- $(P2)$ *A is **monotone**. That is, $Ax \geq 0$ implies $x \geq 0$*
- $(P3)$ *Every **regular splitting** of A is **convergent**. That is, for every representation $A = M - N$, where $M^{-1} \geq 0$, $N \geq 0$, then $\rho(M^{-1}N) < 1$.*
- $(S1)$ *There exists a positive diagonal matrix D such that $AD + DA^t$ is positive definite.*
- $(S2)$ *A is **positive stable**. That is, the real part of each eigenvalue of A is positive.*
- $(S3)$ *There exists a symmetric positive definite matrix W such that $AW + WA^t$ is positive definite.*
- $(SP1)$ *A is **semi-positive**. That is, there exists $x > 0$ with $Ax > 0$.*
- $(SP2)$ *There exists $x \geq 0$ with $Ax > 0$.*
- $(SP3)$ *There exists a positive diagonal matrix D such that AD has all positive row sums.*

6.2.3 A square matrix that is not reducible is said to be **irreducible**. Associated to a square $n \times n$-matrix A there is a digraph $D(A)$ formed by n vertices and edges

(i, j) if $A_{ij} \neq 0$. We say that $D(A)$ is strongly connected if there is a path between every pair of vertices.

Lemma *The following statements are equivalent:*

(1) A is an irreducible matrix.
(2) The digraph $D(A)$ associated to A is strongly connected, that is, every vertex is reachable from any other vertex.
(3) For each i and j, there exists some k such that $(A^k)_{ij} > 0$.
(4) The transitive closure $T(A)$ of A contains all the vertices $\{1, \ldots, n\}$.

6.2.4 We can deduce the existence of the Perron-Frobenius eigenvector from the **Brouwer fixed point theorem**. This latter fundamental result from topology asserts that any continuous self-map of the unit ball B^n (or equivalently, any compact convex set in $\mathbb{R}^n$) admits a fixed point.

Lemma *Let M be an M-matrix such that $M + M^t$ is irreducible but not positive definite. Then $-M^{-1}M^t$ has a real eigenvalue λ. If moreover, $M+M^t$ is not positive semidefinite (resp. it is positive semidefinite) then $\lambda \geq 1$ (resp. all eigenvalues of $-M^{-1}M^t$ have modulus one, 1 is a repeated eigenvalue and the matrix $-M^{-1}M^t$ is not diagonalizable).*

Proof Let D be the diagonal of M, and E a positive diagonal matrix with $E^2 = D^{-1}$. Define for $0 \leq \mu \leq 1$,

$$P(\mu) = (1 + \mu)\mathbf{Id} - EM^tE - \mu EME.$$

Since M is an M-matrix so is EME, and for $\mu > 0$ the (i, j) entry of $P(\mu) = 0$ if and only if the (i, j) and (j, i) entries of M are both zero. But $M+M^t$ is irreducible and then $P(\mu)$ is irreducible for $\mu > 0$. Moreover, $P(\mu)$ has non-negative entries (for $0 \leq \mu \leq 1$), and so by the Frobenius-Perron theorem the spectral radius $\rho(\mu)$ is a simple eigenvalue of $P(\mu)$. It is obvious that $\rho(\mu)$ is a continuous function of μ.

Since EM^tE is an M-matrix its real eigenvalues are all positive (for, the signs of the coefficients of the powers of x in $\det(EM^tE - x\mathbf{Id})$ alternate, and so it must be positive when x is negative). Hence $\rho(0)$ (an eigenvalue of $I - EM^tE$) must be less than 1. However, $E(M + M^t)E$ has an eigenvalue which is non-positive (since $M + M^t$ is not positive definite) and consequently $2I - EM^tE - EME$ has an eigenvalue greater than or equal to 2. So $\rho(1) \geq 2$. By continuity there exists μ with $0 < \mu \leq 1$ and $\rho(\mu) = 1 + \mu$. Moreover, if $M + M^t$ is not positive semidefinite, then $\mu < 1$. Now $\det(-EM^tE - \mu EME) = 0$, and (transposing) $\det(-EME - \mu EM^tE) = 0$. Thus $\det(\mu^{-1}\mathbf{Id} + M^{-1}M^t) = 0$, and $\lambda = \mu^{-1}$ is an eigenvalue of $-M^{-1}M^t$ with the required properties.

Assume that $M+M^t$ is positive semidefinite. By 6.2.1 and a continuity argument, all eigenvalues of $-M^{-1}M^t$ have modulus one. Since $M + M^t$ is singular, 1 is an eigenvalue of $-M^{-1}M^t$. But the polynomial $p(x) = \det(x\mathbf{Id} + M^{-1}M^t)$ has the form

$$(1 + x^n) + a_1(x + x^{n-1}) + a_2(x^2 + x^{n-2}) + \cdots$$

since $x^n p(1/x) = p(x)$, and since $(x-1)^2$ divides $(x^i - 1)(x^{n-i} - 1)$, we see that $x^i + x^{n-i} = 1 + x^n \mod (x-1)^2$. Thus

$$p(x) = (1 + a_1 + a_2 + \cdots)(1 + x^n) \mod (x-1)^2,$$

$$p(x) = 2(1 + a_1 + a_2 + \cdots) \mod (x-1).$$

It follows that $1 + a_1 + a_2 + \cdots = 0$, and $(x-1)^2$ divides $p(x)$. So 1 is a repeated eigenvalue of $-M^{-1}M^t$.

Finally, observe that since $\rho(1) = 2$ is a simple eigenvalue of $2\mathbf{Id}_n - EM^tE - EME$, hence the null space of $-EM^tE - EME$ has dimension 1. Thus the null space of $-M^{-1}M^t - \mathbf{Id}_n$ has dimension 1 despite the fact that 1 is a repeated eigenvalue of $-M^{-1}M^t$. Hence $-M^{-1}M^t$ is not diagonalizable. □

6.3 Coxeter Matrices and Weyl Group

6.3.1 Let $A = (a_{ij})$ be a $n \times n$ **Cartan matrix**, that is, $a_{ii} = 2$ for $1 \le i \le n$, $a_{ij} = 0$ if and only if $a_{ji} = 0$, and a_{ij} is a non-positive integer for $i \ne j$. Associated with A there is a **Coxeter (valued) graph** Δ that we assume to be connected.

Let $E = \mathbb{R}^n$ and $e_1, \ldots, e_n$ be a set of linearly independent vectors in E. Define $(-,-)$ the (non-symmetric) bilinear form in E given by $(e_i, e_j) = a_{ij}$. The **reflections** $s_i \in \mathrm{G}\ell(E)$ are defined by $v \cdot s_i = v - (v, e_i)e_i$. The subgroup W of $\mathrm{G}\ell(E)$ generated by $s_1, \ldots, s_n$ is called the **Weyl group** of A. The Weyl group together with the set of reflections $S = \{s_1, \cdots, s_n\}$ form a **Coxeter system** for A.

Formally, a **Coxeter group** can be defined as a group with the presentation $\langle s_1, s_2, \ldots, s_n \mid (s_i s_j)^{m_{ij}} = 1 \rangle$ where $m_{ii} = 1$ and $m_{ij} \ge 2$ for $i \ne j$. The condition $m_{ij} = \infty$ means that no relations of the form $(s_i s_j)^m$ should be imposed. A number of conclusions can be drawn immediately from the above definition.

The Weyl group W associated to a quadratic form q is generated by the reflections s_i with inner product given as

$$(e_i, e_j) = -cos(\pi/m_{ij}),$$

where m_{ij} are integers with $m_{ii} = 1$ and $m_{ij} > 1$, for $i \ne j$. The **Coxeter element** in W is $c = s_1 \cdots s_n$. The relation $m_{ii} = 1$ means that $(s_i s_i)^1 = (s_i)^2 = 1$, for all i; as such, the generators are involutions. If $m_{ij} = 2$, then the generators s_i and s_j commute. This follows by observing that $xx = yy = 1$, and $xyxy = 1$ implies that $xy = x(xyxy)y = (xx)yx(yy) = yx$.

6.3.2 Theorem *The following are equivalent:*

(1) W is infinite;
(2) c has infinite order;
(3) c has a real eigenvalue greater or equal to 1;
(4) q is not positive definite.

Furthermore q is positive semidefinite and not positive definite if and only if 1 is an eigenvalue of c and all other eigenvalues have modulus 1.

Proof That (1) implies (4) is shown in Bourbaki [[2], Chapter V, §4, No. 8, Theorem 2]. That (4) implies (3) follows from Theorem 2.1 and Corollary 3.3, since U is obviously an M-matrix. If (3) holds and (4) does not then by Lemma 3.1 we see that 1 is an eigenvalue of c. Now Theorem 2.1 shows that $A = U + U^t$ is singular, and hence (3) implies (4). Using (4) and Corollary 3.3, the matrix $-U^{-l}U^t$ either has a real eigenvalue greater than 1 or is not diagonalizable. Hence it has infinite order, that is, (2) holds. That (2) implies (1) is obvious. □

6.3.3 The size of the Coxeter groups. If A is of finite type, all Coxeter transformations are conjugate and therefore have the same order h (=**Coxeter number** of A). The characteristic polynomial $P(x)$ of c has the form

$$P(x) = \Pi_{j=1}^{n}(x - \exp(2i\pi m_j/h)),$$

where $m_1, \dots, m_n$ are integers such that $1 = m_1 < m_2 \leq \cdots \leq m_{n-1} < m_n = h - 1$. Moreover, the order of W is $(m_1 + 1)(m_2 + 1)\cdots(m_n + 1)$.

If A is of affine type, then c has infinite order and $P(x) = (x - 1)^2\Pi_{j=1}^{m_n}(x - \exp(2i\pi m_j/h))$, where $1 \leq m_1 \leq \cdots \leq m_{n-2} \leq h - 1$ are integers and h is the Coxeter number of a Cartan matrix of finite type (canonically associated with A).

In case A is of finite or affine type, we have seen that $\rho(c) = 1$ for a Coxeter element $c \in W$. An element $g \in G\ell(E)$ is said to be **hyperbolic** if there are two eigenvalues λ, μ of g with $|\lambda| \neq |\mu|$. Since an element $w \in W$ has $\det(w) = \pm 1$, then w is hyperbolic if and only if $\rho(w) > 1$.

Following [3], we say that $g \in Gl(E)$ is **sharp** if g has a simple eigenvalue λ such that $\|\mu\| < \|\lambda\| \quad (= \rho(g))$ for any other eigenvalue μ of g. We say that g is **very sharp** if both g and g^{-1} are sharp.

Proposition *Let A be a Cartan matrix and c be a Coxeter transformation. Then*

(a) If A is not of finite type, then $\rho(c)$ is an eigenvalue of c.
(b) If A is of indefinite type, then c is hyperbolic.

6.3.4 If the Coxeter graph Δ associated with A is a **tree** of wild type then c is very sharp [1]. It is an *open problem* if this property holds for any Coxeter graph of wild type.

We say that a Cartan matrix A is **bipartite** if it is **symmetrizable** (that is, there exists a diagonal matrix $D = \text{diag}(f_1, \ldots, f_n)$ with f_i positive integers, $1 \leq i \leq n$, such that $A \cdot D$ is symmetric) and the Coxeter graph Δ of A admits an orientation $\vec{\Delta}$ such that every vertex is either a source or a sink.

If A is bipartite, we will always assume that the vertices of Δ are ordered in such a way that $\{1, \ldots, m\}$ are the sinks of $\vec{\Delta}$ and $\{m+1, \ldots, n\}$ are the sources. Let $c = s_1, \ldots, s_n$ be the corresponding Coxeter transformation.

Proposition (**[2, 5, 8]**) *Let A be a bipartite Cartan matrix of indefinite type. Then:*

(a) The Coxeter transformation is very sharp.
(b) There exists a vector $y^+ \gg 0$ *such that* $y^+ \cdot c = \rho(c)y^+$.

6.3.5 Representing matrices for Coxeter elements. Let W be a Coxeter group with generators $s_1, \ldots, s_n$. Let $c = s_1 s_2 \cdots s_n$ be the Coxeter element corresponding to the given presentation; c obviously depends on the ordering of the generators. Let V be the real inner product space with basis $\{e_1, \cdots, e_n\}$ and inner product given by $(e_i, e_i) = 1$; $(e_i, e_j) = -cos\pi/n_{ij}$, for certain pairs $(i, j) \in K \subset \{1, 2, \cdots, n\}^2$ and $(e_i, e_j) = 0$, if $(i, j) \notin K$.

Observe that V is isomorphic to the Euclidean n-space if and only if the Cartan matrix A, with (i, j) entry $2(e_i, e_j)$, is positive definite. The action of W on V is determined by $s_i(e_j) = e_j - 2(e_i, e_j)e_i$, for $1 \leq i, j \leq n$. That is, the matrix representing s_i, relative to the basis $\{e_1, \cdots, e_n\}$ has the form

$$S_i = \begin{pmatrix} \mathbf{Id}_1 & 0 & 0 \\ a_i & -1 & b_i \\ 0 & 0 & \mathbf{Id}_2 \end{pmatrix},$$

where $\mathbf{Id}_1$ is an $(i-1) \times (i-1)$ identity matrix, $\mathbf{Id}_2$ an $(n-i) \times (n-i)$ identity, the noughts stand for zero matrices of appropriate sizes, and a_i, b_i are row vectors such that the j-th entry of $(a_i, 2, b_i)$ is $2(e_i, e_j)$.

Theorem *Let U be the matrix with* (i, j) *entry* U_{ij} *given by*

$$U_{ij} = \begin{cases} 1, & \textit{if } i = j, \\ 0, & \textit{if } i > j, \\ 2(e_i, e_j), & \textit{if } i < j. \end{cases}$$

Then $-U^{-1}U^t$ *is the matrix of c relative to* $\{e_1, e_2, \cdots, e_n\}$.

Proof For each i define L_i, and U_i such that $U^t = \begin{pmatrix} L_i & 0 \\ X_i^t & U_i^t \end{pmatrix}$ where L_i has size $i \times i$. By induction on i it is easy to show that

$$US_1 \cdots S_i = \begin{pmatrix} -L_i & 0 \\ 0 & U_i \end{pmatrix}.$$

Indeed, this is obviously true for $i = 0$. Assuming it for $i - 1$ and replacing i we have

$$\begin{aligned} US_1 \dots S_i &= \begin{pmatrix} -L_{i-1} & 0 \\ 0 & U_{i-1} \end{pmatrix} S_i \\ &= \begin{pmatrix} -L_{i-1} & 0 & 0 \\ 0 & 1 & -b_i \\ 0 & 0 & U_i \end{pmatrix} \begin{pmatrix} I_1 & 0 & 0 \\ a_i & -1 & b_i \\ 0 & 0 & I_2 \end{pmatrix} \\ &= \begin{pmatrix} -L_{i-1} & 0 & 0 \\ a_i & -1 & 0 \\ 0 & 0 & U_i \end{pmatrix}, \end{aligned}$$

so that the result holds for i also. But the case $i = n$ reduces to $UC = -U^t$ (where C is the matrix of c), and the theorem is proved. □

6.3.6 Let $g, h \in \mathrm{G}\ell(E)$ be two very sharp elements. Let y^+ (resp. y^-, z^+, z^-) be an eigenvector of g (resp. g^{-1}, h, h^{-1}) with eigenvalue λ of norm $|\lambda| = \rho(g)$ (resp. $\rho(g^{-1}), \rho(h), \rho(h^{-1})$). Then g and h are said to be *in general position* if y^+ and y^- (resp. z^+ and z^-) are not eigenvectors of h (resp. of g).

Proposition *If $g, h \in G\ell(E)$ are two very sharp elements in general position, there exists a number N such that for all $m \geq N$, the subgroup of $G\ell(E)$ generated by g^m and h^m is free.*

Proof *Sketch of Proof.* Let y^+, y^-, z^+, z^- be as above and assume that these vectors have euclidean norm 1. Consider the images $\overline{g}, \overline{h}$ of g, h in $\mathrm{PG}\ell(E) = \mathrm{G}\ell(E)/\mathbb{R}^*$. Hence $y^+, y^- \in s^n$ are fixed points of $\overline{g}$. Moreover y^+ is an attracting point and y^- is a repulsing point of $\overline{g}$. That is, for any neighborhood U of y^+ in s^n and for any compact subset K of $s^n \setminus \{y^-\}$, we have $\overline{g}^m(K) \subset U$ for any integer m with $|m|$ big enough. By symmetry, we may find disjoint compact sets K and K' with $y^+, y^- \in K$, $z^+, z^- \in K'$ and such that $\overline{g}^m(K') \subset K, \overline{h}^m(K) \subset K'$ for $|m| \geq N$.

Let $m \in \mathbb{Z}$ be such that $|m| \geq N$ and let $g_1 = g^m$, $h_1 = h^m$. We show that the group generated by g_1 and h_1 is free. This is an application of an argument of Klein that de la Harpe names properly "ping-pong criterium". Assume that

$$g_1^{a_1} m \cdot h_1^{b_1} m \cdot g_1^{a_2} m \cdots h_1^{b_{s-1}} m \cdot g_1^{a_s} m = 1.$$

If $a_1 m \neq 0 \neq a_s$, then

$$\overline{g}_1^{a_1} m \cdot \overline{h}_1^{b_1} m \cdots \overline{h}_1^{b_{s-1}} m \cdot \overline{g}_1^{a_s} m(K') \subset \overline{g}_1^{a_1} m \cdot \overline{h}_1^{b_1} m \cdots \overline{h}_1^{b_{s-1}} m(K) \subset \cdots \subset \overline{g}_1^{a_1} m(K') \subset K,$$

a contradiction. If $a_1 m = 0 < a_s$, then $g_1^{-1} m \cdot h_1^{b_1} m \cdots h_1^{b_{s-1}} m \cdots g_1^{a_s} m \cdot g_1 m = 1$ and we are in the first situation. The remaining cases can be treated similarly. □

6.3.7 When Δ is a Dynkin or affine diagram, the structure of the group W is well known. If Δ is a **wild graph** (that is, Δ is neither Dynkin nor affine), then there is a beautiful result of J. Tits [6].

Theorem *If Δ is a wild graph and $n \geq 3$, then W has a non-abelian free subgroup.*

One of the purposes of this work is to present a self-contained elementary proof of Theorem 6.3.7. In certain cases we indicate how to construct the generators of a free non-abelian subgroup of W.

A matrix $g \in G\ell(V)$ is **hyperbolic** if it has two eigenvalues λ, μ with $|\lambda| \neq |\mu|$. We show the following result.

6.3.8 Theorem *Let A be a $n \times n$ Cartan matrix with Coxeter diagram Δ and $n \geq 3$. The following conditions are equivalent:*

(i) Δ is a wild diagram.
(ii) For every faithful complex representation $\sigma : W \to G\ell(V)$, there exists a locally compact field K, a homomorphism $\mathbb{C} \to K$ and an element $w \in W$ such that $\tilde{\sigma}(w)$ is hyperbolic, where $\tilde{\sigma} : W \to G\ell(V) \otimes_{\mathbb{C}} K$ is the induced representation.

A proof of Theorems 6.3.7 and 6.3.8 will be sketched below in 6.4.4 and 6.4.7.

6.4 Very Sharp Reflections

6.4.1 Let A be a symmetrizable Cartan matrix. Let $D = \text{diag}(f_1, \ldots, f_n)$ be a diagonal matrix with f_i positive integers ($i = 1, \ldots, n$) and $A \cdot D$ be symmetric. Let $\vec{\Delta}$ be an oriented graph whose underlying graph is the Coxeter valued graph Δ of A. We assume that the vertices of $\vec{\Delta}$ form a $(+)$-*accessible sequence* (that is, 1 is a sink of $\vec{\Delta}$, 2 is a sink of $\vec{\Delta} \setminus \{1\} \ldots$).

Fix division rings F_i of dimension f_i over a central field k and for $i < j$, fix $F_i - F_j$-bimodules ${}_iM_j$ of dimension $\dim_{F_i}({}_iM_j) = |a_{ji}|$, dim $({}_iM_j)_{F_j} = |a_{ij}|$. Recall that a *representation* X of the valued graph Δ associates to each vertex i a right F_i-module $X(i)$ and to each arrow $\alpha : i \to j$, a F_j-map $X(\alpha) : X(i) \otimes_{F_i} {}_iM_j \to X(j)$. The *dimension vector* of X is $\mathbf{dim}\, X = (\dim\, X(i)_{F_i})_i \in \mathbb{Z}^n$.

For each vertex i, we denote by P_i (resp. I_i) the indecomposable projective (resp. injective) representation associated with i. Then $\mathbf{dim}\, P_i$ has j-th coordinate

$\sum_s |a_{ix_{s1}} a_{x_{s1}x_{x2}} \cdots a_{x_{sm_s}j}|$, with $\{i \to x_{s1} \to x_{s2} \to \cdots \to x_{sm_s} \to j\}_s$ the set of all paths from i to j in $\vec{\Delta}$. Similarly we may describe **dim** I_i.

Let P (resp. I) be the matrix whose i-th row is the vector **dim** P_i (resp. **dim** I_i). Then P is invertible and we may form the matrix $-P^{-1}I$. The following is well known.

6.4.2 Lemma *Let c be the Coxeter transformation corresponding to the symmetrizable Cartan matrix A. Then $c = -P^{-1}I$.*

Proof Check that (**dim** P_i) $\cdot\, c = -$**dim** I_i. □

Proposition *Let A be a bipartite Cartan matrix of indefinite type and with $n > 2$. Let $c \in W$ be the corresponding Coxeter transformation. Then there exists $i \in \{1, \ldots, n\}$ such that c and $s_i c s_i$ are two very sharp elements in general position.*

Proof By 6.3.4, c and $s_i c s_i$ $(i = 1, \ldots, n)$ are very sharp. Let ρ be the spectral radius of c and $y^+, y^- \gg 0$ be eigenvectors of c with $y^+ \cdot c = \rho y^+$ and $y^- \cdot c = \rho^{-1} y^-$ (observe that in this case $\rho(c^{-1}) = \rho(c)$).

We distinguish two cases.

(a) There are at least two sinks in $\vec{\Delta}$.
Assume that y^+ is an eigenvector of $s_1 c s_1$. There exists a number $\mu \in C$ such that $y^+ \cdot s_1 c = \mu y^+ \cdot s_1$. First we remark that the vectors y^+ and $y^+ \cdot s_1$ are different. Otherwise, ρ is an eigenvalue of $c' = s_2 \ldots s_n$. Let $\bar{A}$ be the Cartan matrix obtained by deleting the first column and the first row of A, then the Coxeter transformation $\bar{c}$ of $\bar{A}$ satisfies

$$c' = \begin{pmatrix} 1 & * \\ 0 & \bar{c} \end{pmatrix}.$$

Hence ρ is an eigenvalue of $\bar{c}$. But [5, 8] show that $\rho(\bar{c}) < \rho(c)$, a contradiction. Therefore $y^+ \cdot s_1 = y^+ + \alpha e_1$ with $\alpha \neq 0$. Since $e_1 =$ **dim** P_1, $(\rho - \mu) y^+ = \alpha$**dim** $I_1 + \mu \alpha e_1$.
Thus $\mu \neq 0$. Since $y^+ \gg 0$, I_1 has all its coordinates different from 0. This implies that 1 is the unique sink of $\vec{\Delta}$. A contradiction.
Similar arguments complete the proof in this case.

(b) The vertex 1 is the unique sink of $\vec{\Delta}$. Then Δ is a valued graph of the form

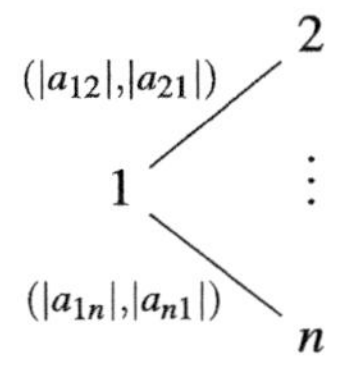

Simple calculations show the following:

$$c = \begin{pmatrix} -1 & a_{12} & \cdots & & a_{1n} \\ -a_{21} & a_{12}a_{21} - 1 & & & \\ \vdots & & \ddots & a_{1i}a_{j1} & \vdots \\ & a_{1i}a_{j1} & & & \\ -a_{n1} & & \cdots & & a_{1n}a_{n1} - 1 \end{pmatrix}$$

$$\rho = \frac{1}{2}(d - 2 + \sqrt{(d-2)^2 - 4}), \quad \text{where } d = \sum a_{1i}a_{i1}$$

and

$$y^+ = \rho/(\rho + 1)(1 + 1/\rho, -a_{12}, \ldots, -a_{1n}).$$

We show that c, s_2cs_2 are in general position. Indeed, $y^+ \cdot s_2 = y^+ + \alpha e_2$, where $\alpha = (\rho - 1)a_{12}/(1 + \rho)$. If y^+ is an eigenvector of s_2cs_2, then $y^+ \cdot s_2c = \mu y^+ \cdot s_2$ for some $\mu \in C$. Hence $(\rho - \mu)y^+ = -\alpha e_2 \cdot c + \alpha\mu e_2$, where $e_2 \cdot c = (-a_{21}, a_{12}a_{21} - 1, a_{13}a_{31}, \ldots, a_{1n}a_{n1})$. Evaluating in the first two coordinates, we get: $a_{12}a_{21} = (1 + \rho)^2/\rho = d$, which implies $n = 2$, against the hypothesis. Similar arguments complete the proof.

□

6.4.3 A Cartan matrix A is said to be **minimal of indefinite type** if A is of indefinite type and every restriction A_i of A, obtained by deleting the i-th column and the i-th row of A, is of finite or affine type.

Lemma *Let A be a Cartan matrix minimal of indefinite type with $n > 3$. Then there is a bijection $\sigma : \{1, \ldots, n\} \to \{1, \ldots, n\}$ such that the Coxeter elements $c = s_{\sigma(1)} \ldots s_{\sigma(n)}$ and s_1cs_1 are very sharp and in general position.*

Proof If A is bipartite the result follows from 6.4.2. It is enough to check the claim for the non bipartite diagrams appearing in the list [7], a task that can be done with the help of a computer. □

6.4.4 Proof of Theorem 6.3.8: Let A be an indefinite $n \times n$ Cartan matrix with $n \geq 3$. If A is bipartite, by 6.3.6, 6.4.2, the Weyl group W of A has a free non-abelian subgroup. If A is non bipartite, the Coxeter graph Δ of A has an induced valued subgraph Δ' such that the corresponding $m \times m$ Cartan matrix A' is of indefinite type, $m \geq 3$ and m is minimal with these properties. Let W' be the Weyl group of A', then W' is a subgroup of W. It is enough to show the result for A'.

It A' is minimal of indefinite type, by 6.4.3, 6.3.6 we get the result. Otherwise, A' has one of the following shapes

1 —— 2 —— 3 or 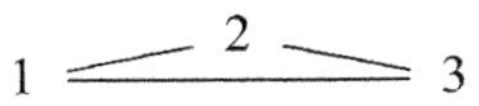

with $a_{12}a_{21} > 4$. Then W' has a quotient isomorphic to a Weyl group $\bar{W}$ of a 3×3 Cartan matrix $\bar{A}$ minimal of indefinite type. Since $\bar{W}$ has a free non-abelian subgroup, therefore W' has also one.

Of course, 6.4.3 can be shown for all minimal indefinite Cartan matrices by a direct calculation. Nevertheless, when it applies 6.4.2 yields a more explicit construction.

6.4.5 Representation of Weyl groups. Let W be a group and k be a field. A **representation** $\sigma : W \to G\ell(V)$, where V is a k-vector space, is called **completely irreducible** if for every subgroup of finite index W_1 of W the restriction $\sigma : W_1 \to G\ell(V)$ is irreducible.

Let A be a $n \times n$ Cartan matrix. Let $E = \mathbb{C}^n$ and $(e_i)_i$ be the canonical basis. Denote by $B \colon E \times E \to \mathbb{C}$ the symmetric bilinear form given by

$$B(e_i, e_j) = \begin{cases} -\cos(\pi/m_{ij}), & \text{if } a_{ij}a_{ji} < 4, \\ -1, & \text{if } a_{ij}a_{ji} \geq 4. \end{cases}$$

Let $\sigma_0 \colon W \to G\ell(E)$ be the canonical inclusion. If B is not degenerated, then σ_0 is irreducible. If B is degenerated, set $\bar{E} = E/E^0$, where E^0 is the radical of B. Then the induced representation $\bar{\sigma}_0 \colon W \to G\ell(\bar{E})$ is irreducible.

Proposition ([4])

(a) If A is not of finite type and B is not degenerated, then σ_0 is completely irreducible.
(b) If A is not of affine type and B is degenerated, then $\bar{\sigma}_0$ is completely irreducible.

6.4.6 If A is a Cartan matrix of indefinite type, we know 6.3.3 that the Coxeter element $c \in W$ is hyperbolic. The proof of Theorem 6.3.7 as given in [4] consists in using the hyperbolic element c to construct two very sharp elements in general position.

Proposition ([4, 6]) *Let A be a $n \times n$ Cartan matrix with $n \geq 3$ and let W be its Weyl group. Assume that $w \in W$ is such that $\sigma_0(w) \in G\ell(E)$ is a hyperbolic matrix. Then W has a non-abelian free subgroup.*

Sketch of Proof Assume that B is not degenerated (the other case is similar).

(a) There is a subgroup of finite index W_1 of W such that $\sigma_0(W_1)$ is contained in $S\ell(E)$. □

Since $\sigma_0(w)$ is hyperbolic, it has infinite order, hence there is some number k such that $w^k \in W_1$. Moreover, $\sigma_0(w^k)$ is hyperbolic. Let $\mu_1, \ldots, \mu_n$ be the set of eigenvalues (with multiplicities) of $\sigma_0(w^k)$. Assume $\|\mu_1\| = \|\mu_2\| = \cdots = \|\mu_m\| > \|\mu_{m+1}\| \geq \cdots \geq \|\mu_n\|$. Consider the induced representation $\tilde{\sigma}: W_1 \to S\ell(\Lambda^m E)$. Then $\tilde{\sigma}(w^k)$ is sharp with spectral radius $\tilde{\rho} = \|\mu_1\|^m$. Moreover, the Zariski closure of $\sigma_0(W_1)$ in $S\ell(E)$ is semisimple. Therefore $\tilde{\sigma}$ is completely reducible. Let $\Lambda^m E = \oplus_1^s F_i$ be a decomposition in irreducible W_1-spaces. An eigenvector of $\tilde{\sigma}(w^k)$ with eigenvalue $\tilde{\rho}$ belong to some F_j. Hence $\tilde{\sigma}_j: W_1 \to S\ell(F_j)$ is an irreducible representation such that $\tilde{\sigma}_j(w^k)$ is sharp.

(b) Let $\sigma: W_1 \to G\ell(F)$ be an irreducible representation such that the Zariski closure of $\sigma(W_1)$ is connected. Let $w_1 \in W_1$ be such that $\sigma(w_1)$ is sharp. Arguments of Tits [6] or Dixon [4] show that there exist elements $x, y \in W_1$ such that $\sigma(x \cdot w_1^s \cdot y \cdot w_1^{-s})$ is very sharp for some $s > 0$.

(c) Let $w_2 \in W_1$ be an element such that $\sigma(w_2)$ is very sharp. An easy topological argument shows that there is some $g \in W_1$ such that $\sigma(w_2)$ and $\sigma(g \cdot w_2 \cdot g^{-1})$ are very sharp in general position. The result then follows from 6.3.6.

6.4.7 Proof of Theorem 6.3.8: Assume that (ii) is satisfied. Then 6.4.6 implies that A is of indefinite type. Alternatively, if A is of finite or affine type, it is easy to see that for any $w \in W$, the eigenvalues of $\sigma_0(w)$ are roots of unity.

Assume that (i) holds. Let $\sigma: W \to G\ell(V)$ be a complex faithful representation. Suppose that $\sigma(w)$ is not hyperbolic for all $w \in W$.

Since $\det \sigma(w) = \pm 1$ then $\|\lambda\| = \pm 1$ for any

$$\lambda \in \mathrm{Spec}\sigma(w) = \{\lambda \in \mathbb{C}: \lambda \text{ eigenvalue of } \sigma(w)\}.$$

(a) Let U be the set of roots of unity in $\mathbb{C}$. Assume that $\mathrm{Spec}\sigma(w) \subset U$ for every $w \in W$.

Let $0 = V_0 \subset V_1 \subset \cdots \subset V_s = V$ be a Jordan-Holder sequence of W-modules that is, the induced representations $\sigma_i: W \to G\ell(V_i/V_{i-1})$ are irreducible, $i = 1, \ldots, s$.

We claim that for all $1 \leq i \leq s$, $w \in W$, we have $\mathrm{Spec}\sigma_i(w) \subset U$ and $\sigma_i(W)$ is a finite group. By induction on i. For $i = 1$, $\sigma_1(w)$ acts as w on V_1, hence $\mathrm{Spec}\sigma_1(w) \subset U$ and by Theorem 6.4.8 below, $\sigma_1(W)$ is a finite group, say of order m_1. Let $w \in W$ and suppose $\mu \in \mathrm{Spec}\sigma_2(w)$. Let $0 \neq v_frm - e \in V_2$, $v_1 \in V_1$ be such that $v_frm - e \cdot w = \mu v_2 + v_1$. Then $(v_2 \cdot w^{m_1} - v_2) \cdot w = \mu(v_2 \cdot w^{m_1} - v_2)$. Hence either $\mu \in \mathrm{Spec}\sigma(w) \subset U$ or $v_2 = v_2 \cdot w^{m_1} = \mu^{m_1} v_2 + (\mu^{m_1 - 1m} v_1 + \mu^{m_1 - 2m} v_1 \cdot w + \cdots + v_1 \cdot w^{m_1 - 1m})$. In the last case, passing to V_1/V_2 we get $\mu^{m_1} = 1$. Again 6.4.8 implies that $\sigma_2(W)$ is a finite group. The induction can be continued in this way.

On the other hand, the representation $\bar{\sigma} = (\sigma_i)_i: W \to \Pi_i G\ell(V_i/V_{i-1})$ has nilpotent kernel (of index $\leq s$). Then W has a nilpotent subgroup of finite index. By Theorem 6.3.7, A is of finite or affine type a contradiction.

(b) Assume that there is some $w \in W$ and some $\lambda \in \mathrm{Spec}\sigma(w)$ such that λ is not a root of unity. By [6],4.1, there is a locally compact field k endowed with an absolute value ω and a homomorphism $\alpha : \mathbb{C} \to k$ such that $\omega(\alpha(\lambda))) \neq 1$.

The induced representation $\tilde{\sigma} : W \to G\ell(V \otimes_{\mathbb{C}} k)$ has an eigenvalue $\alpha(\lambda)$. Since $\det \sigma(w) = \pm 1$, then $\tilde{\sigma}(w)$ is hyperbolic.

6.4.8 Theorem (Schur's First Theorem) *Let $\sigma : W \to G\ell(V)$ be an irreducible complex representation. If for every $w \in W$, $\mathrm{Spec}\sigma(w)$ is formed by roots of unity, then $\sigma(W)$ is finite.*

Proof Let $n = \dim V$. If $\lambda \in \mathrm{Spec}\sigma(w)$, for some $w \in W$, then λ is a root of unity and a root of a polynomial of degree n. There are only finitely many numbers with this property. Hence the traces of $\sigma(w)$, for $w \in W$ take values in a finite set $\{\alpha_1, \ldots, \alpha_t\} \subset \mathbb{C}$. By an argument of Burnside, $\sigma(W)$ is finite of order at most $t^{n^f rm-e}$. □

6.5 On the Decomposition of the Coxeter Polynomial of an Algebra of Cyclotomic Type

Let $A = KQ/I$ be a finite dimensional triangular K-algebra. Consider the Cartan matrix C_A and the Coxeter matrix $\Phi_Q = -C_A^{-t} C_A$. Let $\varphi_A(T) = \det(T\mathbf{Id} - \Phi_A)$ be the Coxeter polynomial of A. The roots of $\varphi(T)$ form the set $\mathrm{Spec}\Phi_A$ of eigenvalues of Φ_A.

6.5.1 Theorem (Theorem 1.1 in [9]) *The following conditions are equivalent for an algebra A.*

(1) Φ_A is a periodic matrix.
(2) A is of cyclotomic type, and Φ_A is diagonalizable.
(3) All eigenvalues of Φ_A are roots of unity, and Φ_A is diagonalizable.

6.5.2 Let A be an algebra with n vertices whose Coxeter polynomial decomposes as $\varphi_A = \prod_{m \in M} F_m(T)^{e(m)}$ for some subset M of $\mathbb{N}$ and integers $e(m) \geq 1$. Several conditions on the coefficients of $\varphi_A = \sum_{i=0}^{n} a_i T^i$ and of the polynomials F_m follow, for instance:

(1) $\sum_{m \in M} e(m)\phi(m) = n$, which follows from the degree calculation, where $\phi(-)$ is *Euler's totient function*;
(2) if $1 \in M$ then $e(1)$ is an even number, which follows from the fact that F_m for $m \neq 1$, as well as $\varphi_A(T)$, are self-reciprocal polynomials, but $F_1(T) = T - 1$ is not so, while $F_1(T)^2 = T^2 - 2t + 1$ is self-reciprocal;

(3) $\sum_{m\in M} e(m)\mu(m) = -a_1$, which follows from the fact that the linear coefficient of $F_m(T)$ is

$$a_1(F_m(T)) = -\mu(m).$$

Use that $a_0(F_m(T)) = 1$ for $m \neq 1$ and $a_0(F_1(T)^{e(1)}) = 1$.

6.5.3 Theorem *Let A be an algebra whose Coxeter polynomial factorizes as $\varphi_A(T) = \prod_{m\in M} F_m(T)^{e(m)}$ for some subset M of $\mathbb{N}$ and integers $e(m) \geq 1$. Set $\epsilon(m) = 0$ if $m \in M$ and $\epsilon(m) = 1$ if $m \notin M$. For any prime number p, let $M(p)$ (resp. $M'(p)$) be the set of those $m \in M$ of the form $m = p^s$ (resp. $m = 2p^s$) for some $1 \leq s$. Let $f(p) = \sum_{p^s=m\in M(p)} e(m)$ and $f'(p) = \sum_{2p^s=m\in M'(p)} e(m)$. Then*

(a) $\varphi_A(1) = \epsilon(1)\prod_{m\in M(p)} p^{e(m)} = \epsilon(1)\prod_{p\ prime} p^{f(p)}$.
(b) $\varphi_A(-1) = \epsilon(2)\,2^{e(1)}\prod_{m\in M'(p)} p^{e(m)} = \epsilon(2)\,2^{e(1)}\prod_{p\ prime} p^{f'(p)}$.

Moreover, if this number is not zero then, for each prime $2 \leq p$, the number $f'(p)$ is even.

Proof

(a) According to [10], the evaluation $F_m(1) = 1$ if $m \neq 1$ and $m \neq p^s$, for all prime p. Otherwise, $F_1(1) = 0$ or $F_{p^s}(1) = p$. Then the evaluation

$$\varphi_A(1) = \prod_{m\in M} F_m(1)^{e(m)} = \epsilon(1) \prod_{m\in M(p)} p^{e(m)}.$$

(b) According to [10], the evaluation $F_m(-1)$ yields $F_m(-1) = -2$ if $m = 1$, $F_m(-1) = 0$ if $m = 2$, $F_m(-1) = p$ if $m = 2p^s$ with p a prime number and $s \geq 1$ and $F_m(-1) = 1$ otherwise. Therefore

$$\varphi_A(-1) = \prod_{m\in M} F_m(-1)^{e(m)} = \epsilon(2)\,2^{e(1)} \prod_{m\in M'(p)} p^{e(m)}.$$

Assume $\varphi_A(-1) = r^2 > 0$ for some integer r, then for any prime $2 < p$

$$\prod_{m\in M'(p)} p^{e(m)} = p^{\sum_{m\in M'(p)} e(m)},$$

is an even power of p. For $p = 2$, use additionally that $e(1)$ is an even number. The claim follows.

□

References

1. A'Campo, N. *Sur les valeurs propres de la transformation de Coxeter.* Inventiones Math. **33**, 61–67 (1976)
2. Anderson, F. W. and Fuller, K. R., *Rings and categories of modules*. Graduate texts in mathematics **13**, Springer-Verlag, New York 1974
3. Assem, I. and Skowroński, A.: *Indecomposable modules over multicoil algebras*. Math. Scand. **71** (1992) 31–61.
4. I. Assem and A. Skowroński, *On tame repetitive algebras*, Fund. Math. **142** (1993), 59–84.
5. M. Auslander, *The representation dimension of artin algebras,* Queen Mary College Mathematics Notes (1971)
6. Auslander M. (1982) *A functorial approach to representation theory*. In: Auslander M., Lluis E. (eds) Representations of Algebras. Lecture Notes in Mathematics, vol 944. Springer, Berlin, Heidelberg. https://doi.org/10.1007/BFb0094058
7. Auslander, M. and Reiten, I. *Representation theory of Artin algebras. III. Almost split sequences*, Communications in Algebra **3** (3) (1975): 239–294
8. Auslander, M. and Reiten, I. and Smalø, S. *Representation theory of Artin algebras*, Cambridge University Press **36** Cambridge Studies in Advanced Mathematics (1995)
9. de la Peña, J.A. *Periodic Coxeter matrices*, Linear Algebra Appl. 365 (2003) 135–142.
10. V. Prasolov, Polynomials. Algorithms and computation in Mathematics vol. 11. Springer Verlag, Berlin, 2001 (second edition).

Chapter 7
Simply Connected Algebras

7.1 The Fundamental Group of a Triangular Algebra

Let K be an algebraically closed field, and A a finite dimensional associative K-algebra with identity, which we assume to be basic.

7.1.1 We recall that a **quiver** Q is defined by its set of vertices Q_0 and its set of arrows Q_1. A **relation** from vertices x and y is a linear combination $\rho = \sum_{i=1}^{m} \lambda_i w_i$ where, for each $1 \leq i \leq m$, λ_i is a non-zero scalar and w_i is a path of length at least two from x to y. A set of relations on Q generates an ideal I, said to be **admissible** in the path algebra KQ of Q. The pair (Q, I) is then called a **bound quiver**. For an algebra A, we denote by Q_A the ordinary quiver of A. It is well-known that, for every basic algebra A, there exists a surjective K algebra morphism $\nu : KQ_A \to A$ (induced by the choice of a set of representatives of basis vectors in the K-vector space $\mathrm{rad}A/\mathrm{rad}^2A$) whose kernel I_ν is admissible. We thus have $A \cong KQ_A/I_\nu$, and the bound quiver (Q_A, I_ν) is called a **presentation** of A. An algebra $A = KQ/I$ can equivalently be considered as a K-linear category, of which the object class is the set Q_0, and the set of morphisms from x to y is the K-vector space $KQ(x, y)$ of all linear combinations of paths from x to y module the subspace $I(x, y) = I \cap KQ(x, y)$. A full subcategory B of A is called **convex** if any path in A with source and target in B lies entirely in B. An algebra A is called **triangular** if Q_A has no oriented cycles. This section is devoted exclusively to triangular algebras.

By an A-module is meant a finitely generated left A-module. We denote by A-mod their category. It is well known that, if $A = KQ/I$, then A-mod is equivalent to the category of all bound representations of (Q, I), we may thus identify a module M with the corresponding representation $(M(x), M(\alpha))_{x \in Q_0, \alpha \in Q_1}$. For each $x \in Q_0$, we denote by S_x the corresponding simple A-module, and by P_x the projective cover of S_x. The algebra A is called **schurian** if $\dim_k \mathrm{Hom}_A(P_x, P_y) \leq 1$ for all $x, y \in Q_0$.

J.-A. de la Peña, *Representations of Algebras*, Algebra and Applications 30,
https://doi.org/10.1007/978-3-031-12288-0_7

7.1.2 The fundamental group. Let (Q, I) be a connected bound quiver. A relation $\rho = \sum_{i=1}^{m} \lambda_i w_i \in I(x, y)$ is called **minimal** if $m \geq 2$ and, for every non-empty proper subset $J \subset \{1, \ldots, m\}$, we have $\sum_{j \in J}^{m} \lambda_j w_j \notin I(x, y)$. For an arrow $\alpha \in Q_1$ we denote by α^{-1} its formal inverse. A **walk** in Q from x to y is a formal composition $\alpha_1^{\epsilon_1} \cdots \alpha_k^{\epsilon_k}$ where $\alpha_i \in Q_1$ and $\epsilon_i = \pm 1$ for $1 \leq i \leq k$, starting at x and ending at y. We denote by e_x the **trivial path** at x.

Let $\sim$ be the smallest equivalence relation on the set of all walks in Q such that:

(a) If $\alpha : x \to y$ is an arrow, then $\alpha^{-1}\alpha \sim e_x$ and $\alpha\alpha^{-1} \sim e_y$.
(b) If $\rho = \sum_{i=1}^{m} \lambda_i w_i$ is a minimal relation, then $w_i \sim w_j$ for all $1 \leq i, j \leq m$.
(c) If $u \sim v$, then $wuw' \sim wvw'$ whenever these compositions make sense.

We denote by $[u]$ the equivalence class of a walk u.

Let $x_0 \in Q_0$ be arbitrary. The set $\pi_1(Q, I, x_0)$ of equivalence classes of all the closed walks starting and ending at x_0 has a group structure defined by the operation $[u][v] = [vu]$. Clearly, the group $\pi_1(Q, I, x_0)$ is independent of the choice of base vertex x_0. We denote it simply by $\pi_1(Q, I)$ and call it the **fundamental group** of (Q, I).

7.1.3 Simple connectedness. Let A be a triangular algebra, and let (Q_A, I_ν) be a presentation of A. It follows from the above description that the fundamental group $\pi_1(Q_A, I_\nu)$ depends essentially on I_ν, thus it is not an invariant of A. For example, consider $A = KQ/I$ where Q is the quiver

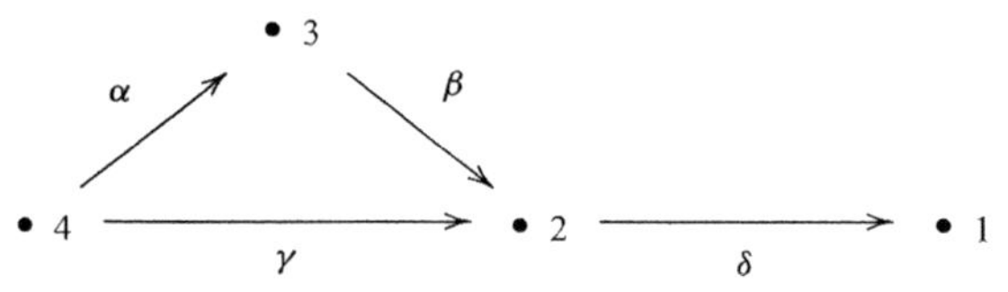

and I is generated by $\delta\beta\alpha - \delta\gamma$. Then $\pi_1(Q, I) = 0$. The choice of the presentation may be modified as follows: take I' generated by $\delta\gamma$, and notice that $A \cong KQ/I'$. Then $0 \neq [\gamma^{-1}\beta\alpha] \in \pi_1(Q, I')$ and $\pi_1(Q, I') \cong \mathbb{Z}$.

Definition Let A be a connected basic triangular algebra.

(a) We say that A is **simply connected** if, for any presentation (Q_A, I_ν) of A, the fundamental group $\pi_1(Q_A, I_\nu)$ is trivial.
(b) We say that A is **freely connected** if, for any presentation (Q_A, I_ν) of A, the fundamental group $\pi_1(Q_A, I_\nu)$ is free.

For instance, any tree algebra (that is, an algebra A whose quiver Q_A is a tree) is simply connected. Any algebra satisfying the separation condition of [4] is simply connected, as are the good algebras of [17] and the completely separating algebras of [9]. Moreover, any **monomial algebra** (that is, an algebra A having a presentation (Q_A, I_ν) where I_ν is generated by finitely many paths in Q_A) is freely connected, as are the representation-finite algebras (cf. [14, (4.3) and (4.4)]).

7.1.4 The fundamental group of a one-point extension. Let A be an algebra, and x be a **source** in Q_A. The full convex subcategory B of A consisting of all objects except x, has as quiver Q_B the quiver obtained from Q_A by deleting x and all arrows starting with x. Any presentation (Q_A, I_ν) yields (by restriction) an induced presentation $(Q_B, I_{\nu'})$ of B, and clearly, all presentations of B are obtained in this way. The A-module $M = \mathrm{rad}(P_x)$ has a canonical B-module structure, and A is isomorphic to the **one-point extension algebra**

$$B[M] = \begin{bmatrix} K & M \\ 0 & B \end{bmatrix},$$

where the operations are the usual addition of matrices, and the multiplication induced by the B-module structure of M.

Let (Q_A, I_ν) be a presentation of A, and let $\approx$ denote the smallest equivalence relation on the set $x^{\rightarrow}$ of all arrows starting at x such that $\alpha \approx \beta$ (for $\alpha, \beta \in x^{\rightarrow}$) whenever there exists $y \in (Q_A)_0$ and a minimal relation $\sum_{i=1}^{m} \lambda_i w_i \in I_\nu(x, y)$ with $w_1 = v_1\alpha$ and $w_2 = v_2\beta$. We denote by $[\alpha]_\nu$ the equivalence class of $\alpha \in x^{\rightarrow}$.

7.1.5 Let A be a connected algebra, and x be a source in Q_A. Let $c(x)$ denote the number of connected components of $Q_A - \{x\}$. Given a presentation (Q_A, I_ν) of A, denote by $t(x, \nu)$ the number of equivalence classes $[\alpha]_\nu$ in $x^{\rightarrow}$. Finally, we denote by $t(x)$ the number of indecomposable direct summands of $\mathrm{rad} P_x$.

Corollary (Proposition 2.1, [1]) *Let A be a triangular algebra, and x be a source in Q_A.*

(a) For any presentation (Q_A, I_ν) of A, we have $c(x) \le t(x, \nu) \le t(x)$.
(b) There exists a presentation (Q_A, I_μ) of Q such that $t(x, \mu) = t(x)$.

7.1.6 Separating vertices. Following [4], we say that a source x in Q_A is a **separating source** if $c(x) = t(x)$. In general, a vertex y in Q_A (not necessarily a source) is called a **separating vertex** if y is separating as a source in the full convex subcategory of A with objects all vertices of Q_A except the vertices z such that there exists a non-trivial path from z to y in Q_A. The algebra A is called **separated** if all vertices in Q_A are separating. There are close relations between the separation property and simple connectedness. For instance, a representation-finite algebra is separated if and only if it is simply connected (see [4]). On the other hand, any separated algebra is simply connected.

Lemma (Lemma 2.5, [1]) *Let A be a triangular algebra.*

(a) Let $A = B[M]$, where $B = \mathrm{rad} P_x$. If B is simply connected and x is separating, then A is simply connected.
(b) If A is simply connected, then all sources in Q_A are separating.

7.1.7 Abelianization. Let $P_1(Q_A, I_\nu)$ be the abelianization of the group $\pi_1(Q_A, I_\nu)$ (that is, the quotient of $\pi(Q_A, I_\nu)$ by its commutator subgroup). For any abelian group Z, we have a functorial isomorphism (the universal property of abelianizations),

$$\mathrm{Hom}(\pi_1(Q_A, I_\nu), Z) \cong \mathrm{Hom}(P_1(Q_A, I_\nu), Z).$$

Moreover, let $A = B[M]$, where $M = \mathrm{rad}P_x$, be a one-point extension. Consider $(Q_B, I_{\nu'})$ the induced presentation of B and $Q^{(1)}, \ldots, Q^{(m)}$ be the connected components of Q_B. Then there is a morphism of abelian groups,

$$\bar{c} : \prod_{j=1}^{m} P_1(Q^{(j)}, I_{\nu(j)}) \longrightarrow P_1(Q_A, I_\nu),$$

induced from the morphisms c_j. One can easily compute the cokernel of $\bar{c}$. Indeed, we have an exact sequence of abelian groups (cf. [1, (2.4)]).

$$0 \longrightarrow Z^{t(x,\nu)-c(x)} \longrightarrow \mathrm{Hom}(P_1(Q_A, I_\nu)) \xrightarrow{\mathrm{Hom}(\bar{c},Z)} \prod_{j=1}^{m} \mathrm{Hom}(P_1(Q^{(j)}, I_{\nu(j)}), Z),$$

for any abelian group Z. Consequently, there exists an exact sequence of abelian groups

$$\prod_{j=1}^{m} P_1(Q^{(j)}, I_{\nu(j)}) \xrightarrow{\bar{c}} P_1(Q_A, I_\nu) \longrightarrow Z^{t(x,\nu)-c(x)} \longrightarrow 0,$$

that is, the cokernel of $\bar{c}$ is isomorphic to $Z^{t(x,\nu)-c(x)}$.

Corollary *The following are equivalent for a freely connected algebra A.*

(a) A is simply connected.
(b) For any presentation (Q_A, I_ν) of A, we have $P_1(Q_A, I_\nu) = 0$.
(c) There is a non-trivial abelian group Z such that, for any presentation (Q_A, I_ν) of A, we have $\mathrm{Hom}(P_1(Q_A, I_\nu), Z) = 0$.

7.1.8 Given an algebra A and the bimodule ${}_AA_A$, the **Hochschild complex** $C = (C^i, d^i)_i \in \mathbb{Z}$ is defined as follows: $C^i = 0$, $d^i = 0$ for $i < 0$; $C^0 = {}_AA_A$, $C^i = \mathrm{Hom}_K(A^{\otimes i}, A)$ for $i > 0$, where $A^{\otimes i}$ denotes the i-fold tensor product $A \otimes_K \cdots \otimes_K A$, $d^0 : A \to \mathrm{Hom}_k(A, A)$ with $(d^0x)(a) = ax - xa$ (for $a, x \in A$) and $d^i : C^i \to C^{i+1}$ with

$$\begin{aligned}(d^i f)(a_1 \otimes \cdots \otimes a_{i+1}) &= a_1 f(a_2 \otimes \cdots \otimes a_{i+1}) \\ &\quad + \sum_{j=1}^{i} (-1)^j f(a_1 \otimes \cdots \otimes a_j a_{j+1} \otimes \cdots \otimes a_{i+1}) \\ &\quad + (-1)^{i+1} f(a_1 \otimes \cdots \otimes a_i) a_{i+1},\end{aligned}$$

for $f \in C^i$ and $a_1, \ldots, a_{i+1} \in A$. Then $H^i(A) := H^i(C)$ is the i-th **Hochschild cohomology group** of A (with coefficients in the bimodule ${}_AA_A$), see [7]. The following theorem of Happel [15, (5.3)] is useful for the calculation of the Hochschild cohomology groups of triangular algebras:

Theorem *Let $A = B[M]$. Then there exists a long exact sequence*

$$0 \to H^0(A) \to H^0(B) \to \mathrm{Hom}_B(M,M)/K \to H^1(A) \xrightarrow{e} H^1(B) \to \mathrm{Ext}^1_B(M,M) \to \cdots$$

$$\cdots \to \mathrm{Ext}^i_B(M,M) \to H^{i+1}(A) \to H^{i+1}(B) \to \mathrm{Ext}^{i+1}_B(M,M) \to \cdots$$

7.1.9 It follows from [11, 22] that there is a close relation between the first Hochschild cohomology group $H^1(A)$ and the fundamental groups $\pi_1(Q_A, I_\nu)$ of A. The following result makes the relation explicit:

Theorem (Theorem 3.2, [1]) *Let A be a triangular algebra, and (Q_A, I_ν) be a presentation of A. There exists an injective group morphism*

$$s : \mathrm{Hom}(\pi_1(Q_A, I_\nu), K^+) \to H^1(A),$$

where K^+ denotes the underlying additive group of the field K.

As consequence we obtain the following criteria for simple connectedness.

Corollary *Let A be a freely connected algebra with $H^1(A) = 0$. Then A is simply connected.*

7.1.10 The converse of this corollary is not true. We now give a procedure for constructing examples of simply connected algebras A such that $H^1(A) \neq 0$. Let B be a simply connected triangular algebra with $H^1(B) = 0$, and M be an indecomposable B-module such that $\mathrm{End}_B(M) \neq K$. Then $A = B[M]$ is simply connected and $H^1(A) \neq 0$. Indeed, the exact sequence

$$0 \to H^0(A) \to H^0(B) \to \mathrm{End}_B(M)/K \to H^1(A) \to H^1(B) \to 0,$$

already shows that $H^1(A) \neq 0$. On the other hand, since M is indecomposable, the extension vertex x such that $M = \mathrm{rad} P_x$ is separating. Then Lemma 7.1.6(a) implies that A is simply connected.

An explicit version of the example is the algebra $A = KQ/I$, where Q is the quiver

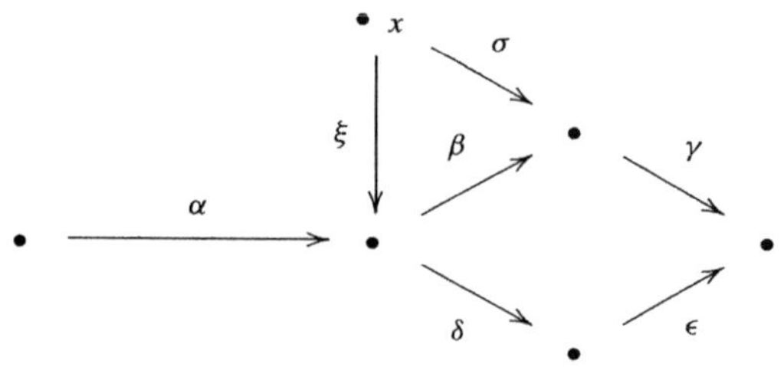

and I is generated by $\gamma\beta\alpha - \epsilon\delta\alpha$, $\gamma\sigma - \epsilon\delta\xi$, $\gamma\beta\xi - \gamma\sigma$. Taking $M = \mathrm{rad} P_x$, we get $A = B[M]$, with B simply connected and $H^1(B) = 0$. Moreover, $\dim_K \mathrm{End}_B(M) = 2$.

Corollary *Let A be triangular algebra with $H^1(A) = 0$. Then all sources of Q_A are separating.*

7.1.11 Strongly simply connected algebras. Recall that, dually, if x is a **sink** in Q_A, then A is isomorphic to the **one-point coextension** $[M]B$ of a full convex subcategory B by a B-module M.

Following [22], we say that a connected triangular algebra is **strongly simply connected** if every full convex subcategory of A is simply connected. We recall the following (cf. [26, (4.1)]):

Theorem *The following are equivalent for a connected triangular algebra A,*

(a) A is strongly simply connected.
(b) Every full convex subcategory of A is separated.
(c) For every full convex subcategory C of A, we have $H^1(C) = 0$.

7.1.12 We now provide a simple method allowing to construct all strongly simply connected algebras.

Proposition *A triangular algebra A is strongly simply connected if and only if there exists a sequence of connected algebras $A = A_0, A_1, \ldots, A_s = K$ and of indecomposable modules ${}_{A_1}M_1, \ldots, {}_{A_s}M_s$ such that either $A_{i-1} = A_i[M_i]$ or $A_{i-1} = [M_i]A_i$ for all $1 \le i \le s$.*

7.2 A Separation Property

7.2.1 Let K be an algebraically closed field and A be a finite dimensional K-algebra. We denote by mod-A the category of finitely generated right A-modules.

The algebra A is said to be **tame** if for each dimension, the indecomposable modules occur in a finite number of discrete and one-parameter families. For tame algebras, a description of some components of the Auslander-Reiten quiver Γ_A of A is known. For instance, in every dimension, almost every indecomposable module lies in a homogeneous tube of Γ_A [15]; if A is a tubular extension of a tame concealed algebra, then almost every component of Γ_A is a ray-inserted or coray-inserted tube [20]; in more generality, for certain polynomial growth algebras, every component of Γ_A is a multicoil [2, 10].

The notion of admissible operations in components of Γ_A was introduced in [2]. There, a **coil** is defined as a translation quiver that is obtained from a stable tube by a sequence of admissible operations. We say that a family $\mathscr{T} = (\mathscr{C}_i)_{i \in I}$ of components of Γ_A is a **weakly separating family of coils** in mod-A if the remaining indecomposable A-modules split into two classes $\mathscr{P}$ and $\mathscr{I}$ satisfying:

(i) The components $\mathscr{C}_i$, $i \in I$, are standard, pairwise orthogonal coils.
(ii) $\mathrm{Hom}_A(\mathscr{I}, \mathscr{P}) = \mathrm{Hom}_A(\mathscr{I}, \mathscr{T}) = \mathrm{Hom}_A(\mathscr{T}, \mathscr{P}) = 0$.
(iii) Any morphism from $\mathscr{P}$ to $\mathscr{I}$ factors through the additive closure $\mathscr{T}$.
Here we will only consider weakly separating families of coils satisfying:
(iv) Any indecomposable projective (resp. injective) module lies in $\mathscr{P}$ or $\mathscr{T}$ (resp. $\mathscr{T}$ or $\mathscr{I}$).

7.2.2 There are important examples of tame algebras A with a weakly separating family of coils in mod-A: tame concealed algebras and tubular extensions of tame concealed algebras [20]; more generally, the tame coil enlargements of tame concealed algebras [3]. Our purpose here is to show the following characterization.

Theorem *Let A be a strongly simply connected tame algebra. Assume that* mod-A *has a weakly separating family of coils. Then A is a coil enlargement of a tame concealed algebra.*

This work is closely related to results on separating tubular families in [6]. The last version of [6] implies some of the results at the end of the chapter, although our proofs for the tame situation are simpler. The chapter is organized as follows: in Sect. 7.3 we recall the definitions and constructions needed in the statement and proof of the theorem. In Sect. 7.4 we consider an important special case of the theorem. We recall from [12] that an algebra A is **quasi-tilted** if gl.dim $A \leq 2$ and every indecomposable A module X has either $\mathrm{p.dim}_A X \leq 1$ or $\mathrm{i.dim}_A X \leq 1$. We show the following:

7.2.3 Theorem *Let A be a strongly simply connected quasi-tilted algebra. If A is tame, then A is of polynomial growth.*

In Sect. 7.5 we give the proof of Theorem 7.2.2, which relies on recent results on polynomial growth tame algebras [18, 21, 23].

7.3 Strongly Simply Connected Algebras

7.3.1 Let us start with some preliminary results. Let A be a basic and connected finite dimensional K-algebra. Thus, A can be written as $A = KQ/I$, where Q is the quiver of A and I is an admissible ideal of the path algebra KQ of Q. We say that A is **triangular** whenever Q has no oriented cycle. We shall consider A as a K-category whose set of objects is the set of vertices Q_0 of Q, and the set of morphisms from i to j is the vector space $KQ(i, j)$ of all linear combinations of paths from i to j in Q modulo the subspace $I(i, j) = I \cap KQ(i, j)$. A full subcategory B of A is **convex** in A if any path with source and sink in B lies completely in B. See [5, 20].

For each vertex $i \in Q_0$, we denote by S_i the corresponding simple A-module, and by P_i (resp. I_i) we denote the projective cover (resp. injective envelope) of S_i. The **dimension vector** of a module X is the vector $\mathbf{dim}X = (\dim_k \operatorname{Hom}_A(P_i, X))_{i\in Q_0}$.

7.3.2 For a triangular algebra A, the **Tits form** $q_A\colon \mathbb{Z}^{Q_0} \to \mathbb{Z}$ is the quadratic form given by

$$q_A(v) = \sum_{i\in Q_0} v(i)^2 - \sum_{i,j\in Q_0} v(i)v(j)\dim_k \operatorname{Ext}^1_A(S_i, S_j) + \sum_{i,j\in Q_0} v(i)v(j)\dim_k \operatorname{Ext}^2_A(S_i, S_j).$$

In this (triangular) situation, the **Cartan matrix** C_A of A is invertible, and we may define the **Euler characteristic** by

$$\langle x, y\rangle = xC_A^{-t}y^t.$$

It has the following homological expression:

$$\langle \mathbf{dim}X, \mathbf{dim}Y\rangle = \sum_{i=0}^{\infty} (-1)^i \dim_K \operatorname{Ext}^i_A(X, Y),$$

for any two A-modules X, Y. Observe that in case that gl.dim $A \le 2$, then $q_A(v) = \langle v, v\rangle$. See [14, 16, 20].

7.3.3 The **one-point extension** of the algebra A by the A-module X is the algebra $A[X] = \begin{pmatrix} A & X \\ 0 & K \end{pmatrix}$ with the usual addition and multiplication of matrices. The quiver of $A[X]$ contains that of A as a full subquiver, and an additional vertex w which is a source. The indecomposable projective $A[X]$-module P_w has $\operatorname{rad} P_w = X$. The $A[X]$-modules are identified with triples (V, M, γ), where V is a K-vector space, M is an A-module and $\gamma\colon V \to \operatorname{Hom}_A(X, M)$ is a K-linear map. An $A[X]$-homomorphism $(V, M, \gamma) \to (V', M', \gamma')$ is a pair (f, g), where $f\colon V \to V'$ is K-linear and $g\colon M \to M'$ is an A-homomorphism satisfying $\gamma' f = \operatorname{Hom}_A(X, g)\gamma$. For an A-module Y, we define the $A[X]$-module $\overline{Y}$ as the triple $(\operatorname{Hom}_A(X, Y), Y, \mathbf{Id}_{\operatorname{Hom}_A(X,Y)})$. One defines dually the **one-point coextension** $[X]A$ of A by X and the modules $\underline{Y}$.

7.3.4 Components of the Auslander-Reiten quiver. For the algebra A, we denote by Γ_A its Auslander-Reiten quiver and by $\tau_A = D\,\mathrm{Tr}$ and $\tau_A^- = \mathrm{Tr}\,D$ the Auslander-Reiten translations. We identify the vertices of Γ_A with the corresponding indecomposable A-modules.

Let $\mathscr{C}$ be a component of Γ_A. We denote by $\operatorname{Ind}\mathscr{C}$ the full subcategory of mod-A with objects the vertices of $\mathscr{C}$. We recall that $\mathscr{C}$ is **standard** if $\operatorname{Ind}\mathscr{C}$ is equivalent to the mesh category $K(\mathscr{C})$ of $\mathscr{C}$, see [20]. We denote by add($\mathscr{C}$) the additive full subcategory of mod-A whose indecomposable modules are those in $\mathscr{C}$.

A translation quiver without multiple arrows is called a **tube** if it contains a cyclic path and its associated topological space is homeomorphic to $\mathbb{S} \times \mathbb{R}^+$.

A tube has two types of arrows: arrows pointing to infinity and arrows pointing to the mouth. An infinite sectional path consisting of arrows pointing to infinity (resp. to the mouth) is called a **ray** (resp. a **coray**). Tubes containing no projective or injective are called **stable**, they are of the form $\mathbb{Z}A_\infty/(\tau^r)$, where r is the rank of the tube.

Let $\mathscr{C}$ be a standard component of Γ_A. For an indecomposable module X in $\mathscr{C}$, called the **pivot**, three types of *admissible operations* are defined, yielding in each case a modified algebra A' of A, and a modified component $\mathscr{C}'$ of $\mathscr{C}$, see [2, 10]:

(1) If the support of $\operatorname{Hom}_A(X, -)|_{\mathscr{C}}$ is of the form:

$$X = X_0 \longrightarrow X_1 \longrightarrow X_2 \longrightarrow \cdots$$

we set $A' = (A \times D)[X \oplus Y_1]$, where D is the full $t \times t$ upper triangular matrix algebra and Y_1 is the unique projective-injective indecomposable D-module. In this case, $\mathscr{C}'$ is obtained by inserting in $\mathscr{C}$ a rectangle consisting of the modules $Z_{ij} = (k, X_i \oplus Y_j, 1)$ for $i \geq 0$, $1 \leq j \leq t$, and $\overline{X_i} = (k, X_i, 1)$ for $i \geq 0$, where Y_j, $1 \leq j \leq t$, denote the indecomposable injective D-modules. If $t = 0$, we set $A' = A[X]$, and the rectangle reduces to the ray formed by the modules of the form $\overline{X_i}$.

(2) If the support of $\operatorname{Hom}_A(X, -)|_{\mathscr{C}}$ is of the form:

$$Y_t \longleftarrow \cdots \longleftarrow Y_1 \longleftarrow X = X_0 \longrightarrow X_1 \longrightarrow X_2 \longrightarrow \cdots$$

with $t \geq 1$ (so that X is injective), we set $A' = A[X]$. In this case, $\mathscr{C}'$ is obtained by inserting in $\mathscr{C}$ a rectangle consisting of the modules $Z_{ij} = (K, X_i \oplus Y_j, \left(\begin{smallmatrix}1\\1\end{smallmatrix}\right))$ for $i \geq 1$, $1 \leq j \leq t$, and $\overline{X_i} = (K, X_i, 1)$ for $i \geq 0$.

(3) If the support of $\operatorname{Hom}_A(X, -)|_{\mathscr{C}}$ is of the form:

$$\begin{array}{ccccccccc} Y_1 & \longrightarrow & Y_2 & \longrightarrow & \cdots & \longrightarrow & Y_t & & \\ \uparrow & & \uparrow & & & & \uparrow & & \\ X = X_0 & \longrightarrow & X_1 & \longrightarrow & \cdots & \longrightarrow & X_{t-1} & \longrightarrow & X_t \longrightarrow \cdots \end{array}$$

with $t \geq 2$ (so that X_{t-1} is injective), we set $A' = A[X]$. In this case, $\mathscr{C}'$ is obtained by inserting in $\mathscr{C}$ a rectangle consisting of the modules $Z_{ij} = (K, X_i \oplus Y_j, 1)$ for $1 \leq j \leq i \leq t$ or $1 \leq j \leq t < i$, and $\overline{X_i} = (K, X_i, 1)$ for $i \geq 0$.

It was shown in [2] that the component of $\Gamma_{A'}$ containing X is $\mathscr{C}'$. The dual operations (1$*$), (2$*$) and (3$*$) are also called admissible. The above admissible operations can be regarded as operations on translation quivers rather than on Auslander-Reiten components, see [10, (2.1)].

Following [2, 10], a translation quiver $\mathscr{C}$ is called a **coil** if there exists a sequence of translation quivers $\Gamma_0, \Gamma_1, \ldots, \Gamma_m = \mathscr{C}$ such that Γ_0 is a stable tube and for each $0 \leq i < m$, Γ_{i+1} is obtained from Γ_i by an admissible operation.

7.3.5 We say that the algebra A is **tame** if for any dimension d, there is a finite number of $A - K[t]$-bimodules M_i which are finitely generated and free as right $K[t]$-modules and such that almost every indecomposable A-module of dimension d is isomorphic to $M_i \otimes_{K[t]} K[t]/(t - \lambda)$ for some $\lambda \in K$ and some i. Let $\mu_A(d)$ be the minimal number of bimodules M_i satisfying the above conditions. Then A is said to be of **polynomial growth** if there is a natural number m such that $\mu_A(d) \leq d^m$ for all $d \geq 1$.

Important examples of tame algebras are the following:

(a) Let $H = K\Delta$ be a hereditary algebra and let T be a (multiplicity-free) tilting H-module, that is, $\operatorname{Ext}^1_H(T, T) = 0$ and T is a direct sum of n pairwise non-isomorphic indecomposable H-modules, where n is the number of vertices of Δ. Then $B = \operatorname{End}_A(T)$ is called a **tilted algebra** of type Δ. If Δ is of Euclidean type $\widetilde{\mathbb{A}}_p$, $\widetilde{\mathbb{D}}_q$, $\widetilde{\mathbb{E}}_6$, $\widetilde{\mathbb{E}}_7$ or $\widetilde{\mathbb{E}}_8$ and T is postprojective, then B is called **tame concealed**. In this case, Γ_B consists of a postprojective component $\mathscr{P}$, a preinjective component $\mathscr{I}$ and a family of tubes $\mathscr{T} = (T_\lambda)_{\lambda \in P_1(K)}$. The family

$\mathscr{T}$ is a **separating family of tubes** in mod-B, that is, $\mathscr{T}$ satisfies conditions (i), (ii), (iv) in the introduction and moreover:

(iii') Any morphism from $\mathscr{P}$ to $\mathscr{I}$ factorizes through add(T_λ), for every $\lambda \in P_1(K)$.

(b) Let C be a tame concealed algebra and $\mathscr{T} = (T_\lambda)_{\lambda \in P_1(K)}$ be the stable separating tubular family of mod-C. An algebra B is called a **coil enlargement** of C if there is a sequence of algebras $C = B_0, B_1, \ldots, B_m = B$ such that for each $0 \leq j < m$, B_{j+1} is obtained from B_j by an admissible operation with the pivot in a stable tube of $\mathscr{T}$ or in a component of Γ_{B_j} obtained from a stable tube of $\mathscr{T}$ by means of the sequence of admissible operations done so far. Observe that, for each $\lambda \in P_1(K)$, almost all modules in $\mathscr{T}_\lambda$ are contained in one component $\mathscr{C}_\lambda$ of Γ_B which is a standard coil [10]. It follows that coil enlargements of C using only the operations of type (1) (resp. of type (1∗)) are **tubular extensions** (resp. **tubular coextensions**) of C in the sense of [20]. By a **coil algebra** we mean a tame coil enlargement of a tame concealed algebra. The following facts were shown in [3, (3.5),(4.2)].

7.3.6 Proposition *Let B be a coil enlargement of a tame concealed algebra C. The following holds:*

(a) There exists a unique maximal tubular extension B^+ of C which is a convex subcategory of B such that B is obtained from B^+ by a sequence of admissible operations of types (1∗), (2∗) *and* (3∗).
(b) There exists a unique maximal tubular coextension B^- of C which is a convex subcategory of B such that B is obtained from B^- by a sequence of admissible operations of types (1), (2) *and* (3).
(c) The family of coils $\mathscr{C} = (\mathscr{C}_\lambda)_{\lambda \in P_1(K)}$ obtained from the tubular family $\mathscr{T} = (T_\lambda)_{\lambda \in P_1(K)}$, is a weakly separating family of coils in mod-B.
(d) B is tame (thus a coil algebra) if and only if B^- and B^+ are tame.

7.3.7 Strongly simply connected algebras. Let A be a triangular algebra with quiver Q. For each vertex i of Q, denote by $Q(i)$ the full subquiver of Q obtained by deleting all vertices of Q being a source of a path ending at i. Then A has the **separation property** if, for each vertex i of Q, the radical radP_i of the indecomposable projective module P_i is a direct sum of pairwise nonisomorphic indecomposable modules whose supports are contained in pairwise different components of $Q(i)$. It was shown in [22] that an algebra A with the separation property is **simply connected**, that is, for any presentation $A \simeq kQ/I$ of A as a bound quiver algebra, the fundamental group $\pi_1(Q, I)$ of (Q, I) is trivial. Also, every convex subcategory of A has the separation property if and only if every convex subcategory of A is simply connected. If this is the case, A is said to be **strongly simply connected**.

We recall the following result from [19] (see [1] for related results).

Proposition *Let B be a convex subcategory of A. If A is strongly simply connected, then B is strongly simply connected and there is a sequence $B = A_0, A_1, \ldots, A_t = A$ of convex subcategories of A such that, for each $0 \leq i < t$, A_{i+1} is a one-point extension or coextension of A_i by an indecomposable A_i-module.*

7.4 Tame Quasi-Tilted Algebras

7.4.1 Following [12], we say that A is a **quasi-tilted algebra** if gl.dim $A \leq 2$ and for every indecomposable A-module X, either $\text{p.dim}_A X \leq 1$ or $\text{i.dim}_A X \leq 1$. (In fact, this is Theorem (2.3) in [12]; by definition A is quasi-tilted if $A = \text{End}_{\mathscr{A}}(T)$ for a tilting object T in an abelian category $\mathscr{A}$).

For example, tilted algebras are quasi-tilted. In particular, the tame concealed algebras are quasi-tilted.

Lemma *Let $\mathscr{T} = (T_i)_{i \in I}$ be a weakly separating family of stable tubes in* mod-A. *Then A is quasi-tilted.*

Proof Assume $\mathscr{T}$ separates $\mathscr{P}$ from $\mathscr{I}$ in mod-A. Let X be a module in $\mathscr{P}$ or in $\mathscr{T}$. By (ii), either X is projective (and $\tau_A X = 0$) or $\tau_A X$ lies also in $\mathscr{P}$ or $\mathscr{T}$. Consider an indecomposable injective A-module I. By (iv), I lies in $\mathscr{T}$ or $\mathscr{I}$, but since $\mathscr{T}$ is formed by stable tubes, then $I \in \mathscr{I}$. Again (ii) implies that $\text{Hom}_A(I, \tau_A X) = 0$. Therefore $\text{p.dim}_A X \leq 1$. Since all projectives are in $\mathscr{P}$, this also implies that $\text{gl.dim}_A \leq 2$. Dually, for any module Y in $\mathscr{T}$ or $\mathscr{I}$, we have $\text{i.dim}_A Y \leq 1$. □

This latter example is an important instance of the algebras considered in our Main Theorem. The results shown in this section for quasi-tilted algebras will be essential for the proof in sect. 7.5.

7.4.2 In [13], it was suggested to consider tame quasi-tilted algebras. The following is the main result of this section:

Theorem *Let A be a strongly simply connected quasi-tilted algebra. If A is tame, then A is of polynomial growth.*

The proof will easily follow in 7.4.5 from the preparatory lemmas given in 7.4.3 and 7.4.4.

7.4.3 There are only four families of tame concealed algebras of type $\widetilde{\mathbb{D}}_n$, given by the following quivers (see for example [20]):

(1)

(2)

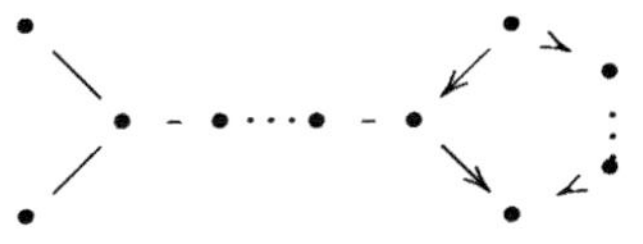

(3)

(4)

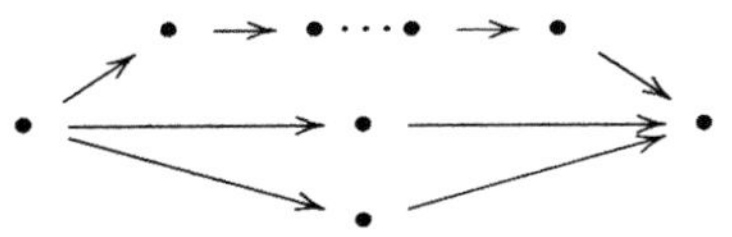

where edges may have any orientation, the algebras (2) and (3) are bound by the commutativity relations, and those of type (4) by the sum of all paths from the unique source to the unique sink.

Following [21], we say that an algebra is pg-**critical** if it is of one of the following forms or their opposites:

$$B_1 = \begin{pmatrix} C & M & 0 & K \end{pmatrix}, \qquad B_2 = \begin{pmatrix} C & N & N \\ 0 & K & 0 \\ 0 & 0 & K \end{pmatrix}, \qquad B_3 = \begin{pmatrix} C & N \\ 0 & D \end{pmatrix}$$

where C is a tame concealed algebra of type $\widetilde{\mathbb{D}}_n$, M is an indecomposable C-module of regular length two lying in a tube of rank $n-2$, N is a simple regular C-module lying in a tube of rank $n-2$, D is given by one of the following quivers:

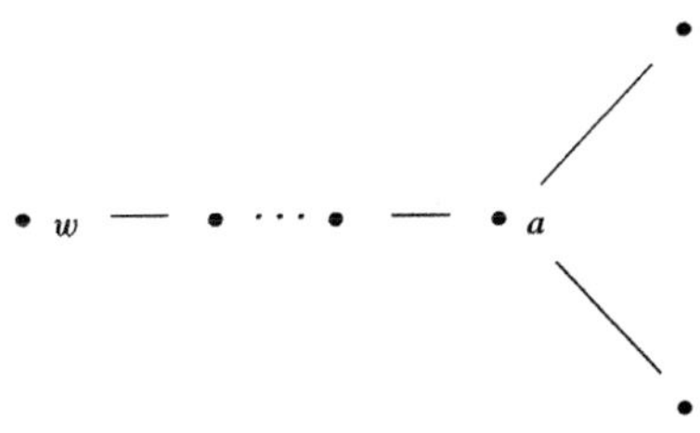

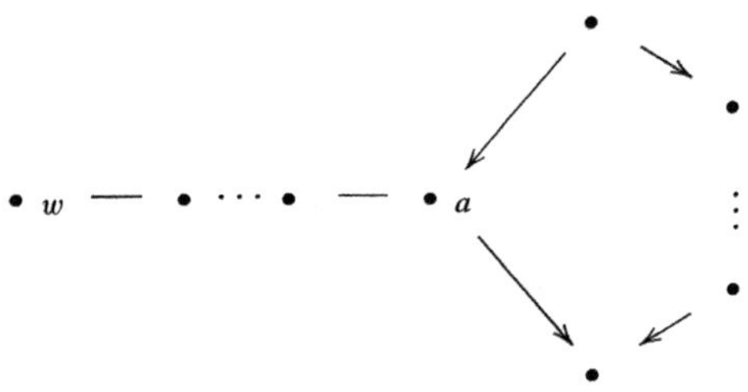

(possibly $w = a$); the right quiver is bound by the commutativity relation, and $B_3(i, j) = N(j) \otimes_K D(i, w)$ for any vertices i in D and j in C. It is known that the pg-critical algebras are tame but not of polynomial growth. The following result from [21] is important.

Theorem *Let A be a strongly simply connected tame algebra. Then A is of polynomial growth if and only if A does not contain any pg-critical convex subcategory.*

7.4.4 We show that pg-critical algebras are not quasi-tilted. In fact, we show the following more precise statement.

Lemma *Let B be a pg-critical algebra. Then there exists an indecomposable B-module X satisfying* $\mathrm{Ext}_B^2(X, X) \neq 0$.

Proof We consider the different possible situations.

(a) Assume that $B = C[M]$, where C is a tame concealed algebra of type $\widetilde{\mathbb{D}}_n$ with $n \geq 5$, and M is a regular C-module of regular length two lying in the tube of rank $n-2$. Let C_0 be the path algebra of the quiver

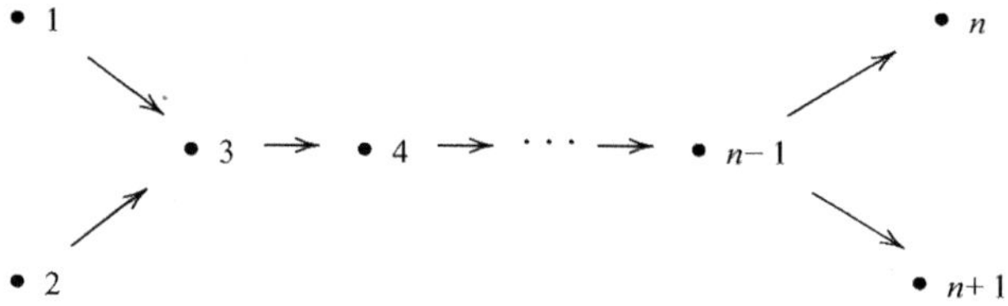

and M_0 be the regular C_0-module with regular length two lying in the tube of rank $n-2$ and with dimension vector

$$\mathbf{dim} M_0 = \begin{array}{ccc} 0 & & 0 \\ & 0\cdots 0 1 1 & \\ 0 & & 0 \end{array}$$

Hence there exists a post-projective tilting C_0-module T such that $C = \mathrm{End}_{C_0}(T)$ and $\mathrm{Hom}_{C_0}(T, M_0) = M$. Let w be the extension vertex in $B_0 = C_0[M_0]$, such that $\mathrm{rad} P_w = M_0$. Then $T' = T \oplus P_w$ is a tilting B_0-module with $\mathrm{End}_{B_0}(T') = B$.

Denote by $F_1[1], F_2[1], \ldots, F_{n-2}[1]$ the simple regular modules in the tube of rank $n-2$ in Γ_{C_0} with $F_1[1] = S_{n-1}$, $F_2[1] = S_{n-2}$ and in general, $\tau_{C_0}^{-1} F_i[1] = F_{i+1}[1]$, $1 \leq i \leq n-3$. Denote by $E_1[1], E_2[1], \ldots, E_{n-3}[1]$ the simple regular modules in the tube of rank $n-3$ in the Auslander-Reiten quiver of the path algebra C_1

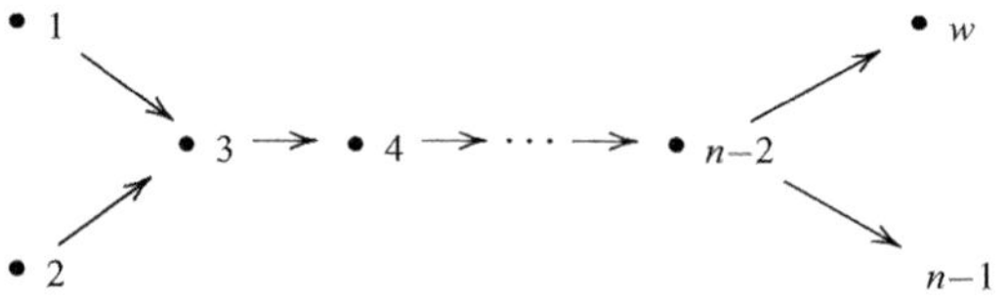

with $\mathbf{dim}E_1[1] = \begin{smallmatrix}1&&0\\&1\cdots 111&\\1&&1\end{smallmatrix}$ and $\tau_{C_1}^{1} E_i[1] = E_{i+1}[1]$, $1 \le i \le n-4$. Hence

$$\mathbf{dim}E_2[1] = \begin{smallmatrix}0&&1\\&0\cdots 001&\\0&&0\end{smallmatrix}$$

which coincides with $\mathbf{dim}\overline{F_2[1]}$ after completing by zeros. Thus, $E_2[1] \cong \overline{F_2[1]}$ as B_0-modules.

The component $\mathscr{C}$ of Γ_{B_0} where M_0 lies, contains the following modules and irreducible maps:

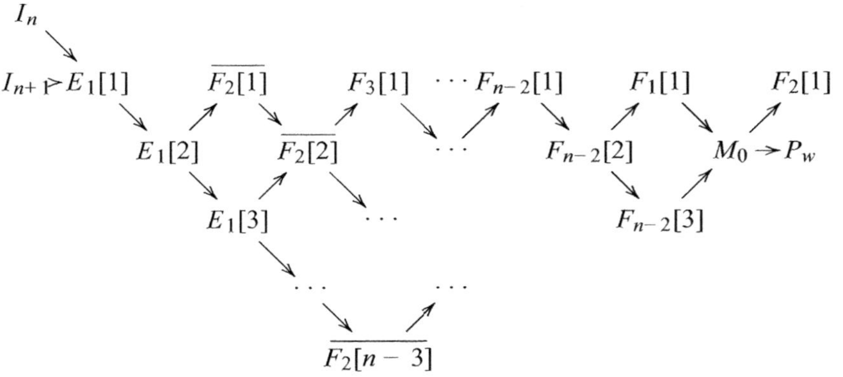

Consider the module $Y = \overline{F_2[n-3]}$. We calculate the value of the Tits form q_{B_0} at $\mathbf{dim}Y$:

$$\begin{aligned} q_{B_0}(\mathbf{dim}Y) &= \langle \mathbf{dim}F_2[n-3] + e_w, \mathbf{dim}F_2[n-3] + e_w\rangle_{B_0} \\ &= q_{C_0}(\mathbf{dim}F_2[n-3]) + 1 + \langle e_w, \mathbf{dim}F_2[n-3]\rangle_{B_0} \\ &\quad + \langle \mathbf{dim}F_2[n-3], e_w\rangle_{B_0} \\ &= 2 - \langle \mathbf{dim}F_1[2], \mathbf{dim}F_2[n-3]\rangle_{C_0} \qquad (7.1) \\ &= 2. \qquad (7.2) \end{aligned}$$

Step (7.1) is justified as follows:

$$\begin{aligned} \langle \mathbf{dim}F_2[n-3], e_w\rangle_{B_0} &= \dim_k \operatorname{Hom}_{B_0}(F_2[n-3], I_w) = 0, \\ \langle e_w, \mathbf{dim}F_2[n-3]\rangle_{B_0} &= \dim_k \operatorname{Hom}_{B_0}(P_w, F_2[n-3]) \\ &\quad - \langle \mathbf{dim}M_0, \mathbf{dim}F_2[n-3]\rangle_{C_0}. \end{aligned}$$

Step (7.2) follows from considerations in the stable tubes of Γ_{C_0}, see [20, (3.1)]. On the other hand,

$$2 = q_{B_0}(\mathbf{dim}Y) = \sum_{i=0}^{2} (-1)^i \dim_K \mathrm{Ext}^i_{B_0}(Y, Y),$$

and $\mathrm{Hom}_{B_0}(Y, Y) = \mathrm{Hom}_{C_0}(F_2[n-3], F_2[n-3]) = K$. Thus $\mathrm{Ext}^2_{B_0}(Y, Y) \neq 0$.

Since $\mathrm{Ext}^1_{B_0}(T', Y) = 0$, the B-module $X = \mathrm{Hom}_{B_0}(T', Y)$ is indecomposable and satisfies

$$q_B(\mathbf{dim}X) = q_{B_0}(\mathbf{dim}Y) = 2, \qquad \mathrm{Hom}_B(X, X) \cong \mathrm{Hom}_{B_0}(Y, Y).$$

Therefore $\mathrm{Ext}^2_B(X, X) \neq 0$.

(*b*) We consider the algebra $B = \begin{bmatrix} C & N \\ 0 & D \end{bmatrix}$, where C is a tame concealed algebra of type $\widetilde{\mathbb{D}}_n$ with $n \geq 5$, N is a simple regular module lying in the tube of Γ_C of rank $n-2$, D is given by the quiver with m vertices

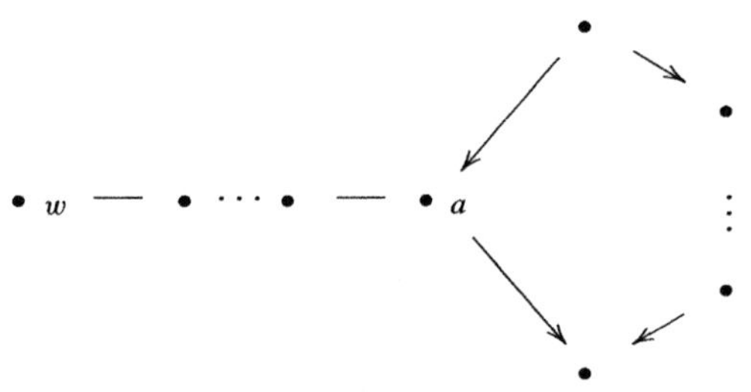

with the commutativity relation, and $B(i, j) = N(j) \otimes_K D(i, w)$, for i in D and j in C.

Let C_0 be the path algebra with quiver

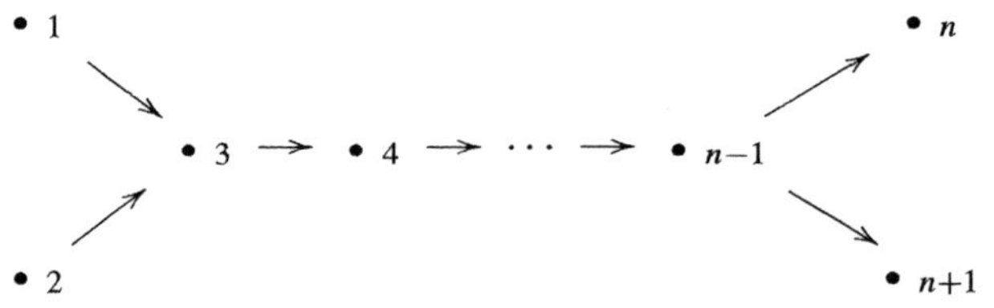

and let N_0 be the simple regular C_0-module S_{n-1}, which lies in the tube of rank $n-2$. There exists a postprojective tilting C_0-module T with $\mathrm{End}_{C_0}(T) = C$ and $\mathrm{Hom}_{C_0}(T, N_0) = N$. Consider the algebra $B_0 = \begin{bmatrix} C_0 & N_0 \\ 0 & D \end{bmatrix}$ with the product defined as above. Then the B_0-module $T' = T \oplus \left(\bigoplus_{x \in D} P_x \right)$ is tilting and $\mathrm{End}_{B_0}(T') = B$.

Denote by $F_1[1], F_2[1], \ldots, F_{n-2}[1]$ the simple regular modules in the tube of rank $n-2$ in Γ_{C_0} with $F_1[1] = S_{n-1}$ and $\tau_{C_0}^{-1} F_i[1] = F_{i+1}[1]$, $1 \le i \le n-3$. In particular,

$$\mathbf{dim} F_{n-2}[1] = \begin{smallmatrix} 1 & & & & 1 \\ & 1 & 1 \cdots 1 & 1 & \\ 1 & & & & 1 \end{smallmatrix}$$

Denote by $E_1[1], E_2[1], \ldots, E_{m+n-4}[1]$ the simple regular modules in the tube of rank $m+n-4$ of the algebra C_1 given by the quiver

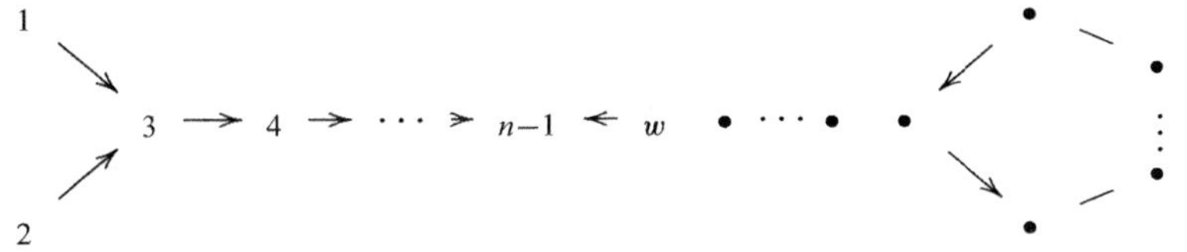

bound by the commutativity relation, with $E_{m+n-4}[1] = S_3$ and $E_i[1] = \tau_{C_1} E_{i+1}[1]$, $1 \le i \le m+n-5$. In particular,

$$\mathbf{dim} E_1[1] = \begin{smallmatrix} & & & & & 0 & \\ 1 & & & & & & 0 \\ & 1\,1 \cdots 1\,0 \cdots 0\,0 & & & & & \cdots \\ 1 & & \scriptstyle w & & & & 0 \\ & & & & & 0 & \end{smallmatrix}$$

and $\tau_{B_0}^{-1} E_{m+n-4}[1] = F_{n-2}[1]$. Therefore, the component of Γ_{B_0} where $N_0 = S_{n-1}$ lies, contains the following modules and irreducible maps:

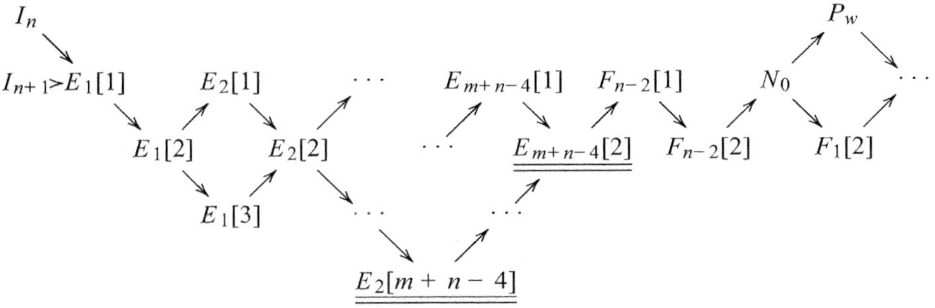

where $\underline{\underline{N}}$ denotes the unique indecomposable B_0-module with $\dim_K \underline{\underline{N}}(n) = 1 = \dim_K \underline{\underline{N}}(n+1)$ and restriction to C_1 equal to N. Let $Y = \underline{\underline{E_2[m+n-4]}}$, we calculate the value of q_{B_0} at $\mathbf{dim}Y$:

$$\begin{aligned} q_{B_0}(\mathbf{dim}Y) &= q_{C_1}(\mathbf{dim}E_2[m+n-4]) + 2 \\ &\quad + \sum_{i=n}^{n+1} (\langle \mathbf{dim}E_2[m+n-4], e_i\rangle_{B_0} \\ &\quad + \langle e_i, \mathbf{dim}E_2[m+n-4]\rangle_{B_0}) \\ &= 2 - 2\langle \mathbf{dim}E_2[m+n-4], \mathbf{dim}E_1[1]\rangle_{C_1} = 2. \end{aligned}$$

As in (a), this implies that $\mathrm{Ext}^2_{B_0}(Y, Y) \neq 0$. Defining the B-module

$$X = \mathrm{Hom}_{B_0}(T', Y),$$

we get $\mathrm{Ext}^2_B(X, X) \neq 0$ as desired. The other cases are similar. □

The fact that the algebras of type B_1 in 7.4.3 are not quasi-tilted was also proved in [12, III(3.9)].

7.4.5 Proof of Theorem 7.4.2

Let A be a strongly simply connected quasi-tilted algebra. Assume that A is not of polynomial growth. By 7.4.3, there is a pg-critical convex subcategory B of A. By 7.4.4, there exists an indecomposable B-module X with $\mathrm{Ext}^2_B(X, X) \neq 0$. Hence $\mathrm{Ext}^2_A(X, X) \neq 0$, which contradicts that A is quasi-tilted.

7.5 Weakly Separating Families of Coils

7.5.1 For the proof of the Main Theorem we will need the following important particular case:

Theorem *Let A be a strongly simply connected tame algebra. Assume that $\mathscr{T}$ is a weakly separating family of stable tubes in* mod-A. *Then A is either a tame concealed algebra or a tubular algebra.*

In 7.5.2, we show how to reduce the proof of the Main Theorem to the above special case. The proof of the above Theorem will be given in 7.5.4 after some preparation.

7.5.2 Theorem *Let A be a strongly simply connected tame algebra. Assume that $\mathscr{T}$ is a weakly separating family of coils in* mod-A. *Then A is a coil algebra.*

Proof Let $P_1, \ldots, P_s$ be all the indecomposable projective modules in $\mathscr{T} = (\mathscr{C}_i)_{i\in I}$.

(a) First assume that $s = 0$. Let B be the full convex subcategory of A in the vertices i such that $I_i \notin \mathscr{T}$ (indeed, if $I_i \notin \mathscr{T}$ and j is a predecessor of i in the path order of Q, then also $I_j \notin \mathscr{T}$). By 7.3.4, each component $\mathscr{C}_i$ is obtained from a stable tube T_i of Γ_B by a sequence of admissible operations of type $(1*)$. In almost all cases $T_i = \mathscr{C}_i$. We shall prove that $\mathscr{T}' = (T_i)_{i \in I}$ is a weakly separating family of stable tubes in mod-B.

Assume that $\mathscr{P}$ and $\mathscr{I}$ are the families of indecomposable A-modules satisfying the conditions (i) to (iv) in the definition. Let $\mathscr{P}'$ be the class of indecomposable B-modules not in $\mathscr{T}'$ or $\mathscr{I}$.

(i) Since $\mathscr{C}_i$ is standard, [10, (2.1)] implies that T_i is also standard, $i \in I$. Clearly, the tubes T_i, $i \in I$, are pairwise orthogonal.
(ii) Clearly, $\mathscr{P}' \subset \mathscr{P}$. Since $\mathscr{T}' \subset \mathscr{T}$, then $\operatorname{Hom}_B(\mathscr{I}, \mathscr{P}') = \operatorname{Hom}_B(\mathscr{T}', \mathscr{P}') = \operatorname{Hom}_B(\mathscr{I}, \mathscr{T}') = 0$ follows.
(iii) Let $f : M \to N$ be a morphism with $M \in \mathscr{P}'$ and $N \in \mathscr{I}$. Let $f = (M \xrightarrow{g} X \xrightarrow{h} N)$ be a factorization of f with $X \in \operatorname{add}(\mathscr{T})$.

Consider the restriction $X' = X|_B$ and an indecomposable factorization $X' = \bigoplus_{j=1}^{m} Y_j$ in mod-B. Then by [20, (4.5)], $Y_j \in \mathscr{T}'$, $1 \le j \le m$. Since h factors through X', f factorizes through $\operatorname{add}(\mathscr{T}')$.
(iv) By construction, all projective B-modules lie in $\mathscr{P}'$ and all injective B-modules in $\mathscr{I}$.

Moreover, since A is strongly simply connected, then B is strongly simply connected 7.3.7. Hence 7.5.1 implies that B is either a tame concealed algebra or a tubular algebra. By construction, A is a coil enlargement of B. Hence A is a coil algebra.

(b) Assume that $s > 0$. Consider a coil $\mathscr{C}$ in $\mathscr{T}$ containing some projectives. Since by [10, (4.5)] the mesh category $K(\mathscr{C})$ has no oriented cycle of projectives, there is a projective P_a in $\mathscr{C}$ such that P_a is a sink in the full subcategory of $k(\mathscr{C})$ consisting of projectives.

We consider two situations:

- P_a is *ordinary*, that is, for each indecomposable direct summand R of $\operatorname{rad} P_a$, the Auslander-Reiten sequence

$$0 \to R \to P_a \oplus R' \to \tau_A^- R \to 0,$$

has either $R' = 0$ or R' indecomposable. In this case, we consider the **wing** $W(P_a)$ formed by all modules X in $\mathscr{C}$ with a sectional path $X \to \cdots \to Y$ for some module Y in the sectional path from P_a to the mouth of $\mathscr{C}$. Then define B as the full convex subcategory of A with vertices i such that $P_i \notin W(P_a)$.

- P_a is *exceptional*, that is, the radical $R = \mathrm{rad} P_a$ is indecomposable and the Auslander-Reiten sequence

$$0 \to R \to P_a \oplus R' \to \tau_A^{-} R \to 0,$$

has $R' = R_1' \oplus R_2'$ an indecomposable decomposition. Then define B as the full convex subcategory of A with vertices $Q_0 \setminus \{a\}$.

In both cases, it is shown in [24, (4.2)] that there is a standard coil $\mathscr{C}'$ in Γ_B such that $\mathscr{C}$ is obtained from $\mathscr{C}'$ by a sequence of admissible operations of type (1), or an admissible operation of type (2) or (3). Assume $\mathscr{C} = \mathscr{C}_{i_0}, \mathscr{C}' := \mathscr{C}_{i_0}'$ and $\mathscr{C}_i' := \mathscr{C}_i$ for $i \in I \setminus \{i_0\}$. Then as in (a), $\mathscr{T}' = (\mathscr{C}_i')_{i \in I}$ is a weakly separating family of coils in mod-B. By induction hypothesis, B is a coil enlargement of a tame concealed algebra C. Then so is A, and we are done. □

7.5.3 For the proof of 7.5.1, we use the following characterization of the class of algebras formed by the tame concealed algebras and the tubular algebras. This strongly relies on the results in [18].

Theorem *Let A be a strongly simply connected algebra. Then the following conditions are equivalent:*

(a) A is either tame concealed or a tubular algebra.
(b) A is of polynomial growth and there is a sincere vector $z \in \mathbb{N}^{Q_0}$ (that is $z(i) > 0$, $i \in Q_0$) such that $q_A(z) = 0$.

Proof It is well known that (a) implies (b), see [20].

Assume that (b) is satisfied. We divide the proof in several steps:

(1) q_A is a non-negative quadratic form. Indeed, let $(x, y) = \frac{1}{2}[q_A(x + y) - q_A(x-y)]$ be the bilinear form associated with q_A. Consider $e_i, i \in Q_0$, the elements of the canonical basis.

Since A is tame, then q_A is weakly non-negative (see for instance [16, IV]). Then for $i \in Q_0$, we have

$$0 \leq q_A(2z + e_i) = 2(z, e_i) + 1 \quad \text{and} \quad (z, e_i) \geq 0.$$

Therefore, $0 = q_A(z) = \sum_{i \in Q_0} z(i)(z, e_i)$, and using that z is sincere, we get that $(z, e_i) = 0$ for every $i \in Q_0$. Now, let $v \in \mathbb{Z}^{Q_0}$ and choose $a \in \mathbb{N}$ with $az + v \in \mathbb{N}^{Q_0}$. Then

$$0 \leq q_A(az + v) = a^2 q_A(z) + a \sum_{i \in Q_0} v(i)(z, e_i) + q_A(v) = q_A(v).$$

(2) The set of vectors $\mathrm{rad} q_A = \{v \in \mathbb{Q}^{Q_0} : q_A(v) = 0\}$ is a $\mathbb{Q}$-vector space. We may choose a set of **critical vectors** of q_A, $z_1, \ldots, z_s \in \mathbb{N}^{Q_0}$ generating $\mathrm{rad} q_A$. We

recall that a vector $v \in \mathbb{N}^{Q_0}$ is a critical vector of q_A if the restriction $q_A^{I(v)}$ of q_A to the support $I(v) = \{i \in Q_0 : v(i) \neq 0\}$ of v is a critical form and v is minimal in $\mathbb{N}^{Q_0}$ generating $\mathrm{rad} q_A^{I(v)}$. See [18] or [8].

Since A is strongly simply connected, then $q_A^{I(z_i)} = q_{A_i}$ for some convex subcategory A_i of A, [18, (2.4)]. We may choose s minimal in (2). Since A is of polynomial growth, then [18, (2.5)] implies that $s \leq 2$. We distinguish two cases.

(3) Assume that $s = 1$. Then z_1 is a sincere vector in $\mathbb{N}^{Q_0}$ and q_A is a critical form. Since A is simply connected, there is a postprojective component $\mathscr{P}_0$ in Γ_A. Then $\mathscr{P}_0$ does not contain injective modules. Otherwise, let I_j be an indecomposable injective module minimal in the path order of $\mathscr{P}_0$. Consider the full convex subcategory B of A with vertices $Q_0 \setminus \{j\}$. Then q_B, as a proper restriction of q_A, is positive and Γ_B has a postprojective component. Thus, [14] implies that B is representation-finite. Moreover, any indecomposable representation X with $X(j) \neq 0$, is a predecessor of I_j in $\mathscr{P}_0$. Therefore A is representation-finite. Again [14] implies that q_A is weakly positive, a contradiction. Hence $\mathscr{P}_0$ does not contain injective modules. Then [20, 4.3 (8)] implies that A is a tame concealed algebra. (For a somewhat different argument the reader may see [16, III]).

(4) Assume that $s = 2$. Then A_1 and A_2 are full convex subcategories of A which jointly contain all the vertices of A. We show that for the two critical vectors z_1, z_2 we have $(z_1, z_2) = 0$. Indeed, for any $i \in Q_0$, $(z_1, e_i) \geq 0$ as in (1). Then $0 = (z, z_1) = \sum_{i \in Q_0} z(i)(z_1, e_i)$ and therefore $(z_1, e_i) = 0$ for all $i \in Q_0$. Hence $(z_1, z_2) = \sum_{i \in Q_0} z_2(i)(z_1, e_i) = 0$. Moreover, A being strongly simply connected of polynomial growth, [21, (4.3)] implies that A is **cycle-finite** (that is, for every cycle $X_0 \xrightarrow{f_1} X_1 \xrightarrow{f_2} \cdots \xrightarrow{f_t} X_t$ of non-zero non-isomorphisms between indecomposable A-modules, $f_i \notin \mathrm{rad}_A^\infty(X_{i-1}, X_i)$, $1 \leq i \leq t$). Therefore, the hypothesis of [18, (2.7)] are fulfilled and in consequence, A is a tubular algebra. We are done. □

7.5.4 Proof of Theorem 7.5.1: Let $\mathscr{T} = (T_i)_{i \in I}$ be a weakly separating family of stable tubes in mod-A. By 7.4.1, A is a quasi-tilted algebra. Therefore, Theorem 7.4.2 implies that A is of polynomial growth.

To apply 7.5.3, we only need to construct a sincere vector $z \in \mathbb{N}^{Q_0}$ such that $q_A(z) = 0$. For each $i \in Q_0$, there is a map $0 \neq f_i : P_i \to I_i$ which factorizes through add($\mathscr{T}$). Then we can choose an indecomposable module X_i in the mouth of some tube $T_{\sigma(i)}$ with $X_i(i) \neq 0$. We may assume that the modules $X_1, \dots, X_t$ with $\{1, \dots, t\} \subset Q_0$ lie in pairwise different tubes and $\{T_{\sigma(j)} : 1 \leq j \leq t\} = \{T_{\sigma(i)} : i \in Q_0\}$. Let r_j be the rank of the tube $T_{\sigma(j)}$, $1 \leq j \leq t$. Then the vector

$$z = \sum_{j=1}^{t} \sum_{i=0}^{r_j - 1} \mathbf{dim}\, \tau_A^i X_j \in \mathbb{N}^{Q_0},$$

is sincere. We show that $q_A(z) = 0$. Indeed, by [20, (3.1)],

$$\langle \mathbf{dim}\tau_A^i X_j, \mathbf{dim}\tau_A^r X_s\rangle = \begin{cases} 1, & \text{if } j = s, i = r, \\ -1, & \text{if } j = s, i = r - 1(\text{mod} r_s), \\ 0, & \text{otherwise.} \end{cases}$$

for any $0 \le i \le r_j - 1, 0 \le r \le r_s - 1$. Since gl.dim $A \le 2$, then

$$\begin{aligned} q_A(z) &= \sum_{1\le j,s\le t} \sum_{i=0}^{r_j-1} \sum_{r=0}^{r_s-1} \langle \mathbf{dim}\tau_A^i X_j, \mathbf{dim}\tau_A^r X_s\rangle \\ &= \sum_{j=1}^{t} \sum_{i=0}^{r_j-1} \Big(\langle \mathbf{dim}\tau_A^i X_j, \mathbf{dim}\tau_A^i X_j\rangle \langle \mathbf{dim}\tau_A^i X_j, \mathbf{dim}\tau_A^{i+1} X_j\rangle \Big) = 0. \end{aligned}$$

7.5.5 Some historic notes on Chap. 7.

(1) The notion of translation quiver and additive functions on translation quivers were introduced by Riedtmann, then a doctoral student of Gabriel. The calculation of additive functions was an essential tool to develop the "knitting procedure" in the construction of Auslander-Reiten quivers, which seems to have started in the middle of the seventies in the work of Bautista and, independently, of Todorov (then a doctoral student of Auslander). It was the technique of Todorov which called for an axiomatic treatment using additive and subadditive functions (copying the additivity property of the ordinary length function on Auslander-Reiten sequences). It was M. Auslander who pointed out during his visit to Bielefeld in June 1979 that the methods of Todorov should furnish an interesting combinatorial characterization of the Dynkin diagrams. In fact, such a characterization follows from the investigations of the Russian mathematician Ernest Vinberg: namely, the Dynkin diagrams are the only finite Cartan matrices with subadditive functions which are not additive. Later, there were further developments in the use of additive functions in the representation theory of finite-dimensional algebras, the most prominent ones were the hammock functions introduced by S. Brenner.
(2) The notion of groups acting on quivers and the construction of quiver quotients and coverings was developed by Riedtmann.
(3) Coils appeared as a natural generalization of stable tubes in the book of Ringel (1984). For tame algebras, a description of some components of the Auslander-Reiten quiver Γ_A of A is known. For instance, in every dimension, almost every indecomposable module lies in a homogeneous tube of Γ_A [Crawley-Boevey]; if A is a tubular extension of a tame concealed algebra, then almost every component of Γ_A is a ray-inserted or coray-inserted tube [Ringel]; in more generality, for certain polynomial growth algebras, every component of Γ_A is a multicoil [Assem-Skowronski]. The operations of insertion of rays on components were worked out by Assem and Skowroński.

References

1. Assem, I. and de la Peña, J.A. *The fundamental groups of a triangular algebra*, Communications in Algebra 24 (1996) 187–208
2. Assem, I. and Skowroński, A.: *Indecomposable modules over multicoil algebras*. Math. Scand. **71** (1992) 31–61.
3. Assem, I., Skowroński, A. and Tomé, B.: *Coil enlargements of algebras*. Tsukuba Journal of Mathematics Vol. **19**, No. 2 (December 1995), pp. 453–479
4. R. Bautista, Larrión, F. and L. Salmerón, *On simply connected algebras*. J. London Math. Soc. (2) 27 (1983), No. 2, 212–220
5. M. Beattie. *A generalization of the smash product of a graded ring*. J. Pure Appl. Algebra 52 (1988) 219–229.
6. O. Butscher and P. Gabriel. *The standard form of a representation-finite algebra*. Bull. Soc. Math. France 111 (1983) 21–40.
7. Cartan, H. and Eilenberg, S. *Homological algebra*. Princeton University Press (1956)
8. Dean, A. and de la Peña, J. A.: *Algorithms for weakly non-negative quadratic forms*. Linear Algebra and its Applications Volume **235**, 1 March 1996, Pages 35–46
9. Dräxler, P., *Completely separating algebras*. Preprint No. 91–100, University of Bielefeld (1991).
10. P. Gabriel. *The universal cover of a representation-finite algebra*. In 'Representation of Algebras'. Springer LNM 903 (1981) 68–105.
11. Happel, D. *Hochschild cohomology of finite dimensional algebras*. Séminaire M.-P. Malliavin (Paris, 1987–88), Lecture Notes in Mathematics 1404, Springer-Verlag (1989) 108–126
12. Happel, D., Reiten, I. and Smalø, S.: *Tilting in abelian categories and quasitilted algebras*. Memoirs of the American Mathematical Society (1996) 88.
13. Happel, D., Reiten, I. and Smalø, S.: *Quasitilted algebras*. NATO Advanced Science Institute Series/Klüwer Ac. Pub.
14. R. Martínez-Villa and J. A. de la Peña. *The universal cover of a quiver with relations*. J. Pure and Appl. Algebra 30 (3) (1983) 277–292.
15. R. Martínez-Villa and J.A. de la Peña. *Automorphisms of representation-finite algebras*. Invent. Math. 72 (1983) 359–362.
16. de la Peña, J.A.: *Quadratic forms and the representation type of an algebra*. Engänzungs-reihe 90–103. Bielefeld (1990).
17. de la Peña, J.A. *Algebras with hypercritical Tits form*. Topics in Algebra, Banach Center Publications, Vol. 26, Part 1, PWN-Polish Scientific Publishers, Warsaw (1990) 353–369
18. de la Peña, J. A.: *On the corank of the Tits form of an algebra*. Journal of Pure and Applied Algebra Volume **107**, Issue 1, 26 February 1996, Pages 89–105
19. de la Peña, J. A. and Skowroński, A.: *Forbidden subalgebras of non-polynomial growth tame simply connected algebras*. Canad. J. Math. **48** (5) (1995), 1018–1043.
20. Ringel, C.M. *Tame algebras and integral quadratic forms*, Springer LNM, **1099** (1984)
21. A. Skowroński, *Cycle-finite algebras*, J. Pure Appl. Algebra **103** (1995), 105–116.
22. Skowroński, A. *Simply connected algebras in Hochschild cohomologies*. In Representations of Algebras., Canad. Math. Soc. Conf. Proc. **14**, Amer. Math. Soc., Providence, RI (1993) 431–447.
23. Skowroński, A.: *Criteria for polynomial growth of algebras*. Bull. Polish Acad. Sci., Mathematics **42** (1994) 173–183.
24. Tomé, B.: *Iterated coil enlargements of algebras*. Fundamenta Mathematicae (1995) Volume **146**, Issue: 3, page 251–266

Chapter 8
Degenerations of Algebras

Here, algebras are associative finite dimensional k-algebras with an identity over an algebraically closed field k. We assume that algebras are basic, and $A = kQ/I$ for a quiver with set of vertices Q_0 and set of arrows Q_1, and I an admissible ideal. By A-mod we denote the category of finite dimensional left A-modules. Recall that the set of vertices $Q_0 = \{1, \dots, n\}$ is in correspondence with the isoclasses of simple A-modules, and let S_i be the simple A-module representing the i-th class. Then there are as many arrows from vertices i to j in Q as $\dim_k \mathrm{Ext}_A^1(S_i, S_j)$. If A is basic then there is a surjective morphism $\nu : kQ \to A$ such that the ideal $\mathrm{Ker}(\nu)$ is admissible, that is, $(\mathrm{rad}A)^m \subset \mathrm{Ker}(\nu) \subset (\mathrm{rad}A)^2$ for some $m \geq 2$. The path algebra kQ has as k-basis the oriented paths in Q, including the trivial paths e_i for $i \in Q_0$, with product given by concatenation of the paths. As before, we identify $A = kQ/I$ with a k-category whose objects are the vertices of Q and whose morphism space $A(i, j)$ is $e_j A e_i$.

A module X in kQ-mod is a representation of Q with vector space $X(i) = e_i X$ for each vertex $i \in Q_0$ and a linear map $X(a) : X(i) \to X(j)$ for each arrow $a : i \to j$ in Q_1. More generally, an A-module is a k-linear functor $X : A \to k$-mod. The **dimension vector** of X is $\mathbf{dim}X = (\dim_k X(i))_{i \in Q_0} \in \mathbb{N}^{Q_0}$, and the **support** of X is the set $\mathrm{supp}X = \{i \in Q_0 \mid X(i) \neq 0\}$. We also consider the duality $D : A\text{-mod} \to A^{op}\text{-mod}$ defined as $D(-) = \mathrm{Hom}_k(-, k)$, where A^{op} is the algebra opposite to A.

8.1 Deformation Theory of Algebras: A Geometric Approach

8.1.1 Consider the **affine space** $V = k^n$ with the Zariski topology, that is, closed sets are of the form

$$Z(p_1, \dots, p_s) = \{v \in V \mid p_i(v) = 0, \quad \text{for all } i = 1, \dots, s\},$$

J.-A. de la Peña, *Representations of Algebras*, Algebra and Applications 30,
https://doi.org/10.1007/978-3-031-12288-0_8

where the $p_i \in k[t_1, \ldots, t_n]$ are polynomials in n indeterminates. In the following we collect some facts about affine spaces.

(i) If $S \subset k[t_1, \ldots, t_n]$, then $Z(S)$ is the zero set of S.
(ii) $Z(S) = Z(\langle S \rangle) = Z(\sqrt{\langle S \rangle})$, where $\langle S \rangle$ is the ideal of $k[t_1, \ldots, t_n]$ generated by S, and

$$\sqrt{I} = \text{radical of } I = \{p \in k[t_1, \ldots, t_n] \mid p^i \in I \text{ for some } i \in \mathbb{N}\}.$$

(iii) $Z(\bigcup_{i \in I} S_i) = \bigcap_{i \in I} Z(S_i)$ and $Z(S \cdot S') = Z(S) \cup Z(S')$.
(iv) **Hilbert's basis theorem:** there exist polynomials $p_1, \ldots, p_s \in S$ with $Z(S) = Z(p_1, \ldots, p_s)$.
(v) **Hilbert's Nullstellensatz:**

$$\{p \in k[t_1, \ldots, t_n] \mid p = 0 \text{ on } Z(S)\} = \sqrt{\langle S \rangle}.$$

8.1.2 We say that $Z = Z(S)$ is an affine variety and $k[Z] = k[t_1, \ldots, t_n]/\sqrt{\langle S \rangle}$ is the **coordinate ring** of Z. An affine variety $Z = Z(p_1, \ldots, p_s)$ is **reducible** if $Z = Z_1 \cup Z_2$ with proper closed subsets $Z_i \subset Z$. Otherwise Z is **irreducible**. There is a finite decomposition of any affine variety $Z = \bigcup_{i=1}^{s} Z_i$ into irreducible subsets $Z_i \subset Z$. If the decomposition is irredundant, we say that $Z_1, \ldots, Z_s$ are te irreducible components of Z. If Z is an irreducible variety, then the maximal length of a chain

$$\emptyset \neq Z_0 \subset \ldots \subset Z_s = Z,$$

is called the **dimension** of $Z(=: \dim Z)$.If $Z = \bigcup_{i=1}^{s} Z_i$ is an irreducible decomposition, then

$$\dim Z = \max_i \dim Z_i.$$

Consider for instance the following varieties

(i) $\text{Bil}(n) = \{\text{bilinear maps } k^n \times k^n \to k^n\}$, with the Zariski topology of k^{n^3}.
(ii) $\text{Ass}(n) = \{\text{associative algebras}\}$, it is a closed subset of $\text{Bil}(n)$, therefore it is an affine variety.
(iii) $\text{Alg}(n) = \{\text{associative algebra structures on } k^n \text{ which have a } 1\}$.

8.1.3 Lemma

(1) $\text{Alg}(n)$ *is an open subset of* $\text{Ass}(n)$.
(2) The map $\text{Alg}(n) \to k^n$ *determined by* $A \mapsto 1_A$ *is a regular map.*
(3) $\text{Alg}(n)$ *is an affine variety.*

Proof For a finite dimensional k-algebra A corresponding to the bilinear map m (not necessarily with 1), denote by $L_a^m, R_a^m : A \to A$ the left and right multiplication by $a \in A$. Then A has a 1 exactly when for some $a \in A$ with both L_a and R_a invertible, in case $1 = L_a^{-1}(a)$.

(1) $D(a) := \{m \in \mathrm{Ass}(n) \mid \det(L_a^m)\det(R_a^m) \neq 0\}$ is open in $\mathrm{Ass}(n)$. Then $\mathrm{Alg}(n) = \bigcup_a D(a)$ is open.
(2) On $D(a)$ the map is equal to $m \mapsto (L_a^m)^{-1}(a)$ which is a quotient of polynomial functions on $\mathrm{Bil}(n)$: the denominator is $\det(L_a^m)$ which does not vanish on $D(a)$.
(3) In fact, $\mathrm{Alg}(n) = \{(m, a) \in Ass(n) \times k^n \mid a \text{ is } 1 \text{ for } m\}$.

□

8.1.4 Let G be an algebraic group. An **action** of G on the affine variety Z is a morphism $\mu : G \times Z \to Z$ satisfying:

(i) $\mu(1_G, z) = z$;
(ii) $\mu(g, \mu(h, z)) = \mu(gh, z)$.

We will write $\mu(g, z) =: gz$. The general linear Group $\mathrm{G}\ell(n)$ acts on $\mathrm{Alg}(n)$ by conjugation and the orbits are the isomorphism classes of algebras. The **orbit** of A is denoted $o(A) := \mathrm{G}\ell(n)A$.

The **stabilizer** $\mathrm{Stab}_{\mathrm{G}\ell(n)}(A)$ of an algebra A is the automorphism group $\mathrm{Aut}(A)$. Then

$$\dim o(A) = \dim \mathrm{G}\ell(n) - \dim \mathrm{Aut}(A).$$

Given a variety Z and morphism $\mu : Z \to \mathrm{Alg}(n)$, we say that $(\mu(z))_{z\in Z}$ is an algebraic family. Given two algebras A and B in $\mathrm{Alg}(n)$ we say that B is a **degeneration** of A if there is an algebraic family $(A_z)_{z\in Z}$ such that $A_z = A$ for z in an open and dense subset of Z and $A_{z_0} = B$ for some $z_0 \in Z$. It is a non-trivial fact that each degeneration can be obtained along the affine line $\mathbb{C}$.

8.1.5 Proposition $\mathrm{Alg}(n)$ *is connected and contains exactly one closed orbit, namely that of the commutative algebra* $B_0 = \mathbb{C}[t_2, \dots, t_n]/(t_i t_j \mid 2 \leq i, j \leq n)$.

Proof Let A be any algebra in $\mathrm{Alg}(n)$ with a basis $1 = a_1, a_2, \dots, a_n$ such that $a_i a_j = \sum_{s=1}^n \gamma_{i,j}^s a_s$, and the $\gamma_{i,j}^s$ are called the structure constants. For $t \in \mathbb{C}$ we define the algebra A_t with basis $1 = a_1, a_2, \dots, a_n$ and structure constants

$$\gamma_{i,j}^s(t) = \begin{cases} t\gamma_{i,j}^s, & \text{for } i, j, s \neq 1, \\ t^2\gamma_{i,j}^s, & \text{for } i, j \neq 1, s = 1, \\ \gamma_{i,j}^s, & \text{otherwise.} \end{cases}$$

We get $A_t \cong A$ for $t \neq 0$, and A_0 is the given commutative algebra B_0. □

8.1.6 It is an interesting and difficult problem to determine the number of irreducible components of $\mathrm{Alg}(n)$ and the generic structures of those components, that is, those algebras which are not degenerations of other algebras.

For instance, for $n = 3$ consider the non-commutative algebra $B_3 = \begin{pmatrix} \mathbb{C} & \mathbb{C} \\ 0 & \mathbb{C} \end{pmatrix}$, then the generic structures in $\mathrm{Alg}(3)$ are:

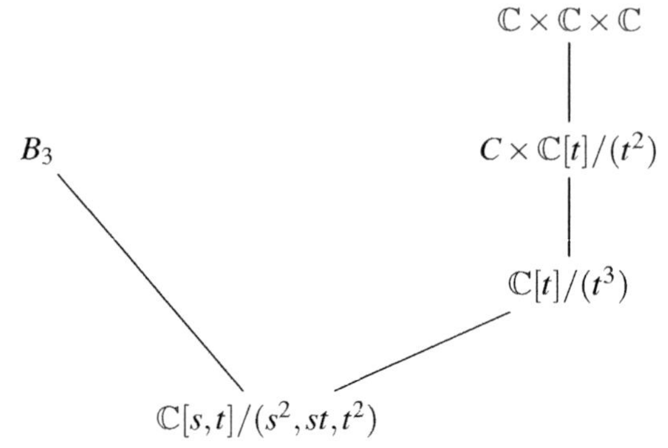

$\mathrm{Alg}(3)$ has two components, one of dimension 9, the closure of the orbit of $\mathbb{C}\times\mathbb{C}\times\mathbb{C}$, and one of dimension 7, the closure of the orbit of B_3.

Few varieties $\mathrm{Alg}(n)$ have been described (Gabriel, Mazzola, Happel). Shafarevich proved in 1990 that $\mathrm{Alg}(n)$ has at least $n - \sqrt{7n}$ irreducible components. One component of $\mathrm{Alg}(n)$ is the closure of the orbit of the semi-simple algebra $\mathbb{C}^n$. Mazzola showed that for $n \geq 7$ this set is formed by the commutative n-dimensional algebras. But in dimension 10 there are commutative algebras which are not degenerations of $\mathbb{C}^n$.

8.1.7 A fundamental result is the following:

Theorem (Chevalley) *Let $\mu : Y \to Z$ be a morphism between affine varieties. Then the function*

$$y \mapsto \dim_y \mu^{-1}(\mu(y)) = \max\{\dim C \mid y \in \mathbb{C} \textit{ irreducible component of } \mu^{-1}(\mu(y))\},$$

*is **upper semicontinuous** (that is, $d : Y \to \mathbb{N}$ has $\{y \in Y \mid d(y) < n\}$ open in Y, for all $n \in \mathbb{N}$).*

As illustration consider $\mu : Y \to Z$ is neither open nor closed, but $\mu(Y)$ is a finite union of locally closed subsets of Z. A finite union of locally closed subsets of a variety Z is called a **constructible subset**.

8.1.8 Proposition *If $\mu : Y \to Z$ is a morphism and $Y' \subset Y$ is a constructible subset, then $\mu(Y')$ is also constructible.*

Few consequences:

(i) The function $A \to \dim_k Z(A)$ where $Z(A)$ is the **center** of an algebra A, is upper semicontinuous. In particular, the commutative algebras form a closed set in $\mathrm{Alg}(n)$

Proof The set $Z := \{(a, A) \mid a \in Z(A)\}$ is closed in $k^n \times \mathrm{Alg}(n)$. Consider the maps: $\pi : Z \to \mathrm{Alg}(n)$ induced by the projection and the section $\sigma : \mathrm{Alg}(n) \to Z$, defined by $A \mapsto (0, A)$. Clearly, $\pi^{-1}(A) = Z(A) \times \{A\}$ and $\dim_k Z(A) = \dim \pi^{-1}(A) = \dim_{\sigma(A)} \pi^{-1}(\pi(\sigma(A)))$. □

(ii) The function $A \to \dim \mathrm{Aut}(A)$ is upper semicontinuous. In particular

(a) the set $\{A \in \mathrm{Alg}(n) \mid \dim o(A) \leq s\}$ is closed for each s;
(b) the set $\{A \in \mathrm{Alg}(n) \mid \dim o(A) = s\}$ is locally closed for each s.
In particular, if B is a degeneration of A in $\mathrm{Alg}(n)$ then:
(c) $\dim_k Z(B) \geq \dim_k Z(A)$;
(d) $\dim \mathrm{Aut}(B) \geq \dim \mathrm{Aut}(A)$ (which we already knew by a dimension of orbits argument).

8.2 Degenerations of Algebras: A Homological Interpretation

8.2.1 Let V be a n-dimensional k-vector space. Recall that the Zariski closed set of associative maps is the affine space $\mathrm{Hom}_k(V \otimes_k V, V)$. Moreover, the associative algebra structures with 1 form an affine open subvariety $\mathrm{Alg}(n)$ of the associative structures. On $\mathrm{Alg}(n)$ operates the algebraic group $\mathrm{G}\ell(n)$ by transport of structure. Thus the orbits of the points of this variety are in one to one correspondence with the isoclasses of the n-dimensional associative k-algebras with 1.

An algebra A defines the **isotropy group** $\mathrm{Aut}(A)$ which is a closed subscheme of $\mathrm{G}\ell(n)$. In general, we use small Greek letters $(\alpha, \beta, \ldots)$ for the points of $\mathrm{Alg}(n)$ and the respective capital roman letters $(A, B, \ldots)$ for the corresponding k-algebras.

Lemma

(a) For given $d, i \in \mathbb{N}$, the function

$$\delta^i : \mathrm{Alg}(n) \to \mathbb{N}, a \mapsto \dim H^i(A),$$

is upper semicontinuous.

(b) If $H^n(A) = 0$, there exists an open neighborhood $\mathcal{U}$ of α in $\mathrm{Alg}(n)$ *and integers c_α, $\tilde{c}_\alpha$ such that for all $\beta \in \mathcal{U}$ we have*

(i) $\dim_k \mathrm{Ker} d_\beta^{n-1} = c_\alpha$.

(ii) $\sum_{i=0}^{n-1} \dim_k H^i(B) = \tilde{c}_\alpha$, *and*

(iii) $H^n(B) = 0$.

Proof Observe that $\dim_k H^0(A) = \dim_k \mathrm{Ker} d_\alpha^0$ and for $i > 0$,

$$\dim_k H^i(A) = \dim_k \mathrm{Ker} d_\alpha^i + \dim_k \mathrm{Ker} d_\alpha^{i-1} - \dim_k C^{i-1}.$$

Since $d^i : \mathrm{Alg}(n) \to \mathrm{Hom}_k(C^i, C^{i+1})$ is a regular map, then $\alpha \mapsto \dim_k \mathrm{Ker} d_\alpha^i$ is an upper semicontinuous function by a simple subdeterminant argument. This shows (a).

Assume now that $H^n(A) = 0$. The upper semicontinuity of δ^n implies the existence of an open connected neighborhood $\mathcal{U}$ of α in $\mathrm{Alg}(n)$ where $\delta^n(\beta) = 0$ for $\beta \in \mathcal{U}$. In particular, for $\beta \in \mathcal{U}$ we have

$$\dim_k \mathrm{Ker} d_\beta^{n-1} = -\dim_k \mathrm{Ker} d_\beta^n + \dim_k C^{n-1},$$

which is a constant function in $\mathcal{U}$ (the left side and the additive inverse of the right side being upper semicontinuous). Finally, observe that

$$\sum_{i=0}^{n-1} (-1)^i \dim_k H^i(B) = \sum_{i=0}^{n-2} (-1)^i \dim_k C^i + (-1)^{n-1} \dim_k \mathrm{Ker} d_\beta^{n-1},$$

is a constant for $\beta \in \mathcal{U}$. In particular, if β is a degeneration of A, for every $n \in \mathbb{N}$ we have

$$\dim_k H^n(B) \geq \dim_k H^n(A).$$

□

8.2.2 Suppose $V \subset k^n$ is defined by certain polynomials $f(t_1, \ldots, t_n)$. For $x \in V$, define the derivative of f at the point x. Then the **tangent space** of V of x is the linear variety $T_x(V)$ in the k^n defined by the vanishing of all $d_x f$ as $f(t)$ ranges over the polynomials in the radical ideal $I(V)$ defining V.

There are more algebraic ways to define tangent spaces: let $R = k[V]$ be the affine algebra associated with V and M_x be the maximal ideal of R vanishing at x. Since R/M_x can be identified with k and M_x is a finitely generated R-module, then the R/M_x-module M_x/M_x^2 is a finite dimensional k-vector space. Then $(M_x/M_x^2)^*$ the dual space over k may be identified with $T_x(V)$.

For $\alpha \in \mathrm{Alg}(n)$ we have canonical inclusions:

$$T_{\mathrm{Alg}(n),\alpha} \to \mathrm{Ker} d^2_\alpha, \quad \text{and} \quad T_{\mathrm{Aut}(A),\mathbf{Id}} \to \mathrm{Ker} d^1_\alpha.$$

Moreover, we denote $T^0_{\mathrm{Alg}(n),\alpha}$ the tangent space to the orbit $o(\alpha)$ of α at α, then the image of d^1_α is included in $T^0_{\mathrm{Alg}(n),\alpha}$.

8.2.3 Proposition *Assuming that $H^1(A) = 0$ then the following holds:*

(1) the maps $T_{\mathrm{Aut}(A),\mathbf{Id}} \to \mathrm{Ker} d^1_\alpha$ and $\mathrm{Im} d^1_\alpha \to T^0_{\mathrm{Alg}(n),\alpha}$ are isomorphisms;
(2) there is a canonical inclusion

$$T_{\mathrm{Alg}(n),\alpha}/T^0_{\mathrm{Alg}(n),\alpha} \to H^2(A);$$

(3) In case $H^3(A) = 0$, then the above inclusion is an isomorphism. Moreover, the point α is smooth in the variety $\mathrm{Alg}(n)$.

Corollary

(a) If $H^2(A) = 0$ then the orbit $o(A)$ is open in $\mathrm{Alg}(n)$.
(b) There are (up to isomorphism) only finitely many algebras A with dimension n and $H^2(A) = 0$.
(c) If $H^1(A) = 0 = H^3(A)$, then $o(A)$ is open in $\mathrm{Alg}(n)$ *if and only if $H^2(A) = 0$.*

8.2.4 Proposition *Let A be a n-dimensional k-algebra. If $H^1(A) = 0$, there is an open neighborhood $\mathcal{U}$ of α in* $\mathrm{Alg}(n)$ *such that the dimension of the* $\mathrm{G}\ell(n)$*-orbits of points in $\mathcal{U}$ is constant (in fact, equal to $n^2 - n + \dim_k H^0(A)$). In particular, if A is a degeneration of B, then A and B are isomorphic.*

Proof There is an open neighborhood $\mathcal{U}$ of α such that $H^1(B) = 0$ and $\dim_k H^0(B) = \dim_k H^0(A)$, for every $\beta \in \mathcal{U}$. Moreover, all algebras in $\mathcal{U}$ have smooth automorphism groups of constant dimension $n - \dim_k H^0(A)$. Therefore, for $\alpha \in \mathrm{Alg}(n)$, the $\mathrm{G}\ell(n)$-orbit of α in $\mathrm{Alg}(n)$ has dimension $n^2 - n + \dim_k H^0(A)$.

Let A be a degeneration of B, that is, α belongs to the closure of the $\mathrm{G}\ell(n)$-orbit of β. Therefore $\mathcal{U}$ contains a point corresponding to an algebra B' isomorphic to B. Since the orbits of A of B' have the same dimension (and are irreducible), they coincide. Hence, B is isomorphic to A. □

8.2.5 We recall that A is a **one-point extension** of B by M if

$$A = \begin{pmatrix} B & M \\ 0 & k \end{pmatrix},$$

with the usual matrix operations. For $A = B[M]$ with M a B-module, **Happel's long exact sequence** relates the Hochschild cohomology groups $H^i(A)$ and $H^i(B)$

in the following way:

$$0 \to H^0(A) \to H^0(B) \to \mathrm{End}_B(B)/k \to H^1(A) \to H^1(B) \to \mathrm{Ext}^1_B(M, M) \to H^2(A) \to \cdots$$

Many applications arise:

(i) Inductive calculation of $\dim_k H^i(A)$;
(ii) if M is exceptional, that is $\mathrm{Ext}^1(M, M) = 0$, we get $H^i(A) = H^i(B)$, for $i \geq 0$, and moreover, the cohomology rings $H^*(A)$ and $H^*B)$ are isomorphic;
(iii) if A is representation finite with a preprojective component and $H^1(A) = 0$, then for all $n \geq 1$ we have $H^n(A) = 0$.

Let $C_\lambda = kQ/I_\lambda$, for $\lambda \in k^*$, be the algebra given by the following quiver:

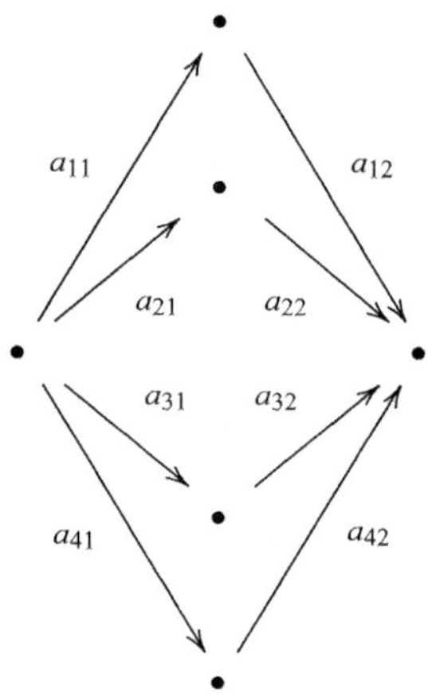

and ideal I_λ generated by the relations $\sum_{i=1}^{3} a_{i2}a_{i1}$ and $a_{12}a_{11} + \lambda a_{22}a_{21} + a_{42}a_{41}$.

The following holds:

(i) The algebras C_λ are isomorphic in Alg(16);
(ii) $\dim\mathrm{Aut}(C_1) = 15$, hence $\dim o(C_1) = \dim \mathrm{G}\ell(16) - 14 = 242$;
(iii) we have $C_1 = H[M]$ for a hereditary algebra of type $\widetilde{\mathbb{D}}_4$ and M an indecomposable module with $\dim_k \mathrm{End}_{C_1}(M) = 1 = \dim_k \mathrm{Ext}^1_{C_1}(M, M)$, then

$$H^i(C_1) = 0, \quad \text{for } i \neq 0, 2,$$

and $H^0(C_1) = k = H^2(C_1)$;
(iv) The algebra C_0 is a degeneration of C_1, that is, $C_0 = \lim_{\lambda \to 0} C_\lambda$. In this case, $C_0 = H[N_1 \oplus N_2]$ and there exists an exact sequence

$$0 \to N_1 \to M \to N_2 \to 0;$$

(v) $H^0(C_0) = H^1(C_0) = H^2(C_0) = k$, all others $H^n(C_0) = 0$.

8.2.6 Let $k[[t]]$ be the algebra of formal power series and $\tilde{p} : k[[t]] \to k$ be the canonical projection. A **formal deformation** of $\alpha \in \text{Alg}(n)$ is an element

$$\tilde{\alpha} \in \text{Alg}(n)_{k[[t]]}(n), \qquad \text{such that } \text{Alg}(n)(\tilde{p})(\tilde{\alpha}) = \alpha.$$

Two formal deformations α_1 and α_2 of α are **equivalent** if they are conjugate in $\text{Alg}_{k[[t]]}(n)$ under some $g \in G\ell_{k[[t]]}(n)$ of the form $g = E_n + tg_1 + t^2 g_2 + \ldots$ where for all i, g_i is a $n \times n$ matrix over k. Moreover, a deformation of α is **trivial** if it is equivalent to α. An **infinitesimal deformation** of $\alpha \in \text{Alg}(n)$ is an element $\tau \in \text{Alg}_{k[\epsilon]}(n)$ such that $\text{Alg}(p)(\tau) = \alpha$. Thus the infinitesimal deformations of α may be identified with the tangent space $T_{\text{Alg}(n),\alpha}$. The equivalence classes of the infinitesimal deformations of α may be identified with $H^2(A)$. An infinitesimal deformation τ of α is **integrable** if there exists a formal deformation $\tilde{\alpha}$ such that the projection $\text{Alg}_{k[[t]]}(n) \to \text{Alg}_{k[\epsilon]}(n)$ sends $\tilde{\alpha}$ to τ. Let $\tilde{\alpha} \in \text{Alg}_{k[[t]]}(n)$ be a formal deformation of $\alpha \in \text{Alg}(n)$. We can take

$$\tilde{\alpha} = \alpha + \alpha_1 t + \alpha_2 t^2 + \ldots$$

for k-linear maps $\alpha_i \in \text{Hom}_k(k^n \otimes k^n, k^n)$, for $i \geq 1$.

8.2.7 Let $s \in \mathbb{N}$ and let J_s be the canonical nilpotent $s \times s$-Jordan block. Consider the ring of truncated polynomials $R_s = k[t]/(t^2)$ and let $p_s : \text{Alg}_{k[[t]]}(n) \to \text{Alg}_{R_s}(n)$ be the map induced by the canonical quotient. Then

$$A_s^\alpha := p_s(\tilde{\alpha})(J_s) = \alpha + \alpha_1 J_s + \ldots \alpha_{s-1} J_s^{s-1},$$

is an algebra in $\text{Alg}(n)$. The following holds:

(1) Equivalent deformations α and β yield isomorphic algebras A_s^α and A_s^β.
(2) A is a degeneration of A_s^α.

For (2) consider the algebraic family $B_\lambda := \alpha + \alpha_1 \lambda J_s + \ldots + \alpha_{s-1}(\lambda J_s)^{s-1}$ which lies in $o(A^{\alpha_s})$ for all $\lambda \neq 0$ and $B_0 = A$.

Proposition

(a) If $H^3(A) = 0$, every infinitesimal deformation $\tau \in T_{\text{Alg}(n),\alpha}$ can be lifted to a formal deformation $\tilde{\alpha}_\tau$ of α.
(b) If $H^2(A) = 0$ every formal deformation is trivial.

Proof (*b*): Let $\alpha_t = \alpha + \alpha_1 t \alpha_2 t^2 + \ldots$ be a formal deformation with $a_1 = a_2 = \ldots = a_{n-1} = 0$ and $a_n \neq 0$. Then $a_n \in \text{Ker} d_\alpha^2 = \text{Im} d_\alpha^2$, thus there exists $g_n \in C^1 = \text{Hom}_k(k^n, k^n)$ such that $\alpha_n = d_\alpha(g_n)$ and with $g = \mathbf{Id} + g_n t^n$ we get $\alpha_t^g = \alpha_0 + \alpha'_{n+1} t^{n+1} + \ldots$. □

8.2.8 The algebra A is said to be **absolutely rigid** if $H^2(A) = 0$. It is said to be **analytically rigid** if every formal deformation of α is trivial. Finally, A is said to be **geometrically rigid** if the orbit of α in $\mathrm{Alg}(n)$ is open. We have the chain of implications:

$$\text{absolutely rigid} \implies \text{analytically rigid} \implies \text{geometrically rigid}.$$

The converse of the first implication is known to be false for positive characteristic, while it is true if $H^3(A) = 0$. If k has characteristic zero, the converse of the second implication holds.

8.3 Tame and Wild Algebras: Definitions and Degeneration Property

8.3.1 A fundamental problem in the representation theory of algebras is the classification of all indecomposable A-modules (up to isomorphism). We say that A is of finite representation type if there are only finitely many indecomposable A-modules up to isomorphism. One of the first successes of modern representation theory was the identification by Gabriel of the Dynkin diagrams as the underlying graphs of quivers Q such that kQ is representation-finite. But representation-infinite algebras are common.

Already in the nineteenth century, Kronecker completed work of Weierstrass to classify all indecomposable pencils by means of infinite families of pairwise non-isomorphic normal forms, which in modern terminology corresponds to the classification of the indecomposable modules over the Kronecker algebra.

8.3.2 Let A be a n-dimensional algebra with basis $1 = e_1, e_2, \ldots, e_n$ with structure constants α_{ij}^*, that is, $e_i e_j = \sum_{s=1}^{n} a_{ij}^s e_s$. The **module variety** $\mathrm{mod}_A(r)$ is the closed subset of the affine space $\mathrm{mat}_k(r)^n$ formed by the matrices $(\mathbf{Id}_r = E_1, E_2, \ldots, E_n)$ satisfying $E_i E_j = \sum_{s=1}^{n} a_{ij}^s E_s$. On $\mathrm{mod}_A(r)$ acts by conjugation the algebraic group $\mathrm{G}\ell(r)$ in such a way that the orbit of a point μ identifies with the isomorphism class of modules M corresponding to μ. We write $o(\mu) = \mathrm{G}\ell(r)M$, and get:

(i) $\mathrm{Stab}_{\mathrm{G}\ell(r)}(M) = \mathrm{Aut}_A(M)$ which is open and dense in the variety $\mathrm{End}_A(M)$;
(ii) $\mathrm{dim} o(\mu) = \mathrm{dim}\mathrm{G}\ell(r) - \mathrm{dim}\mathrm{Stab}_{\mathrm{G}\ell(r)}M = r^2 - \mathrm{dim}_k\mathrm{End}_A(M)$;
(iii) the tangent space $T_{(\mathrm{mod}_A(r),\mu)}$ to $\mathrm{mod}_A(r)$ at μ has as subspace $T_{(o(\mu),\mu)}$ and te quotient $T_{(\mathrm{mod}_A(r),\mu)}/T_{(o(\mu),\mu)}$ is a subspace of $\mathrm{Ext}_A^1(M, M)$;
(iv) $\mathrm{dim}\mathrm{mod}_A(r) \le \mathrm{dim}T_{(\mathrm{mod}_A(r),\mu)} \le \mathrm{dim}_k\mathrm{Ext}_A^1(M, M) + \mathrm{dim}T_{(o(\mu),\mu)}$

$$= \mathrm{dim}_k\mathrm{Ext}_A^1(M, M) + \mathrm{dim}\, o(\mu) = \mathrm{dim}_k\mathrm{Ext}_A^1(M, M) + r^2 - \mathrm{dim}_k\mathrm{End}_A(M);$$

(v) $\dim \mathrm{G}\ell(r) - \dim \mathrm{mod}_A(r) \geq \dim_k \mathrm{End}_A(M) - \dim_k \mathrm{Ext}^1_A(M, M)$;
(vi) there are only finitely many modules M (up to isomorphism) of dimension r satisfying $\mathrm{Ext}^1_A(M, M) = 0$.

Let M be a r-dimensional A module corresponding to the point $\mu \in \mathrm{mod}_A(r)$. The orbit $o(\mu) = \mathrm{G}\ell(r)M$ is locally closed. In particular, $\bar{o}(\mu)\backslash o(\mu)$ is formed by the union of orbits of dimension strictly smaller than $o(\mu)$.

Let $X, Y \in \mathrm{mod}_A$ be modules of dimension r. If the orbit $o(y)$ is contained in $\bar{o}(x)$, we say that Y is a degeneration of X.

8.3.3 Proposition *Let $X \in \mathrm{mod}_A$ of dimension r. We have the following.*

(a) Let $0 \to X' \to X \to X'' \to 0$ be an exact sequence. Then $X' \oplus X''$ is a degeneration of X.
(b) Consider the semi-simple module $\mathrm{gr}\, X = \bigoplus_{i=1}^{s} S_i$, obtained as direct sum of the composition factors S_i of X. Then $\mathrm{gr}\, X$ is a degeneration of X.

Corollary *The orbit $\mathrm{G}\ell(r)X$ is closed if and only if X is semi-simple.*

8.3.4 Let us consider a couple of examples.

(i) Let $F = k\langle T_1, \dots, T_m\rangle$ be the free algebra in m indeterminants. Let M be a $A - F$-bimodule which is free as right F-module. Then the functor $M \otimes_F - : \mathrm{mod}_F \to \mathrm{mod}_A$ induces a family of regular maps $f^n_M : \mathrm{mod}_F(n) \to \mathrm{mod}_A(nr)$ for some number $r \in \mathbb{N}$ and every $n \in \mathbb{N}$. Indeed, set $r = \mathrm{rk}_R(M)$. Since $M \otimes F^s = M^s$, then

$$M \otimes \mathrm{CoKer}(\ F^s \xrightarrow{v} F^t\ ) = \mathrm{CoKer}(\ M \otimes F^s \xrightarrow{M\otimes v} M \otimes F^t\ ) = \mathrm{CoKer}(\ M^s \xrightarrow{M(v)} M^t$$

(ii) The subset $\mathrm{ind}_A(r)$ of $\mathrm{mod}_A(r)$ is constructible. Indeed, the set of pairs

$$\{(X, f) \mid X \in \mathrm{mod}_A(r), f \in \mathrm{End}_A(X) \text{ with } 0 \neq f \neq \mathbf{Id}_X \text{ and } f^2 = \mathbf{Id}_X\},$$

is a locally closed subset of $\mathrm{mod}_A(r) \times k^{r^2}$. The projection $\pi_1 : \mathrm{mod}_A(r) \times k^{r^2} \to \mathrm{mod}_A(r)$ is a regular map with image

$$\mathrm{mod}_A(r)\backslash \mathrm{ind}_A(r).$$

8.3.5 Let $\mathrm{mod}_A(r, s)$ be the set of all modules $\mathrm{mod}_A(r)$ with orbit $\dim o(M) = s$. By upper semi-continuity, the set $\mathrm{mod}_A(r, s)$ is locally closed in $\mathrm{mod}_A(r)$. Let Y be a constructible subset of $\mathrm{mod}_A(r)$ which is closed under the action of $\mathrm{G}\ell(r)$, we set $Y_{(s)} = Y \cap \mathrm{mod}_A(r, s)$ which is constructible. We define the number of parameter of $\mathrm{G}\ell(r)$ on Y as

$$\mu(Y) = \max_s(\dim Y_{(s)} - s).$$

Observe that:

(1) If Z is a constructible subset of Y meeting each orbit, then $\mu(Y) \leq \dim Z$.
(2) Let $f : \mathrm{mod}_B(t) \to \mathrm{mod}_A(r)$ be a regular map and Y be a constructible subset of $\mathrm{mod}_A(r)$ which is closed under the action of $\mathrm{G}\ell(r)$. Assume that Z is a constructible subset of $\mathrm{mod}_B(t)$ restricting to $f : Z \to Y$ such that $\dim f^{-1}(o(y)) \leq d$ for each $y \in Y$. Then $\mu(Y) \geq \dim Z - d$.

8.3.6 Proposition

(1) If A is wild then there is some r such that $\mu(\mathrm{mod}_A(st)) \geq s^2$, for all s.
(2) If A is tame then $\mu(\mathrm{mod}_A(r)) \leq r$, for all r.

Proof

(1) If A is wild we find a number r and regular maps $\mathrm{mod}_{k\langle x,y\rangle} \to \mathrm{mod}_A(sr)$ which has as inverse image of an $\mathrm{G}\ell(sr)$-orbit a $\mathrm{G}\ell(s)$-orbit. Then

$$\mu(\mathrm{mod}_A(st)) \geq \dim \mathrm{mod}_{k\langle x,y\rangle}(s) - \dim \mathrm{G}\ell(s) = 2s^2 - s^2 = s^2.$$

(2) If A is tame for every $n \in \mathbb{N}$ there is a finite family of $A - k[t]$-bimodules $M_{n,1} \ldots, M_{n,\ell(n)}$ with the following properties:

(i) $M_{n,i}$ is finitely generated free as a right $k[t]$-module;
(ii) almost every indecomposable left A-module X with $\dim_k X = n$ is isomorphic to a module of the form $M_{n,i} \otimes_{k[t]} S_\lambda$ for some $\lambda \in k$.

Let r be a positive number and $1 \leq i_1, \ldots, i_s \leq r$ be a sequence with $\sum_{p=1}^{s} \mathrm{rk}(M_{i_p,j_p}) = r$ for some selection of $1 \leq j_p \leq t(i_p)$ for each p. Then $\sum\limits_{p=1}^{s} M_{i_p,j_p} \otimes S_\lambda$ defines a constructible subset of $\mathrm{mod}_A(r)$ of dimension $\leq s \leq r$. Let Z be the union of all these constructible sets for all possible sequences. Since Z meets every orbit, we get $\mu(\mathrm{mod}_A(r)) \leq \dim Z \leq r$.

□

Corollary *An algebra is not simultaneously tame and wild.*

8.3.7 Theorem *A degeneration of a wild algebra is wild.*

Proof The set $\{\alpha \in \mathrm{Alg}(n) \mid \alpha \text{ is wild}\} = \bigcup_r W_r$, where

$$W_r = \{\alpha \in \mathrm{Alg}(n) \mid \mu(\mathrm{mod}_\alpha(r)) > r\}.$$

The sets W_r are closed and $\mathrm{G}\ell(r)$-stable. Hence, if $\beta \in \bar{o}(\alpha)$ and α is wild, then $\alpha \in W_r$ for some r which implies $\beta \in W_r$ and β is wild. □

For instance, take

$$A = k\langle x, y\rangle/(x^2 - yxy, y^2 - xyx, (xy)^2, (yx)^2),$$

degenerated to $B = k\langle x, y\rangle/(x^2, y^2, (xy)^2, (yx)^2)$. Indeed, a point A_t in Alg(7) with basis $1, x, y, xy, yx, xyx, yxy$ and with the multiplication laws given by $x^2 = yxy$, $y^2 = xyx$, $(xy)^2 = 0 = (yx)^2$, satisfies:

$$A_t \equiv A, \quad \text{for } t \neq 0 \text{ and } A_0 \equiv B.$$

Since B is tame, then A is tame. This is the only known proof of the tameness of A.

8.3.8 The following is a central fact about the structure of the **Auslander-Reiten quiver** Γ_A of a tame algebra A.

Theorem *Let A be a tame algebra. Then almost every indecomposable lies in a homogeneous tube. In particular, almost every indecomposable X satisfies $X \cong \tau X$.*

It is an open problem whether it is true that an algebra is of tame type if and only if almost every indecomposable module belongs to a **homogeneous tube**.

Proposition *Let A be an algebra such that almost every indecomposable lies in a standard tube. Then A is tame.*

Proof Our hypothesis implies that almost every indecomposable X satisfies

$$\dim_k \mathrm{End}_A(X) \leq \dim_k X.$$

We will see that this condition implies the tameness of A. Indeed, assume that A is wild and let M be a $A - k\langle u, v\rangle$-bimodule which is finitely generated free as right $k\langle u, v\rangle$-module and the functor $M \otimes_{k\langle u,v\rangle} -$ insets indecomposables. Consider the algebra B given by the quiver with $\mathrm{rad}^2 = 0$:

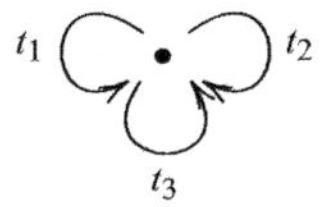

Then there is a $A - B$-bimodule N such that N_B is free and $N \otimes_B - : B\text{-mod} \to A\text{-mod}$ if fully faithful.

Therefore, the composition $F = M \otimes_A (N \otimes_B -)$ is faithful and insets indecomposables. Moreover, $\dim_k FX \leq m\dim_k X$ for any $X \in B$-mod if we set $m = \dim_k(M \otimes_A N)$. Consider also the functor $H : A\text{-mod} \to B\text{-mod}$ sending X to the space $X' = X \oplus X$ with endomorphisms

$$X'(t_1) = \begin{bmatrix} 0 & X(w) \\ 0 & 0 \end{bmatrix}, \quad X'(t_2) = \begin{bmatrix} 0 & X(v) \\ 0 & 0 \end{bmatrix} \quad \text{and} \quad X'(t_3) = \begin{bmatrix} 0 & \mathbf{Id}_X \\ 0 & 0 \end{bmatrix}.$$

This functor insets indecomposables. For the simple A-modules X of dimension n, we get indecomposable A-modules $FH(X)$ with

$$\dim_k FH(X) \leq m\dim_k H(X) = 2mn,$$

and

$$\dim_k \mathrm{End}_A(FH(X)) \geq \dim_k \mathrm{End}_B(H(X)) = n^2 + \dim_k \mathrm{End}_A(X) = n^2 + 1.$$

□

8.3.9 Consider the (infinite dimensional) algebra $A := k[x, y]/(y^2 - x^3 + x)$ with $\mathrm{chr}(k) \neq 2$. We claim that A is neither tame nor wild.

(i) A is not wild. If it was wild, then there is a functor

$$_AM_B \otimes - : B\text{-mod} \to A\text{-mod},$$

which insets indecomposable modules, where B is a finite dimensional wild algebra and M_B is free of finite rank. Choose $m_1, \ldots, m_s$ a k-basis of M and define $f_A \to B^n$ the morphism such that $f(a) = (am_1, \ldots, am_s)$. Then $A' = A/\mathrm{Ker} f$ is a finite dimensional wild algebra. But A' is a quotient of some algebra $A'' = k[x, y]/(y^2 - x^3 + x, p(x))$ for some polynomial $p(x) = k[x]$. One can easily show that A'' is a quotient of $k[x]$ which is representation-finite.

(ii) A is not tame. Assume otherwise, and consider the module variety $\mathrm{mod}_A(1)$. There should exist an open subset U of k and a regular map $U \to \mathrm{mod}_A(1)$. Since A is commutative without nilpotent elements, $A = k[\mathrm{mod}_A(1)] \subset k(t)$. By Lüroth's Theorem we have $A = k(x)$ for a certain transcendental variable x. However, this is not true for A, that is, the curve $y^2 - x^3 + x$ is not rational.

8.4 The Tits Quadratic Form and the Degeneration of Algebras

8.4.1 Let $A = kQ/I$ be a finite dimensional k-algebra and fix a finite set $R \subset \bigcup_{x,y \in Q_0} I(x, y)$ of admissible generators of I. Let $z \in \mathbb{N}^{Q_0}$ be a dimension vector.

The module variety $\mathrm{mod}_A(z)$ is the closed subset, with respect to the Zariski topology, of the affine space $k^z = \prod_{x \to y} k^{z(y)z(x)}$ defined by the polynomial equations given by the entries of the matrices

$$m_r = \sum_{i=1}^{t} \lambda_i m_{\alpha i1} \cdots m_{\alpha i s_i}, \quad \text{where } r = \sum_{i=1}^{t} \lambda_i \alpha_{i1} \cdots \alpha_{is_i} \in R,$$

and for each arrow $x \xrightarrow{\alpha} y$, m_α is the matrix of size $z(y) \times z(x)$

$$m_\alpha = (X_{\alpha ij})_{ij},$$

where $X_{\alpha ij}$ are pairwise different indeterminates. We will identify points in the variety $\mathrm{mod}_A(z)$ with representations X of A with vector dimension $\mathbf{dim}(X) = z$.

For example, let $A = kQ/I$ where $Q = \bullet \xrightarrow{\alpha} \bullet \xrightarrow{\beta} \bullet$ and $I = \langle \alpha\beta \rangle$,

$$\begin{pmatrix} x_{\alpha 11} & x_{\alpha 12} \\ x_{\alpha 21} & x_{\alpha 22} \end{pmatrix} \begin{pmatrix} x_{\beta 11} & x_{\beta 12} \\ x_{\beta 21} & x_{\beta 22} \end{pmatrix} = \begin{pmatrix} x_{\alpha 11}x_{\beta 11} + x_{\alpha 12}x_{\beta 21} & x_{\alpha 11}x_{\beta 12} + x_{\alpha 12}x_{\beta 22} \\ x_{\alpha 21}x_{\beta 11} + x_{\alpha 22}x_{\beta 21} & x_{\alpha 21}x_{\beta 12} + x_{\alpha 22}x_{\beta 22} \end{pmatrix}.$$

Then $\mathrm{mod}_A(2, 2, 2) \subset k^{2\times 2} \times k^{2\times 2}$ is defined by the four equations above.

8.4.2 The group $G(z) = \prod_{i \in Q_0} G\ell_{z(i)}(k)$ acts on k^z by conjugation, that is, for $X \in k^2$, $g \in G(z)$ and $\alpha : x \to y$, then $X^g(\alpha) = g_y X(\alpha) g_x^{-1}$. By restriction of this action, $G(z)$ also acts on $\mathrm{mod}_A(z)$. Moreover, there is a bijection between the isoclasses of A-modules X with $\mathbf{dim}X = z$ and the $G(z)$-orbits in $\mathrm{mod}_A(z)$. Given $X \in \mathrm{mod}_A(z)$, we denote by $G(z)X$ the $G(z)$-orbit of X. Then

$$\mathrm{dim}G(z)X = \mathrm{dim}G(z) - \mathrm{dimStab}_{G(z)}(X).$$

Theorem (Voigt) *Let $X \in \mathrm{mod}_A(z)$. Consider $T_X(G(z)X)$ as a linear subspace of $T_X(\mathrm{mod}_A(X))$. Then there exists a natural linear monomorphism*

$$T_X(\mathrm{mod}_A(X))/T_X(G(z)X) \longrightarrow \mathrm{Ext}_A^1(X, X).$$

Moreover, if X satisfies $\mathrm{Ext}_A^2(X, X) = 0$, then the linear morphism

$$T_X(\mathrm{mod}_A(X))/T_X(G(z)X) \longrightarrow \mathrm{Ext}_A^1(X, X),$$

is an isomorphism.

The inclusion in the theorem above is not always an isomorphism, as the following simple example shows. Let $A = k[T]/(T^2)$. Consider the simple module $S \in \mathrm{mod}_A(1)$. Then $\mathrm{mod}_A(1) = G(1)S = \{S\}$, and $T_S(\mathrm{mod}_A(1))$ is trivial. On the other hand, $\mathrm{Ext}_A^1(S, S)$ has dimension 1.

8.4.3 Let $A = kQ/I$ be a triangular algebra, that is, Q has no oriented cycles. Choose R a minimal set of generators of I, such that $R \subset \bigcup_{i,j \in Q_0} I(i, j)$. We have:

(i) $\mathrm{dim}_k \mathrm{Ext}_A^1(S, S)$ is the number of arrows from i to j.
(ii) $r(i, j) = |R \cap I(i, j)|$ is independent of the choice of R.
(iii) $r(i, j) = \mathrm{dim}_k \mathrm{Ext}_A^2(S_i, S_j)$.

The **Tits form** of A is the quadratic form $q_A : \mathbb{Z}^{Q_0} \to \mathbb{Z}$ given by

$$q_A(v) = \sum_{i \in Q_0} v(i)^2 - \sum_{i \to j} v(i)v(j) + \sum_{i,j \in Q_0} r(i,j)v(i)v(j).$$

For example, taking

$A =$ 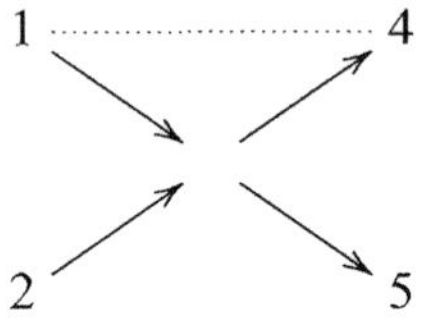

we have

$$\begin{aligned} q_A(x) &= \sum_{i=1}^{5} x_i^2 - x_1x_3 - x_2x_3 - x_3x_4 - x_3x_5 + x_1x_4 \\ &= (x_1 - \frac{1}{2}x_3)^2 + (x_2 - \frac{1}{2}x_3)^2 + \frac{1}{2}(x_3 - \frac{1}{2}x_4 - \frac{1}{2}x_5)^2 + \frac{5}{8}(x_4 + \frac{1}{5}x_5)^2 + \frac{17}{20}x_5^2. \end{aligned}$$

8.4.4 Proposition *Assume that $A = kQ/I$ is triangular, and take $z \in \mathbb{N}^{Q_0}$. Then for any $X \in \mathrm{mod}_A(z)$,*

$$q_A(z) \geq \dim_k \mathrm{End}_A(X) - \dim_k \mathrm{Ext}_A^1(X, X).$$

Proof Let $X \in \mathrm{mod}_A(z)$. The local dimension $\dim_X \mathrm{mod}_A(z)$ is the maximal dimension of the irreducible components of $\mathrm{mod}_A(z)$ containing X. By **Krull's Hauptidealsatz**, we have

$$\dim_X \mathrm{mod}_A(z) \geq \sum_{i \to j} z(i)z(j) - \sum_{i,j \in Q_0} r(i,j)z(i)z(j).$$

Therefore, we get the following inequalities,

$$\begin{aligned} q_A(z) &\geq \dim G(z) - \dim_X \mathrm{mod}_A(z) \geq \dim G(z) - \dim T_X \\ &\geq \dim_k \mathrm{End}_A(X) - \dim_k \mathrm{Ext}_A^1(X, X). \end{aligned}$$

□

8.4.5 In 1975, Brenner observed certain connections between properties of q_A and the representation type of A. She wrote about her observations: '... is written in the spirit of experimental science. It reports some regularities and suggests that there should be a theory to explain them' [2].

Theorem *Let $A = kQ/I$ be a triangular algebra.*

(i) (Bongartz, [1]) If A is representation-finite, then q_A is weakly positive.
(ii) [3] If A is tame, then q_A is weakly non-negative.

Proof In general, for $v \in \mathbb{N}^{Q_0}$ we have

$$\dim \mathrm{mod}_A(z) \geq \sum_{i \to j} v(i)v(j) - \sum_{i,j \in Q_0} r(i,j)v(i)v(j),$$

and $\dim G(v) = \sum_{i \in Q_0} v(i)^2$,

$$q_A(v) \geq \dim G(v) - \dim \mathrm{mod}_A(v).$$

If A is tame, then $q_A(v) \geq 0$. If A is representation finite, $\mathrm{mod}_A(v) = \bigcup_{i=1}^{m} G(v)X_i$, where $X_1, \ldots, X_m$ represent the isoclasses of indecomposable A-modules of $\mathbf{dim} = v$. Hence

$$\dim \mathrm{mod}_A(v) = \dim G(v)X_j = \dim G(v) - \dim \mathrm{Stab}_{G(v)}X_j \leq \dim G(v) - 1,$$

and $q_A(v) \geq 1$. □

8.4.6 An algebra A is **strongly simply connected** if for every convex subalgebra B of A, we have that the first Hochschild cohomology vanishes $H^1(A) = 0$. The Tits form of a strongly simply connected algebra A is weakly non-negative if and only if A does not contain a convex subalgebra (called a hypercritical algebra) which is a preprojective tilt of a wild hereditary algebra of one of the following tree types,

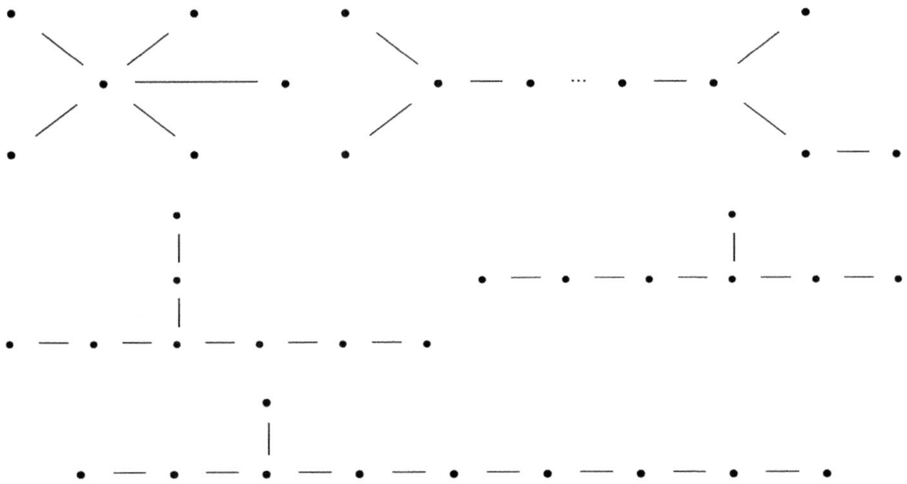

8.4.7 Proposition *Let A be a strongly simply connected algebra. The following are equivalent:*

(i) A is of polynomial growth.
(ii) A does not contain a convex subalgebra which is pg-critical or hypercritical.
(iii) The Tits form q_A if A is weakly non-negative and A does not contain a convex subcategory which is pg-critical.

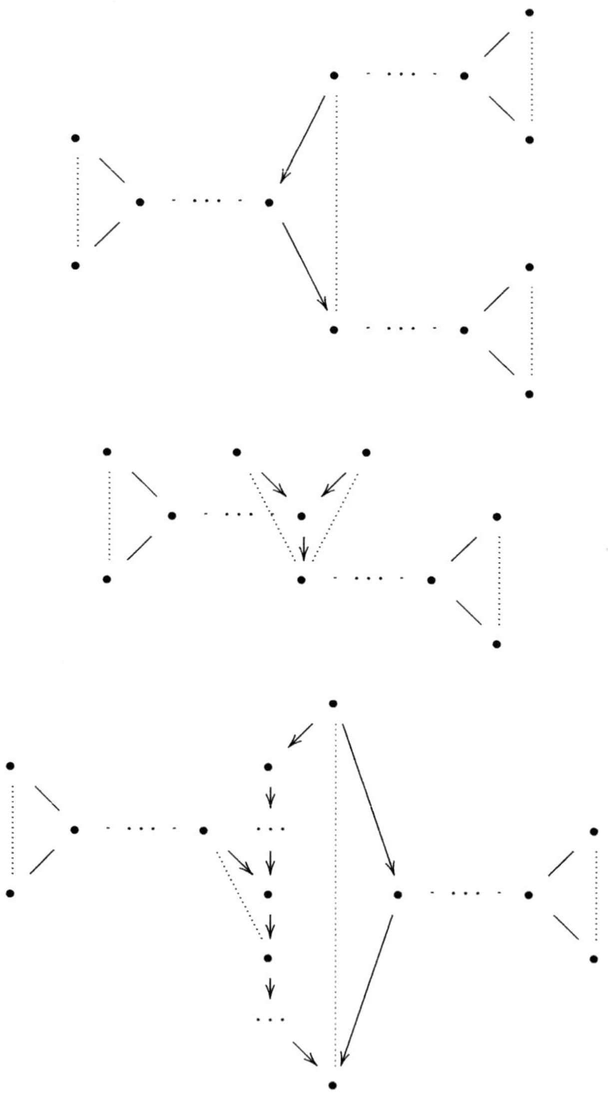

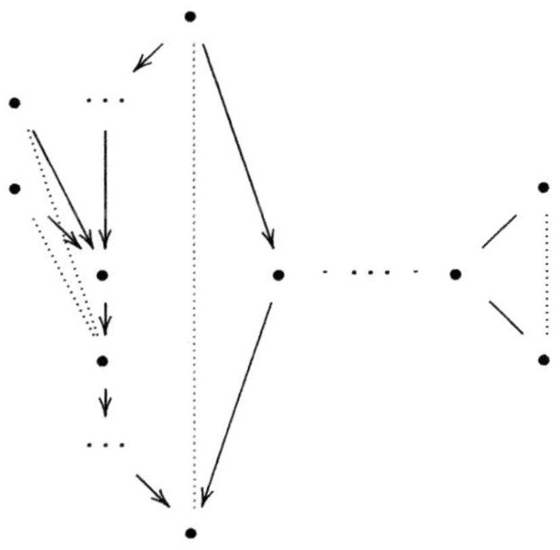

8.4.8 Theorem *Let A be a strongly simply connected algebra. Then A is tame if and only if the Tits form q_A of A is weakly non-negative.*

Corollary *Let A be a strongly simply connected algebra. Then A is tame if and only if A does not contain a convex hypercritical subalgebra.*

Since the quivers of hypercritical algebras have at most 10 vertices, we obtain also the following consequence.

Corollary *Let A be a strongly simply connected algebra. Then A is tame if and only if every convex subcategory of A with at most 10 objects is tame.*

8.4.9 An algebra A is said to be **special biserial** if A is isomorphic to a bound quiver algebra kQ/I, where the bound quiver satisfies the condition:

(a) each vertex of Q is a source and sink of at most two arrows,
(b) for any arrow α of Q there are at most one arrow β and at most one arrow Γ with $\alpha\beta \notin I$ and $\gamma\alpha \notin I$.

For example, take the quiver

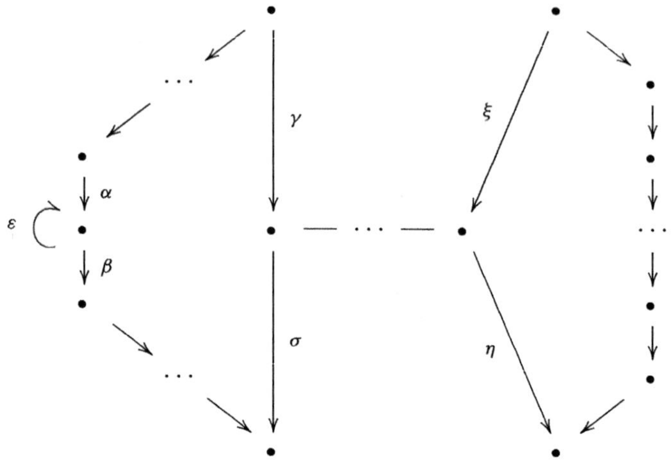

with relations $\varepsilon^2 = 0$, $\alpha\beta = 0$, $\gamma\sigma = 0$ and $\xi\eta = 0$.

8.4.10 Proposition *Every special biserial algebra is tame.*

An essential role in the proof of our main result will be played by the following proposition.

Proposition *Every pg-critical algebra degenerates to a special biserial algebra.*

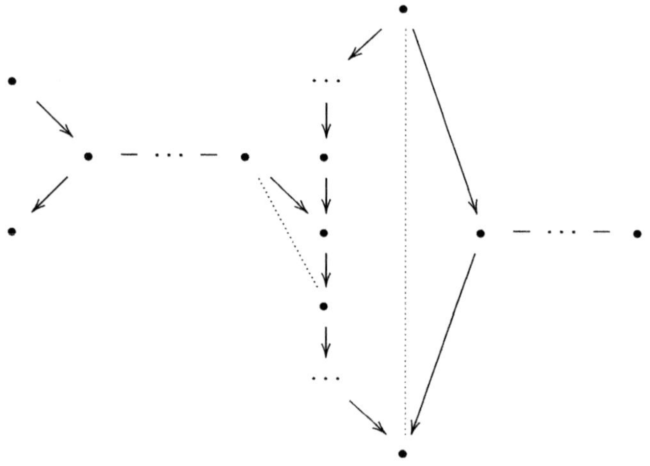

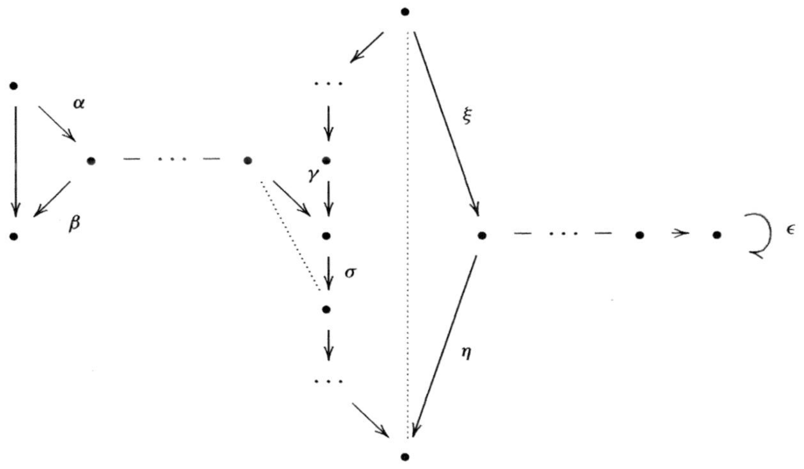

8.4.11 Lemma *Let $A = kQ/I$ be a bound quiver algebra whose quiver Q contains a convex subquiver Q' of the form*

$$x_1 \xrightarrow{\alpha_1} y \xleftarrow{\alpha_2} x_2 \ ,$$

where x_1, x_2 are sources of Q, and α_1 and α_2 are unique arrows starting at x_1 and x_2, respectively. Assume that the ideal I admits a set R of generators of the form

$$R = \{\alpha_1 b_1, \ldots, \alpha_1 b_n, \alpha_2 b_1, \ldots, \alpha_2 b_n, c_1, \ldots, c_m\},$$

with certain elements $b_1, \ldots, b_n \in e_y(kQ)$ and $c_1, \ldots, c_m \in e_z(kQ)$ for $x_1 \neq z \neq x_2$.

Let $\bar{A} = k\bar{Q}/\bar{I}$ be a bound quiver algebra obtained from A as follows: the quiver $\bar{Q}$ is obtained from Q by replacing the subquiver Q' by the subquiver $\bar{Q}'$ of the form

$$\varepsilon \circlearrowright x \xrightarrow{\alpha} y \ ,$$

and $\bar{I}$ is the ideal of $k\bar{Q}$ generated by the set

$$\bar{R} = \{\varepsilon^2, \alpha b_1, \ldots, \alpha b_n, c_1, \ldots, c_m\}.$$

Then A degenerates to $\bar{A}$.

8.4.12 Let A be the algebra given by the following quiver with all commutative relations:

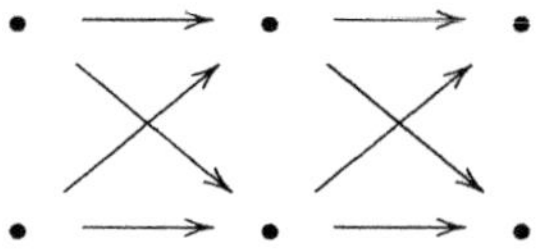

Observe that the group G with two elements acts on A without fixing vertices. Consider the Galois covering functor

$$F : A \to \bar{A} = A/G.$$

We show that $\bar{A}$ is a tame or a wild algebra depending whether or not $\mathrm{chr}(k) \neq 2$. Assume first that $\mathrm{chr}(k) = 2$ and consider the following change of variables:

$$x_0 = \alpha_0 + \beta_0, \quad y_0 = \beta_0, \quad x_1 = \alpha_1 + \beta_1, \quad y_1 = \beta_1.$$

Then $\bar{A}$ is isomorphic to the algebra A' given by the following quiver with relations $x_1 x_0 = 0$ and $y_1 x_0 = x_1 y_0$,

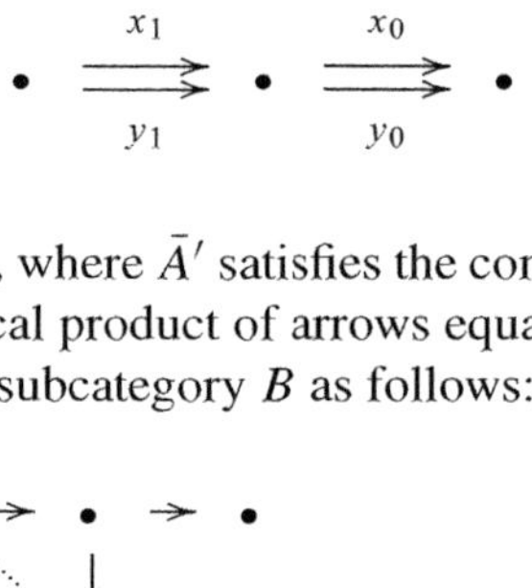

There is a covering $\bar{A}' \to A'$, where $\bar{A}'$ satisfies the commutativity relations marked by dotted lines and the vertical product of arrows equal zero, defined by the action of $\mathbb{Z}$ admitting a full convex subcategory B as follows:

Since $q_B(v) = -1$, then B (and thus $\bar{A}'$) is wild. Then A is also wild.

8.4.13 Assume that $\mathrm{chr}(k) \neq 2$. By Galois covering theory, if A is tame then so is $\bar{A}$. There is an equivalence $F : \mathrm{mod}_A \to \mathrm{mod}_C$ where C is given by:

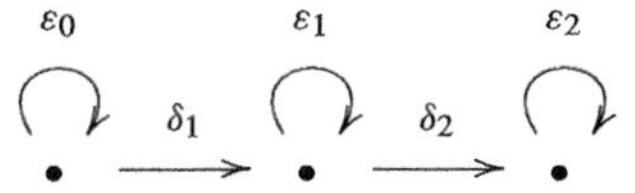

satisfying $\delta_1\delta_2 = 0$ and $\varepsilon_i^2 = \varepsilon_i$, for $i = 0, 1, 2$. The algebra C is isomorphic to C_λ (for $\lambda \neq 0$) with $\varepsilon_i^2 = \lambda\varepsilon_i$. Hence C deforms to C_0, which is a special biserial algebra and hence tame.

References

1. Bongartz K. *Quadratic forms and finite representation type.* In: Malliavin MP. (eds) Séminaire d'Algébre Paul Dubreil et Marie-Paule Malliavin. Lecture Notes in Mathematics, vol. **1146**. Springer, Berlin, Heidelberg. https://doi.org/10.1007/BFb0074545
2. Brenner S. *Quivers with commutativity conditions and some phenomenology of forms.* In: Dlab V., Gabriel P. (eds) Representations of Algebras. Lecture Notes in Mathematics, vol**488**. Springer, Berlin, Heidelberg (1975). https://doi.org/10.1007/BFb0081215
3. de la Peña, J.A. *On the representation type of one point extensions of tame concealed algebras*, Manuscripta Mathematica**61** (2), 183–194 (1988)

Chapter 9
Further Comments

In this brief chapter we collect some comments on dichotomy problems in more general base fields, as well as some historical remarks.

9.1 More on Dichotomy Problems

As stated in the preface, one of our main requirements throughout the book is the base field to be algebraically closed. Therefore, we do not review any of the efforts to generalize both the dichotomy problems or Brauer-Thrall conjectures to more general base fields. Here we comment on some of such generalizations.

9.1.1 A wide generalization was pursued by Crawley-Boevey in [5], independent of the underlying base field K of the algebra A. It depends on the existence of indecomposable A-modules G of infinite dimension, whose length over the ring $\mathrm{End}_A(G)$, called **endo-length** of G, is finite. Such modules are called **generic**, and the algebra A is called **generically tame** if for every $d \in \mathbb{N}$ there is a finite number of isomorphism classes of generic modules of endo-length d.

9.1.2 Theorem ([5]) *The following are equivalent for a finite dimensional algebra A over an algebraically closed field K.*

(a) A is tame.
(b) A is generically tame.
(c) $\mathrm{End}_A(G)$ is a principal ideal ring for every generic A-module G.

9.1.3 Generic modules also provide a description of those indecomposable finite dimensional A-modules parametrized by the field in the classical definition of tameness.

J.-A. de la Peña, *Representations of Algebras*, Algebra and Applications 30,
https://doi.org/10.1007/978-3-031-12288-0_9

Theorem ([5]) *If A is a finite dimensional algebra over an algebraically closed field K, and A is tame, then for each generic A-module G we can choose a A-R_G-bimodule M_G, such that the following hold.*

(a) R_G is a finitely generated localization of $K[x]$. As a right R_G-module, M_G is free of rank equal to the endo-length of G. If K_G is the quotient field of R_G, then $M_G \otimes_{R_G} K_G \cong G$.
(b) The functor $M_G \otimes_{R_G} -$ -from R_G-modules to A-modules preserves isomorphism classes, indecomposability and Auslander-Reiten sequences.
(c) For each $d \in \mathbb{N}$, almost all indecomposable A-modules of dimension d arise as $M_G \otimes_{R_G} R_G/(r)$ for some G and some $O \neq r \in R_G$.

In particular, the algebra A has no generic module (or is **generically trivial**) if and only if it has finite representation type. We also characterize the **domestic algebras** as those with only finitely many generic modules.

Corollary *A finite dimensional algebra over an algebraically closed field has finite representation type if and only if it is generically trivial.*

In view of Theorem 9.1.3, it is natural to say that an algebra A is **generically wild** if there is a generic module whose endomorphism ring is not a principal ideal ring.

9.1.4 A different approach may be traced back to the origins of the theory. After Gabriel's results, Ringel observed that the indecomposable representations of Dynkin quivers may be given via matrices using only the 0 and 1 in the field, and the indecomposability of this matricial presentation is independent of the field. Remarkably, this is no longer true for Euclidean quivers. One may question, how much of the base field is needed to present all indecomposable modules of a certain dimension? This question may be formalized in terms of fields of definition and essential dimensions, as in [8].

9.1.5 Here, F will denote a base field and A a finite-dimensional associative algebra over F. If K/F is a field extension (not necessarily algebraic), we will denote the tensor product $K \otimes_F A$ by A_K. Let M be an A_K-module. Unless otherwise specified, we will always assume that M is finitely generated (or equivalently, finite-dimensional as a K-vector space). If L/K is a field extension, we will write M_L for $L \otimes_K M$.

9.1.6 An intermediate field $F \subset K_0 \subset K$ is called a **field of definition** for M if there exists a K_0-module M_0 such that $M \cong (M_0)_K$. In this case we will also say that M descends to K_0.

A field of definition K_0 of M is said to be **minimal** if whenever M descends to a field L with $F \subset L \subset K$, we have $K_0 \subset L$. A field F is called **quasi-algebraically closed** (or C_1) if every non-constant homogeneous polynomial P over F has a non-trivial zero provided the number of its variables is more than its degree. The idea of quasi-algebraically closed fields was investigated by C. C. Tsen, a student of Emmy

Noether, in a 1936 paper; and later by Serge Lang in his 1951 Princeton University dissertation and in his 1952 paper.

Formally, if P is a non-constant homogeneous polynomial in variables $X_1, \ldots, X_N$, and of degree d satisfying $d < N$ then it has a non-trivial zero over F; that is, for some x_i in F, not all 0, we have

$$P(x_1, \ldots, x_N) = 0.$$

In geometric language, the hypersurface defined by P, in projective space of degree $N - 2$, then it has a point over F.

9.1.7 Theorem *Let F be a C_1-field, A be a finite-dimensional F-algebra, K/F be a separable algebraic field extension and M be an A_K-module. Then M has a minimal field of definition $F \subset K_0 \subset K$ such that $[K_0 : F] < \infty$. The Theorem remains valid if K/F is not assumed to be algebraic.*

9.1.8 Theorem *Let F be a C_1-field, A be a finite-dimensional F-algebra of finite representation type, K/F be a field extension, and M be an A_K-module. Assume further that F is perfectly closed in K. Then M has a minimal field of definition $F \subset K_0 \subset K$ such that $[K_0 : F] < \infty$.*

9.1.9 Given the A_K-module M, the **essential dimension** $\mathrm{ed}(M)$ is defined as the minimal value of the transcendence degree $\mathrm{trdeg}(K_0/F)$, where the minimum is taken over all fields of definition $F \subset K_0 \subset K$. The integer $\mathrm{ed}(M)$ may be viewed as a measure of the complexity of M. Note that $\mathrm{ed}(M)$ is well-defined, irrespective of whether M has a minimal field of definition or not. We also remark that this number implicitly depends on the base field F, which is assumed to be fixed throughout. As a consequence of Theorem 9.1.8, we deduce the following.

Theorem *Let F be a C_1-field, A be finite-dimensional F-algebra of finite representation type, K/F be a field extension, and M be an A_K-module. Then* $\mathrm{ed}(M) = 0$.

9.1.10 Denote by $r_A(n)$ the essential dimension of the functor of representations of A of dimension at most n. By definition, $r_A(n)$ is the smallest integer $m \geq 0$ such that for every field extension K/k and every representation M of $A_K = A \otimes kK$ such that $\dim_K M \leq n$, there exist a subfield $k \subseteq K_0 \subseteq K$ such that its degree of transcendence over k is no greater than m, and a representation N of A_{K_0} such that $N \otimes_{K_0} K \cong M$.

9.1.11 Proposition *Let k be a perfect field and A be a finite-dimensional k-algebra. Assume that A is of finite representation type. Then there exists a constant C such that $r_A(n) \leq C$ for every $n \geq 1$.*

9.1.12 A finite-dimensional F-algebra A is said to be of finite representation type if there are only finitely many indecomposable finitely generated A-modules (up to isomorphism). Recall that, when k is perfect, A is of finite representation type if and only if $A_{\bar{k}}$ is.

9.1.13 Proposition *Let k be an arbitrary field, and let A be a k-algebra. Assume that $A_{\bar{k}}$ is tame. Then there exists a constant $c > 0$ such that*

$$cn - 1 \leq r_A(n) \leq 2n - 1,$$

for every $n \geq 1$.

The rational maps appearing in the proof of by Proposition 9.1.13 have already been constructed and used by the author in [[1], §1.4, 1.5]. Since every indecomposable summand of M is defined over F, Noether-Deuring's Theorem, M is also defined over F, hence $\mathrm{ed}_{\mathrm{k}} M \leq m(m-1)$. Since M was arbitrary, we obtain $r_A(n) \leq m(m-1)$ for every $n \geq 0$.

9.1.14 Proposition *Let k be an arbitrary field, and let A be a k-algebra. Assume that $A_{\bar{k}}$ is wild. Then there exists a constant $c > 0$ such that*

$$r_A(n) \geq cn^2 - 1,$$

for every $n \geq 1$.

9.1.15 Some Conjectures Let B be the free algebra in two indeterminates and W an $A-B$ bimodule which is free as a right B-module. We say that A is wild if $M-B$ preserves indecomposability and isomorphism classes. We say that A is strictly wild if $M\otimes -$ is full. Let $r(A)$ the minimal rank of such an M. By Cohn, B is an Invariant Basis Number (IBN) ring. Then the rank of a free B-module is unique. Hence $r(A)$ is well defined.

For $d \in \mathbb{N} = \{1, 2, 3, \ldots\}$, $A(d)$ denotes the affine variety of associative algebra structures with identity on k^d. We identify algebras with points in $A(d)$. The linear group $GLd(k)$ operates on $A(d)$ by transport of structure. Put $T(d)$ those points in $A(d)$ which are tame and $W(d)$ (resp. wild).

Wild Rank Conjecture There is a function $F : \mathbb{N} \to \mathbb{N}$ such that $r(A)$ is $\leq f(d)$ for A in $W(d)$.

One remarkable result in the geometry of representations is finite representation type is open, i.e., all d-dimensional k-algebras of finite representation type form an open subset. Similarly, we ask:

Tame-Open Conjecture For any $d \in \mathbb{N}$, all tame algebras in $A(d)$ form an open subset of $A(d)$.

Theorem (Y. Han, 2005) *The Wild-Rank Conjecture implies the Tame-Open Conjecture.*

9.1.16 Third Brauer-Thrall Conjecture We note here possible connections of the various embeddings presented between categories of locally finite modules to a conjecture due to D. Simson. In [6], the following is posed as a question.

Conjecture If A is a finite dimensional algebra which is not of finite type, then for any (infinite) cardinality λ, there is an indecomposable A-module M of dimension $\dim(M) \geq \lambda$.

As the classical two Brauer-Thrall conjectures asked whether algebras which are not of finite type have arbitrarily large finite dimensional indecomposable representations, we may call the above a Brauer-Thrall 3 (BT3) question. We remark that Ringel [3, 4] proved that the BT3 statement on existence of arbitrarily large indecomposables holds for the Kronecker quiver with two arrows, and for tame hereditary algebras. This means that it holds also for any algebra A whose Ext quiver Q is not schurian, in the sense that Q contains the Kronecker quiver $\mathbb{K}_2$. One can see this because in this case the quiver coalgebra of $\mathbb{K}_2$ embeds in $C = A^*$, and so there is an exact and full representation embedding of modules over $\mathbb{K}_2$ into C-comodules (equivalently, A-modules); this embedding is then seen to "preserve dimension", as we recall below. Hence, Ringel's work shows that the BT3 statement works for all such algebras.

9.1.17 The above conjecture is also proved to hold for several other classes of algebras in [6], such as fully wild algebras; this is a result of the fact that such algebras are Wild (this statement follows also from the more general embedding of Theorem 3.3.4). We recall here this method: if B is fully wild, let W be a wild algebra and a full faithful exact embedding G : mod-W $\to$ mod-B, where G can be assumed to be of the form $G(X) = P \otimes_W X$ for P finitely generated projective over W. This P is in fact finite dimensional and one can easily argue that G preserves (infinite) dimension, and if indecomposable W-modules of arbitrary dimension exist, then the statement will hold for B. This can be done also by an argument independent of the finite dimensionality of P, which can potentially be used in other situations: if we assume that the indecomposable B-modules have bounded cardinality, then their isomorphism classes form a set I; since W has modules of arbitrarily large cardinality, we can pick a set of non-isomorphic W modules J of cardinality larger than that of I. But since G respects isomorphisms, the map $G : J \to I$, $X \mapsto G(X)$ has to be injective, and so the cardinality of I is at least as large as that of J, a contradiction. By the results in [6, Theorem 3.1 and Corollary 3.2], there are finite dimensional wild algebras which satisfy the above conjecture (namely, the incidence algebra of any finite poset of wild representation type whose Tits form is not positive definite on vectors with entries non-negative integers), and hence such algebras W can be used to show other algebras satisfy the conjecture provided suitable embeddings between large module categories can be found. We refer also to [7] for connections between various other variations of the notion of wild.

9.2 Some Historical Notes

Jordan's Canonical Form Theorem was first proved by Camille Jordan on 1870 and published only on 1930. For the time of publication, the Jordan decomposition was already central in the theory of matrices. In 1874, a strong controversy on the theory of bilinear and quadratic forms opposed Camille Jordan and Leopold Kronecker. The arithmetical ideal of Kronecker faced Jordan's claim for the simplicity of his algebraic canonical form. As the controversy combined mathematical and historical arguments, it gave rise to the writing of a history of the methods used by Lagrange, Laplace and Weierstrass in a century long mathematical discussion (1760–1860) around the "secular equation" (cf. F. Brechenmacher [2]).

Systems of equations of the form $A\frac{dx}{dt} + Bx = f(t)$ were considered by Weierstrass who was able to solve them in case A and B are invertible square matrices. In 1855 Kummer arrived to Berlin to fill the vacancy which occurred when Dirichlet left for Göttingen. Borchardt had lectured at Berlin since 1848 and, in late 1855, he took over the editorship of Crelle's Journal on Crelle's death. In 1856 Weierstrass came to Berlin, so within a year of Kronecker returning to Berlin, the remarkable team of Kummer, Borchardt, Weierstrass and Kronecker was in place in Berlin. In 1890, while working on the problem posed by Weierstrass, Kronecker classified what nowadays are called representations of the **Kronecker quiver** $\mathbb{K}_2$, with two vertices 1 and 2 and two arrows between them as follows,

$$1 \rightrightarrows 2$$

The associated path algebra is denoted $\mathbf{K}_2$, therefore the category of modules $\mathbf{K}_2$-mod is equivalent to the category of representations rep($\mathbb{K}_2$).

In the old times of the new theory there were many mathematicians attracted by the hot topics, we will encounter their names associated (at least sometimes!) to their results. Some produced contributions in representation theory but gave their main work to other subjects. Many of them saw their careers interrupted by the wars:

Hermann Weyl (1885–1955) was a German mathematician. Many years he alternated his life and work between Zürich and Princeton, but he is identified with the mathematical tradition of Göttingen, represented by Hilbert and Minkowski. Weyl invited Richard Brauer to assist him at Princeton's Institute for Advanced Studies in 1934, where together with Nathan Jacobson, he edited Weyl's lectures: *Structure and Representation of Continuous Groups*. As a student of Georg Frobenius (1849–1917), Issai Schur (1875–1941) worked on group representations (the subject with which he is most closely associated), but also worked in combinatorics and number theory. He is perhaps best known today for his result on the existence of the Schur decomposition and for his work on group representations (Schur's lemma). Schur had a number of students, including Richard Brauer, B. H. Neumann, Heinz Prüfer, and Richard Rado.

Emmy Noether (1882–1935) is the founder of modern abstract algebra. She was a German mathematician of Jew ascendancy who worked on invariant theory. Her contributions to theoretical physics and abstract algebra were fundamental. She was considered by Hilbert, Einstein and many others as the most important woman in mathematics in history.

The influence of Schur and the group of his students was strong: Hopf spent the year after his doctorate at the University of Göttingen, where David Hilbert, Richard Courant, Carl Runge, and Emmy Noether were working. Richard Brauer (1901–1977) was a German and American mathematician. He worked mainly in abstract algebra, but made important contributions to number theory. He was the founder of modular representation theory. Brauer (who wrote his thesis under Schur), provided an algebraic approach to irreducible, continuous, finite-dimensional representations of real orthogonal (rotation) groups. He expounded central division algebras over a perfect field while in Königsberg (before going to Princeton); the isomorphism classes of such algebras form the elements of the Brauer group he introduced. Part of scientific community organized to help Brauer and other Jewish scientists. Brauer was offered an assistant professorship at University of Kentucky (see for instance [9]).

Israel Gelfand (1913–2009) was born in Ukraine. He is considered one of the greatest mathematicians of the twentieth century. He made major contributions to many areas of mathematics. His achievements also include well-known work in biology. Among other central concepts in representation theory, he invented the preprojective algebra associated to a quiver. The modules of preprojective algebras are intimately related to representations of the quiver, but it is often the modules for the preprojective algebra which are of relevance in other parts of mathematics. There is beautiful geometry linked to the preprojective algebra, including Kleinian singularities and H. Nakajima's quiver varieties.

Pierre Gabriel (1933–2015), also known as Peter Gabriel, was a French mathematician at the universities of Strasbourg (1962–1970), Bonn (1970–1974) and Zürich (1974–1998) who worked on category theory, algebraic groups, and representation theory of algebras. The subject was started by P. Gabriel in 1972, when he discovered that the quivers with only finitely many indecomposable representations are exactly the $\mathbb{ADE}$ Dynkin diagrams which occur in Lie theory. Gabriel tells the following story of the day he presented this result in Oberwolfach. After claiming there were only few diagrams of finite representation type, he started to state numerical results: $\mathbb{A}_n$ *accepts exactly* $\frac{1}{2}n(n+1)$ *indecomposable modules*, $\mathbb{D}_n$ *accepts exactly* $n(n-1)$ *indecomposable modules*, while saying these numbers, a voice raised in the room to anticipate the right answers. Gabriel looked at the audience and continue with the exceptional cases: *for* $\mathbb{E}_6$ *there are* 36 *indecomposable modules*, and again the same voice had anticipated correctly the number. *Who said this?*, asked Gabriel: it was Jacques Tits. *Did you already proved the Theorem yourself?* asked Gabriel. Tits said: *no, I did not prove the theorem, I was just guessing the numbers, as the number of roots in the corresponding root system for the semi-simple Lie algebra.* Quickly they checked the remaining cases $\mathbb{E}_7$ and $\mathbb{E}_8$ with 63 and 120 indecomposable modules, respectively. The guess was correct but the reasons were obscure for many years. Quivers and their representations now appear in all sorts of areas of mathematics and physics, including representation theory, cluster algebras, geometry (algebraic, differential, symplectic), non-commutative geometry, quantum groups, string theory, and more.

Maurice Auslander (1926–1994) was a Jewish-American mathematician who worked on commutative algebra and homological algebra at Brandeis University. He proved the Auslander–Buchsbaum theorem saying that regular local rings are factorial, the Auslander–Buchsbaum formula, and introduced Auslander–Reiten theory in the representation theory of algebras. Given a representation-finite algebra A with indecomposable modules $M_1, \dots, M_s$, the algebra $E = \mathrm{End}_A(\bigoplus_{i=1}^{s} M_i)$ is called the Auslander algebra of A.

Andrei Roiter (1937–2006) was an Ukrainian mathematician. Roiter was the founder of a Kiev school of representation theory, one of whose traits was an extensive use of methods of linear algebra. He solved two conjectures on representations of finite dimensional algebras, proposed by Brauer and Thrall, which became one of the best-known leitmotivs of the theory. Among Roiter's numerous contributions to the general theory of matrix problems we mention two: (1) in a joint paper with Nazarova (his wife and collaborator) he introduced representations of partially ordered sets, an important class of matrix problems with many applications; (2) in a series of papers he introduced representations of bocses, a very large class of matrix problems that provided the framework for Drozd's tame and wild theorem saying that a finite dimensional algebra over an algebraically closed field is either tame or wild but not both.

The influence of those mathematicians extended also through the work of their students and collaborators. We mention only some of the most relevant of them (at least from the point of view of the results presented in this book) and the schools they created:

In Germany, Claus Ringel had a deep influence through his European students (Happel, Unger, Krause) and opened his teachings to Chinese students. His book *Tame algebras and integral quadratic forms* published in 1984 was very influential in the development of the field for many years.

In Switzerland, and Germany, Gabriel and his students Riedtmann and Bongartz were very influential. The author of this book spent post-doctoral time (1983–1986) working under the supervision of Gabriel in Zürich and then started long-term collaborations with other European mathematicians (Helmut Lenzing in Germany, Andrzej Skowroński in Poland and others).

The influence of Auslander spread to several countries: his longtime collaborator Idun Reiten and Sverre Smalø founded the Norwegian school; Todorov, Igusa, Green, the German born Birge Huisgen-Zimmermann and others supported the growth of the theory in the United States; Bautista and Martínez-Villa introduced the theory to Mexico.

There was already an important tradition in algebra in Japan at the first part of XX century, later Japanese mathematics much adapted to western traditions. Particularly influential was the so called "zen" school in category theory as represented by Nabuo Yoneda (1930–1996) and his lemma. The Yoneda lemma suggests that instead of studying the (locally small) category $\mathscr{C}$, one should study the category of all functors of $\mathscr{C}$ into **Set** (the category of sets with functions as morphisms). There are several immediate benefits of this approach, for instance, **Set** is a category we understand well, and a functor of $\mathscr{C}$ into **Set** can be seen as a "representation" of $\mathscr{C}$ in terms of known structures. The original category $\mathscr{C}$ is contained in this functor category.

Other successful developments happened in Poland (led mainly by Simson and Skowroński), Canada (led by Dlab and, years later by Assem), Brazil (led by Merklen, Coelho and Marcos), Argentina (led by Platzeck and Trepode). Further developments are due to the impetus of students (now in the fourth or fifth generation) and young researchers around the world.

Due to their regularity and temporal proximity, the International Congress of Mathematicians (ICM) serve as a good indicator of the general health of mathematics. Adolf Hurwitz addressed the participants of the first ICM (Zurich, 1897), saying: "It is true that most of the great ideas of our science have risen and matured on the silence of the working studio; no other science, but possibly for Philosophy, presents a character so eremitic, and secluded as mathematics. And yet, at the heart of a mathematician lives the necessity for communicating and expressing himself to his colleagues. And each of us knows, by personal experience, how stimulating personal scientific intercourse can be". The first lecture of the ICM in 1904 was held in commemoration of the centennial Anniversary of Jacob Jacobi's birth. In view of the distressful world situation, due to WW2, the ICM to be held in Cambridge, in September 1940 was postponed. It finally took place in September 1950. Along the years, there were talks on representation theory of algebras given in the ICM by Ringel, Gabriel, Reiten, Geiss, and others.

At the end, I thank the institutions which, along the years, have made possible my work –in particular, for this book. I thank my many collaborators, whose work I have used as source of the results reported. Very specially, I thank Jesús Jiménez for his support during the writing of the text.

References

1. Auslander M. (1966) *Coherent Functors*. In: Eilenberg S., Harrison D.K., MacLane S., Röhrl H. (eds) Proceedings of the Conference on Categorical Algebra. Springer, Berlin, Heidelberg.
2. Brechenmacher, F., *Algebraic generality vs arithmetic generality in the controversy between C. Jordan and L. Kronecker (1874)*. (2008) ffhal-00340071v1
3. Corner, A. L. S.: *Endomorphism algebras of large modules with distinguished submodules*, J. Algebra **11** (1969) 155–185
4. W. Crawley-Boevey, *On tame algebras and bocses*, Proc. London Math. Soc. **56** (1988), 451–483.
5. W. Crawley-Boevey, *Tame algebras and generic modules*, Proc. London Math. Soc. **63** (1991), 241–265.
6. P. Dowbor, *Stabilizer conjecture for representation-tame Galois coverings of algebras*, J. Algebra **239** (2001), 119–149.
7. P. Dowbor and A. Skowroński, *Galois coverings of representation-infinite algebras*, Comment. Math. Helv. **62** (1987), 311–337.
8. Scavia, F.: *Essential dimension of representations of algebras*, arXiv:1804.00642 (2020)
9. Weil, A., *Souvenirs d'apprentissage*. Basel; Boston. Birkhäuser. Vita mathematica, V. 6 (1990).

Index

J.-A. de la Peña, *Representations of Algebras*, Algebra and Applications 30,
https://doi.org/10.1007/978-3-031-12288-0

Printed by Printforce, the Netherlands